METHODS IN MOLECULAR BIOLOGY

For further volumes:
http://www.springer.com/series/7651

For over 35 years, biological scientists have come to rely on the research protocols and methodologies in the critically acclaimed *Methods in Molecular Biology* series. The series was the first to introduce the step-by-step protocols approach that has become the standard in all biomedical protocol publishing. Each protocol is provided in readily-reproducible step-by-step fashion, opening with an introductory overview, a list of the materials and reagents needed to complete the experiment, and followed by a detailed procedure that is supported with a helpful notes section offering tips and tricks of the trade as well as troubleshooting advice. These hallmark features were introduced by series editor Dr. John Walker and constitute the key ingredient in each and every volume of the *Methods in Molecular Biology* series. Tested and trusted, comprehensive and reliable, all protocols from the series are indexed in PubMed.

Rhodopsin

Methods and Protocols

Edited by

Valentin Gordeliy

Institut de Biologie Structurale (IBS), Université Grenoble Alpes, CEA, CNRS, Grenoble, France

Humana Press

Editor
Valentin Gordeliy
Institut de Biologie Structurale (IBS)
Université Grenoble Alpes, CEA, CNRS
Grenoble, France

ISSN 1064-3745 ISSN 1940-6029 (electronic)
Methods in Molecular Biology
ISBN 978-1-0716-2328-2 ISBN 978-1-0716-2329-9 (eBook)
https://doi.org/10.1007/978-1-0716-2329-9

This Humana imprint is published by the registered company Springer Science+Business Media, LLC, part of Springer Nature.
The registered company address is: 1 New York Plaza, New York, NY 10004, U.S.A.

Preface

Since the discovery of the first microbial rhodopsin — a light-driven proton pump bacteriorhodopsin (BR), a seven alpha helical membrane protein comprising the retinal, in 1971, BR was a major driver of groundbreaking developments in membrane protein research. BR studies stimulated important methodological developments in the fields like electron microscopy and membrane protein crystallization. However, before 1999 there were only several and only archaeal known rhodopsins, functioning as proton, chloride pumps, and also sensory rhodopsins. At the end of the twentieth century not many of us, if any at all, would imagine a new era of the rhodopsin world and its contributions to the applications in neuroscience. Not many enthusiasts were eager to continue working with these proteins.

But nature plays tricks on us. A new microbial rhodopsin era started in 1999–2000 when the first bacterial rhodopsin, a proton pump, was discovered. Later it appeared that there are unexpectedly many rhodopsins and they are present in all the domains of life and even in viruses. It turned out (which is amazing) that they execute such a diversity of functions being "nearly the same." However, in the late 1990s nobody could predict groundbreaking discoveries of other rhodopsin functions and among them, probably the most significant one, the discovery of channel rhodopsins (ChRs). For life science, it is the discovery of the proteins which are different from any known channels. For neuroscience, it was the biggest luck. ChRs are probably among two most important proteins in applications and opened a new field—optogenetics. A major core of optogenetics is microbial rhodopsins.

Among over 10,000 currently identified potential rhodopsin genes only a few of the related proteins are characterized. This dramatically limits our knowledge of their molecular functions and mechanisms and biotechnological applications. Another limitation is that high-resolution structures and molecular mechanisms of the recently discovered rhodopsins (proteorhodopsin, sodium pump, inward proton pump) are yet unknown or, in the best case, only structures of the ground states are available. This is also the case of native channel rhodopsins which are the first and still major tool of optogenetics.

The amazing scientific and technological potential of rhodopsins is still to be realized. With the fast expansion of the field and the available numerous publications, a growing number of the scientists involved in the corresponding studies require comprehensive and practical knowledge concentrated in one place which will help to realize this challenge.

The chapters of the book provide up-to-date information on the advances made in the field of rhodopsins. The first three chapters introduce the reader to new discoveries and developments which led to an amazing progress in the field of rhodopsins. The following chapters will introduce the reader into the details of the advanced methods used in the field under discussion. They also provide detailed protocols which, as the authors believe, are requested by a wide extensively growing auditorium of the researchers working with rhodopsins and in the related scientific fields. For the convenience of the readers from different scientific fields the chapters are grouped in accordance with the description of the methods. Two chapters are devoted to the methods which are used to search for rhodopsins with potentially new functions and properties. The following six chapters provide detailed information on the methods used to characterize the function and structure with a focus on those necessary for the studies of rhodopsins at molecular level. Most of the new genes were discovered using metagenomics and in many cases the host is unknown or/and cannot be cultured in the laboratory. This fact poses an additional problem in their studies. Therefore, the methods described in these chapters play a major role in the functional characterization of new rhodopsins. Taking into account a huge number of new (not yet studied) rhodopsins, high-throughput methods are required for the studies. Therefore, some of the chapters describe such methods. New cutting-edge technologies, like nano volume high throughput in meso crystallization, time-resolved serial crystallography at X-ray free electron lasers, and fourth generation X-ray synchrotron sources, are also presented in these chapters. Optogenetics is a diamond produced by rhodopsin studies. It revolutionized neuroscience promising new approaches to the treatment of severe diseases and has a potential for a further big application as a tool to control the physiology of different cells. Therefore, the closing chapters of the book describe major advances in the development of this technology and its major applications.

Last but not least, there is little doubt about the huge potential and further great progress in the rhodopsin field and also biotechnological applications of these proteins. New specific applications may require modifications of the methods and protocols described in this volume. To face this challenge each of the chapters has an introduction aiming to provide a laconic but clear and deep understanding of the principles of the methods. This should help the reader to modify the methods and protocols efficiently.

I thank all the authors of this volume of *Methods in Molecular Biology* for their agreement to share their outstanding experience with others and spend considerable time to provide the reader with general and practical knowledge in so effective and concentrated a form.

Grenoble, France *Valentin Gordeliy*

Contents

Contributors

ALEXEY ALEKSEEV • *Research Center for Molecular Mechanisms of Aging and Age-Related Diseases, Moscow Institute of Physics and Technology (National Research University), Dolgoprudny, Russia*

YURI N. ANTONENKO • *Department of Bioenergetics, Belozersky Institute of Physico-Chemical Biology, Lomonosov Moscow State University, Moscow, Russia*

ROMAN ASTASHKIN • *Institut de Biologie Structurale (IBS), Université Grenoble Alpes, CEA, CNRS, Grenoble, France*

TARAS BALANDIN • *Institute of Biological Information Processing (IBI-7: Structural Biochemistry), Forschungszentrum Jülich GmbH, Jülich, Germany; JuStruct: Jülich Center for Structural Biology, Forschungszentrum Jülich GmbH, Jülich, Germany*

ERNST BAMBERG • *Max Planck Institute of Biophysics, Frankfurt am Main, Germany*

THOMAS R. M. BARENDS • *Max Planck Institute for Medical Research, Heidelberg, Germany*

JOHANNA BECKER-BALDUS • *Institute of Biophysical Chemistry, Goethe University Frankfurt, Frankfurt am Main, Germany; Centre for Biomolecular Magnetic Resonance, Goethe University Frankfurt, Frankfurt am Main, Germany*

ODED BÉJÀ • *Faculty of Biology, Technion - Israel Institute of Technology, Haifa, Israel*

VALENTIN BORSHCHEVSKIY • *Research Center for Molecular Mechanisms of Aging and Age-Related Diseases, Moscow Institute of Physics and Technology (National Research University), Dolgoprudny, Russia; Joint Institute for Nuclear Research, Dubna, Russia*

DMITRY BRATANOV • *Institute of Biological Information Processing (IBI-7: Structural Biochemistry), Forschungszentrum Jülich GmbH, Jülich, Germany; JuStruct: Jülich Center for Structural Biology, Forschungszentrum Jülich GmbH, Jülich, Germany*

VOLKER BUSSKAMP • *Department of Ophthalmology, Universitäts-Augenklinik Bonn, University of Bonn, Bonn, Germany*

I-WEN CHEN • *Sorbonne Université, INSERM, CNRS, Institut de la Vision, Paris, France*

KAI CHEN • *Department of Biomedical Engineering, University of Alabama at Birmingham, Birmingham, AL, USA; Division of Cardiovascular Disease, Department of Medicine, University of Alabama at Birmingham, Birmingham, AL, USA*

VADIM CHEREZOV • *Bridge Institute, Department of Chemistry, University of Southern California, Los Angeles, CA, USA*

IGOR CHIZHOV • *Institute for Biophysical Chemistry, Hannover Medical School, Hannover, Germany*

R. BRUCE DOAK • *Max Planck Institute for Medical Research, Heidelberg, Germany*

MARTIN ENGELHARD • *Department Structural Biochemistry, Max Planck Institute of Molecular Physiology, Dortmund, Germany*

PATRICK ERNST • *Department of Biomedical Engineering, University of Alabama at Birmingham, Birmingham, AL, USA; Division of Cardiovascular Disease, Department of Medicine, University of Alabama at Birmingham, Birmingham, AL, USA*

LIEF E. FENNO • *Departments of Psychiatry and Neuroscience, University of Texas Austin Dell Medical School, Austin, TX, USA*

SHIQIANG GAO • *Department of Neurophysiology, Institute of Physiology, Biocenter, University of Wuerzburg, Wuerzburg, Germany*

CLEMENS GLAUBITZ • *Institute of Biophysical Chemistry, Goethe University Frankfurt, Frankfurt am Main, Germany; Centre for Biomolecular Magnetic Resonance, Goethe University Frankfurt, Frankfurt am Main, Germany*
VALENTIN GORDELIY • *Institut de Biologie Structurale (IBS), Université Grenoble Alpes, CEA, CNRS, Grenoble, France*
ALEXANDER GOREL • *Max Planck Institute for Medical Research, Heidelberg, Germany*
MARIE LUISE GRÜNBEIN • *Max Planck Institute for Medical Research, Heidelberg, Germany*
IVAN GUSHCHIN • *Research Center for Molecular Mechanisms of Aging and Age-Related Diseases, Moscow Institute of Physics and Technology (National Research University), Dolgoprudny, Russia*
ROUHOLLAH HABIBEY • *Department of Ophthalmology, Universitäts-Augenklinik Bonn, University of Bonn, Bonn, Germany*
HIDEKI KANDORI • *Department of Life Science and Applied Chemistry, Nagoya Institute of Technology, Showa-ku, Nagoya, Japan*
MARCO KLOOS • *Max Planck Institute for Medical Research, Heidelberg, Germany*
GABRIELA NASS KOVACS • *Max Planck Institute for Medical Research, Heidelberg, Germany*
KIRILL KOVALEV • *Institut de Biologie Structurale (IBS), Université Grenoble Alpes, CEA, CNRS, Grenoble, France; Institute of Biological Information Processing (IBI-7: Structural Biochemistry), Forschungszentrum Jülich GmbH, Jülich, Germany; JuStruct: Jülich Center for Structural Biology, Forschungszentrum Jülich GmbH, Jülich, Germany; Research Center for Molecular Mechanisms of Aging and Age-Related Diseases, Moscow Institute of Physics and Technology (National Research University), Dolgoprudny, Russia; Institute of Crystallography, University of Aachen (RWTH), Aachen, Germany*
CLARA NASSRIN KRIEBEL • *Institute of Biophysical Chemistry, Goethe University Frankfurt, Frankfurt am Main, Germany; Centre for Biomolecular Magnetic Resonance, Goethe University Frankfurt, Frankfurt am Main, Germany*
RIVKA LEVY • *Department of Neurobiology, Weizmann Institute of Science, Rehovot, Israel*
CHANGYANG LINGHU • *McGovern Institute for Brain Research, MIT, Cambridge, MA, USA*
XIAOGUANG MARGARET LIU • *Department of Biomedical Engineering, University of Alabama at Birmingham, Birmingham, AL, USA*
THOMAS MAGER • *Advanced Optogenes Group, Institute for Auditory Neuroscience and InnerEarLab, University Medical Center Gottinggen, Gottingen, Germany; Cluster of Excellence "Multiscale Bioimaging: From Molecular Machines to Networks of Excitable Cells" (MBExC), University of Goettingen, Goettingen, Germany*
NINA L. MALIAR • *Research Center for Molecular Mechanisms of Aging and Age-Related Diseases, Moscow Institute of Physics and Technology (National Research University, Dolgoprudny, Russia*
YASUHISA MIZUTANI • *Department of Chemistry, Graduate School of Science, Osaka University, Osaka, Japan*
GEORG NAGEL • *Department of Neurophysiology, Institute of Physiology, Biocenter, University of Wuerzburg, Wuerzburg, Germany*
IVAN OKHRIMENKO • *Research Center for Molecular Mechanisms of Aging and Age-Related Diseases, Moscow Institute of Physics and Technology (National Research University), Dolgoprudny, Russia*
ALINA PUSHKAREV • *Faculty of Biology, Technion - Israel Institute of Technology, Haifa, Israel*

FRANCISCO RODRIGUEZ-VALERA • *Research Center for Molecular Mechanisms of Aging and Age-Related Diseases, Moscow Institute of Physics and Technology (National Research University), Dolgoprudny, Russia; Evolutionary Genomics Group, Departamento de Producción Vegetal y Microbiología, Universidad Miguel Hernández, San Juan de Alicante, Alicante, Spain*

TATYANA I. ROKITSKAYA • *Department of Bioenergetics, Belozersky Institute of Physico-Chemical Biology, Lomonosov Moscow State University, Moscow, Russia*

RICCARDO ROSSELLI • *Departamento de Fisiología, Genética y Microbiología, Facultad de Ciencias, Universidad de Alicante, Alicante, Spain*

ILME SCHLICHTING • *Max Planck Institute for Medical Research, Heidelberg, Germany*

KRITIKA SHARMA • *Department of Ophthalmology, Universitäts-Augenklinik Bonn, University of Bonn, Bonn, Germany*

OR A. SHEMESH • *McGovern Institute for Brain Research, MIT, Cambridge, MA, USA; Department of Neurobiology, Pittsburgh Institute for Neurodegenerative Diseases, University of Pittsburgh School of Medicine, Pittsburgh, PA, USA*

ROBERT L. SHOEMAN • *Max Planck Institute for Medical Research, Heidelberg, Germany*

SERGEY A. SILETSKY • *Department of Molecular Energetics of Microorganisms, Belozersky Institute of Physico-Chemical Biology, Lomonosov Moscow State University, Moscow, Russia*

DMYTRO SOLOVIOV • *Institute for Safety Problems of Nuclear Power Plants, NAS of Ukraine, Chornobyl, Ukraine; Faculty of Physics, Adam Mickiewicz University, Poznan, Poland*

JOHANNES STRIEBEL • *Department of Ophthalmology, Universitäts-Augenklinik Bonn, University of Bonn, Bonn, Germany*

DIMITRII TANESE • *Sorbonne Université, INSERM, CNRS, Institut de la Vision, Paris, France*

YUEHUI TIAN • *Department of Neurophysiology, Institute of Physiology, Biocenter, University of Wuerzburg, Wuerzburg, Germany; Environmental Microbiomics Research Center, School of Environmental Science and Engineering, Southern Marine Science and Engineering Guangdong Laboratory (Zhuhai), Sun Yat-sen University, Guangzhou, China*

DMYTRO VOLKOV • *Institute of Biological Information Processing (IBI-7: Structural Biochemistry), Forschungszentrum Jülich GmbH, Jülich, Germany; JuStruct: Jülich Center for Structural Biology, Forschungszentrum Jülich GmbH, Jülich, Germany*

OFER YIZHAR • *Department of Neurobiology, Weizmann Institute of Science, Rehovot, Israel*

DMITRII ZABELSKII • *Research Center for Molecular Mechanisms of Aging and Age-Related Diseases, Moscow Institute of Physics and Technology (National Research University), Dolgoprudny, Russia*

VALERIA ZAMPINI • *Sorbonne Université, INSERM, CNRS, Institut de la Vision, Paris, France*

LUFANG ZHOU • *Department of Biomedical Engineering, University of Alabama at Birmingham, Birmingham, AL, USA; Division of Cardiovascular Disease, Department of Medicine, University of Alabama at Birmingham, Birmingham, AL, USA*

EGOR ZINOVEV • *Research Center for Molecular Mechanisms of Aging and Age-Related Diseases, Moscow Institute of Physics and Technology (National Research University), Dolgoprudny, Russia*

Chapter 1

Microbial Rhodopsins

Valentin Gordeliy, Kirill Kovalev, Ernst Bamberg, Francisco Rodriguez-Valera, Egor Zinovev, Dmitrii Zabelskii, Alexey Alekseev, Riccardo Rosselli, Ivan Gushchin, and Ivan Okhrimenko

Abstract

The first microbial rhodopsin, a light-driven proton pump bacteriorhodopsin from *Halobacterium salinarum* (*Hs*BR), was discovered in 1971. Since then, this seven-α-helical protein, comprising a retinal molecule as a cofactor, became a major driver of groundbreaking developments in membrane protein research. However, until 1999 only a few archaeal rhodopsins, acting as light-driven proton and chloride pumps and also photosensors, were known. A new microbial rhodopsin era started in 2000 when the first bacterial rhodopsin, a proton pump, was discovered. Later it became clear that there are unexpectedly many rhodopsins, and they are present in all the domains of life and even in viruses. It turned out that they execute such a diversity of functions while being "nearly the same." The incredible evolution of the research area of rhodopsins and the scientific and technological potential of the proteins is described in the review with a focus on their function–structure relationships.

Key words Rhodopsin, Retinal, Membrane protein, Ion transport, Optogenetics

1 Introduction

1.1 Historical Overview

The long time period starting from the discovery by W. Stoeckenius and D. Oesterhelt in 1971 of the first microbial rhodopsin, a light-driven proton pump bacteriorhodopsin (*Hs*BR) [1], could be divided into two periods.

The first period lasted until 1999. During that time, only a few rhodopsins were known, which were exclusively archaeal proteins. It was established that these light-sensitive α-helical membrane proteins comprise the retinal molecule as a prosthetic group (chromophore). They function as proton and chloride pumps and also as photosensor components of two-component signaling systems [2–4]. Very successful efforts of several excellent research teams pushed forward the rhodopsin research to an outstanding level in the membrane protein field. Moreover, groundbreaking discoveries in

Valentin Gordeliy (ed.), *Rhodopsin: Methods and Protocols*, Methods in Molecular Biology, vol. 2501,
https://doi.org/10.1007/978-1-0716-2329-9_1,

this field promoted the development of new biophysical and biochemical methods. There were also growing expectations that the light-driven proteins could be used to develop new technologies. Outstanding attempts to obtain the structure of *Hs*BR resulted in the development of the cryogenic electron microscopy (cryoEM) by R. Henderson and colleagues [5] and the invention of lipidic cubic phase (LCP) crystallization in 1996 by J. Rosenbush and E. Landau [6]. Finally, the structure of *Hs*BR was solved in 1997 by E. Pebay-Peyroula and colleagues [7, 8]. Then within the following years, the structures of the intermediate states were solved [9–14]. Seemingly, no questions about bacteriorhodopsin structures were left unanswered. However, there were two serious reasons to revise the *Hs*BR structures.

First of all, the structural data and corresponding conclusions of three research groups, working on molecular mechanisms of proton pumping, were controversial [15–17]. Second, a visible X-ray-radiation-induced damage of *Hs*BR was detected even under low exposure time [18]. This observation became possible only due to the use of high-quality crystals of the protein [18]. Some complications also stemmed from merohedral twinning of *Hs*BR crystals belonging to the $P6_3$ space group. The $P6_3$ crystals were mostly used by researchers since they diffracted to the highest resolution; however, with a rare exception, the crystals were highly twinned, which reduced the quality of the data. A comprehensive study of both problems found ways to overcome this drawback [13, 18–21]. Only when the X-ray data are free from these problems and simultaneously extend to atomic resolution can one obtain a consistent picture of the molecular mechanisms of action of rhodopsins. Nevertheless, some of the authors were still quite optimistic about the quality of the above-mentioned data [22]. Later the problems were again reconsidered to explain the importance of time-resolved serial crystallography studies using X-ray free electron lasers (XFELs) and X-ray synchrotron sources [17, 23–25]. Still, the key problem of the approach is that nearly at any given time of a rhodopsin functional photocycle, the proteins in the crystals are not exactly in the same state. It means that in a general case, time-resolved crystallographic data comprise a mixture of different states (structures), which then must be separated by data treatment. This considerably limits the quality of the crystallographic data of the intermediate states. This becomes a problem even in a conventional approach when an intermediate state is cryo-trapped in crystals. In that case, crystallographic data comprise a mixture of a target (unknown) and the ground (usually known) states. Another problem occurs in time-resolved crystallography of rhodopsins, both at synchrotrons and XFELs, which considerably reduces the quality of crystallographic data, when there is a mixture of at least two unknown states. It means that ultrahigh resolution of time-resolved crystallographic data is required for resolving a reliable molecular picture of the proteins at work. The currently available resolution of time-resolved crystallographic data is about 2 Å, which can be insufficient to resolve the tiny (but key!) details of the structures, as is seen from the history of conventional structural studies of the intermediate states [17, 23, 25]. Several other serious problems of time-resolved crystallography of rhodopsins were recently discussed in detail in the paper by the group of I. Schlichting [25]. We believe that the time-resolved approach has a considerable potential; however, it will take time to fully realize its promises.

Returning to the first period of the rhodopsin era, we might conclude that, in general, a widespread opinion was that the age of the active research of rhodopsins is nearly over. However, nature plays tricks on us. A new microbial rhodopsin era started just at the turn of the century when the first bacterial rhodopsin, a proton pump, was discovered [26]. A little bit later, it appeared that there are unexpectedly many rhodopsins, and they are present in all the domains of life and even in viruses. It appeared that they execute such a diversity of functions while being "nearly the same." However, in the late 1990s, nobody could predict groundbreaking discoveries of other rhodopsin functions, and among them, probably one of the most significant, the discovery of channelrhodopsins (ChRs). For life science, it was the discovery of the proteins which were different from any other known channels. For neuroscience, it was the most significant luck. Cation and anion-conducting ChRs are probably among the two most important types of proteins in neurophysiological applications.

Thus, the second period started in 2000 when the first bacterial rhodopsin, a proton pump, was discovered [26]. Then it appeared that there are unexpectedly many rhodopsins, and as it was already mentioned, they are present in all the domains of life and even in viruses. It was also unexpected that these "seemingly similar" proteins can execute diverse functions. Groundbreaking discoveries and characterization of cation ChRs in 2002–2003 [27], light-driven sodium pump in 2013 [28], enzymerhodopsins, such as rhodopsin-guanylyl cyclase in 2014 [29], anion ChRs in 2015 [30], and inward proton pumps xenorhodopsins in 2016 [31]/2017 [32] support this conclusion. Despite their diversity, all of them are oriented in membranes with their N-termini facing the outside of the cells. Recently Pushkarev et al. discovered a distinct abundant family of rhodopsins named heliorhodopsins (HeRs) [33]. The N-termini of HeRs are exposed to the cytoplasm of the cell. The structure of two proteins belonging to this family was recently solved [34–36].

It is incredible how evolution was using the "minimum" protein modification to engineer rhodopsins with such different functions. Certainly, rhodopsins may provide a unique opportunity to understand better the molecular aspects of evolution, and they may find new important applications. Global abundance and diversity of their function tell us that these proteins are not local but global players, which might influence the ecology and climate of our planet [37].

Unfortunately, little is known on the mechanisms of rhodopsin function. It is surprising, but the fact is that only some of the above-mentioned proteins have been structurally well characterized. Indeed, mostly the structures of the ground state of these rhodopsins were determined. Moreover, the physiological roles of most of these rhodopsins (sodium pumps, inward proton pumps, heliorhodopsins, and viral rhodopsins) are yet unknown.

To summarize, rhodopsins are the most abundant light-harvesting proteins on the Earth, performing very diverse functions and present in all the domains of life, and their genes are also present in viruses [38]. The discovery, in 2000, of the light-driven pump proteorhodopsin in marine sediments triggered an extensive search for metagenomes of light-activated proteins [15]. More than 10,000 rhodopsin genes have already been identified in archaea, bacteria, unicellular eukaryotes, and viruses, although the molecular mechanisms of function, as well as the biological roles of most of these proteins, remain elusive. It is also true for viral rhodopsins that form two distinct groups encoded by Nucleocytoplasmic Large DNA Viruses (NCLDV) that infect algae. The biological role of these rhodopsins discovered by E. Koonin and N. Yutin [38] also remains unknown [39, 40].

In the next sections, we will summarize in more detail the current knowledge about the phylogeny and the search for new rhodopsins, function and structure of the proteins discovered in the twenty-first century (proteorhodopsins, channelrhodopsins, enzymes, sodium pumps, outward proton pumps, heliorhodopsins, and viral rhodopsins). We will also discuss the function–structure relationship between different rhodopsins, which may help to understand better the evolution of rhodopsins and the molecular mechanisms of their function. This is important for understanding the "basic laws" of rhodopsin evolution and their global role in life on Earth. Finally, we will speculate on the future of rhodopsin research and its impact on different scientific fields and applications.

2 Phylogeny and New Rhodopsins

Rhodopsins are extremely widespread convertors of solar energy. They have probably been around for a very long evolutionary time [41]. This protein family has evolved for so long and in so many branches of the microbial phylogenetic tree that sequence similarity between some of the family members has been all but completely lost (Fig. 1). Only the size and the overall 3D structure have been well conserved. The other characteristic that has been conserved

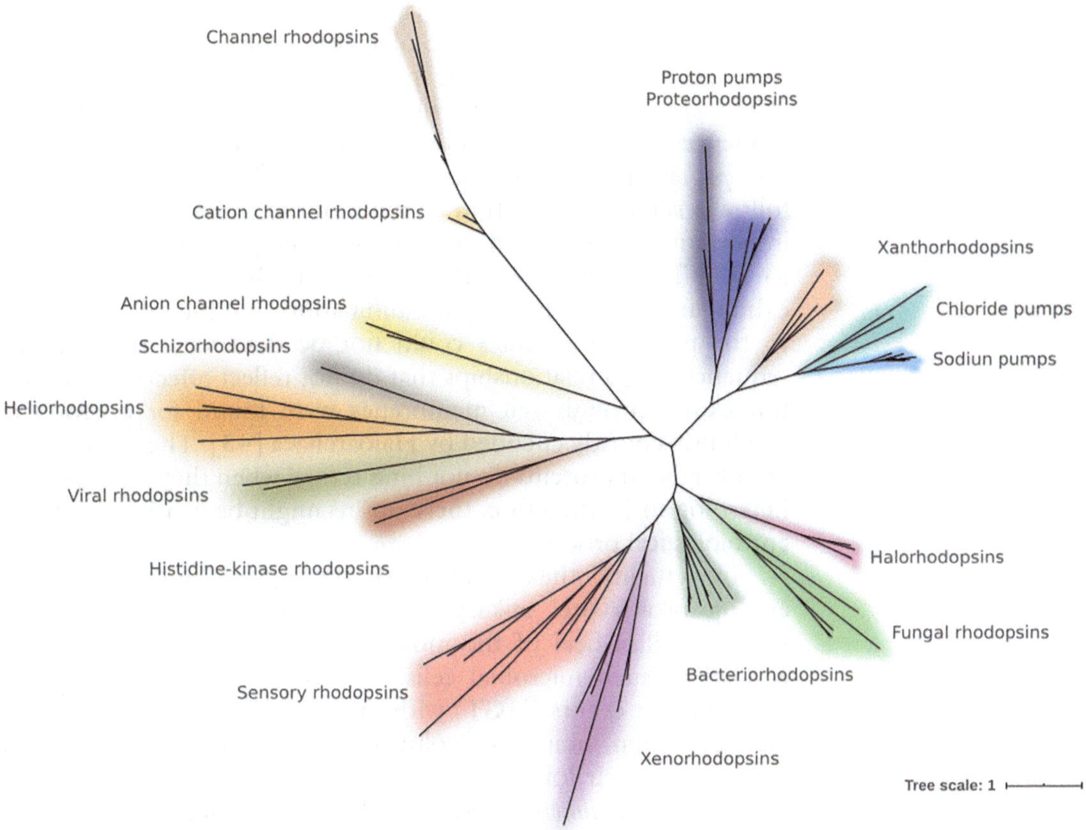

Fig. 1 Phylogenetic tree of type-1 rhodopsins. Obtaining a reliable phylogenetic topology that includes all rhodopsins is not an easy task. Evolutionary divergences and sequence dissimilarities determine loss of the phylogenetic signal and decrease the robustness of the topology in terms of statistical support (i.e., bootstrap). Nevertheless, rhodopsins with similar physicochemical properties tendentially constitute distinct clades, and therefore it is possible to formulate functional hypotheses about new proteins by using the already known information. Type-1 rhodopsins can be roughly divided into seven major clades: (1) Bacteria- and Archaea-rhodopisns, that includes proton pumps and proteorhodopsins (i.e. blue- and green-absorbing proteorhodopsins, BPR and GPR), xanthorhodopsins (XR), sodium- and chloride pumps (i.e., IaNaR, BaClR). (2) Rhodopsins encoded by Halobacteriales archaea. This lineage is constituted by halorhodopsins (HR), bacteriorhodopsins (BR), and sensory rhodopsins (SRI, SRII). Inward-pump xenorhodopsins appear related to Halobacteriales sensory rhodopsins, although they were identified in five different bacterial phyla. (3) Channel rhodopsins encoded by microeukaryotes. The lineage includes mono/divalent cation- (i.e., CrChR2) and anion-channelrhodopsins (ACRs, such as GtACR1). (4) Histidine-kinase rhodopsins (i.e., HKR1). (5) Two types of viral rhodopsins encoded by viruses that prey on unicellular algae. (6) Heliorhodopsins (HeR) that differ from other type-1 rhodopsins (and type-2) for the opposite orientation in the plasma membrane (N-termini facing the cytoplasm, and extracellular C-termini). (7) Schizorhodopsins (SzR), placed between HeR and the other type-1 rhodopsins. These rhodopsins are encoded by a particular group of Archaea (Asgardarchaeaota) that has been hypothesized to occupy a basal position of the eukaryote tree of life, and share common features with both, type-1 and HeR rhodopsins

(till now) is the use of retinal as a cofactor. Even the latter has been challenged recently in the case of a large group of rhodopsins that were identified from marine mesopelagic (dark ocean) samples [42]. The phylogenetic analysis conducted using amino acid sequences placed these rhodopsins within the proteorhodopsins group. The proteins possess the unusual DTT motif (here we follow a standard definition of the motif as three key functionally important amino acids in most of the rhodopsins, that correspond to D85-T89-D96 in *Hs*BR numbering). However, they did not exhibit any activity upon light illumination, and the retinal binding to the protein molecules could not be detected either. Similar sequence-based relationships that do not reflect effective molecule trafficking through cell membranes were found for a group of rhodopsins mostly encoded by Haloarchaea [43]. These phylogenetically related proteins do not bind retinal when they are expressed heterologously; therefore, these opsins might be involved in different biological processes.

In actuality, the evolution of rhodopsins and their spread among prokaryote, eukaryote and virus genomes represent a major Achilles heel of the molecular chronometer theory when applied to such microbial cells. The combination of horizontal gene transfer events (HGT) with molecular tinkering and convergent evolution, blurs the evolutionary trajectories that they have followed in their long histories as species [44–46]. A remarkable example is *Salinibacter ruber* [47], a hyperhalophilic bacterium that lives in the same environment (NaCl saturated brines) as the microbe in which the microbial rhodopsins were discovered: the archaeon *Halobacterium salinarum* [1] (Fig. 2). In spite of the wide phylogenetic chasm between the microbes, both species carry four rhodopsin genes with highly similar functions [50]. The proton pump SRU_1500 of *S. ruber* is related to other bacterial rhodopsins that sometimes use a carotenoid antenna as accessory pigment, the xanthorhodopsins [51, 52]. Xanthorhodopsins are distantly related to the haloarchaeal proton pumps, and therefore it is a case of convergence. The other three rhodopsins appear to have a more complex origin involving HGT and molecular tinkering. The halorhodopsin gene SRU_2780 appears phylogenetically related to the *Halobacterium* one, and keeps a very similar genomic context. Like in haloarchaea, immediately upstream of the halorhodopsin gene, there is an oxidoreductase-encoding gene (possibly a regulator of pump activity) [50]. Also, in both groups a potassium symporter gene *trk*A is found nearby. These observations suggest that *S. ruber* would have acquired these genes through HGT from a haloarchaeon, and that the activity of halorhodopsin might be critical to survive in NaCl saturated waters with a "*salt-in strategy*" [53, 54], reviewed by [55]. The evolutionary history of the other two rhodopsins of *Salinibacter* appears even more remarkable. It involves not only genetic exchanges between two completely

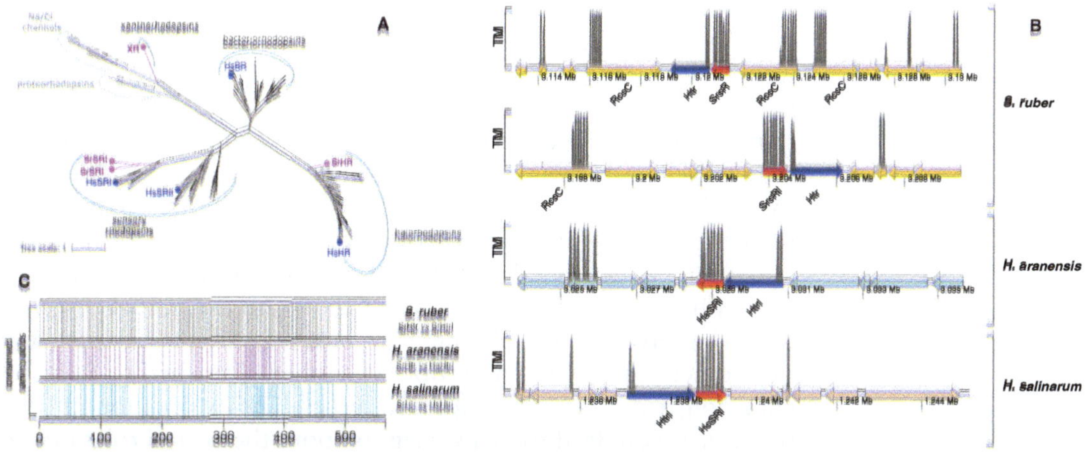

Fig. 2 Comparison between Halobacteriales and *Salinibacter* rhodopsins. (**a**) The phylogenetic reconstruction was inferred by retrieving rhodopsins from complete *Salinibacter ruber* genomes (eight strains) and Halobacteria genomes (75 strains) available at NCBI. The draft genome of *Halovenus aranensis* (RefSeq assembly GCF_900100385.1) was also included in the dataset. The rhodopsins of *H. aranensis* genome are the closest relatives to the *S. ruber* ones. The main groups of halophilic Archaea rhodopsins are bacteriorhodopsins (BR), halorhodopsins (HR), and sensory rhodopsins (SR). Rhodopsins encoded exclusively by *S. ruber* belong to the xanthorhodopsin group (XR). *S. ruber* strains also encode two sensory rhodopsins, *Sr*SR and *Sr*SRI, that are related to the SRI of *Halobacterium salinarum* (*Hs*SRI) and other haloarchaea. No homologs of *Hs*SRII have been identified in the genome of *S. ruber*. The other rhodopsin encoded by *S. ruber*, *Sr*HR, is related to the HR group. (**b**) The genomic contest in which *S. ruber* sensory rhodopsins SrSR and SrSRI are located is compared with that of *H. aranensis* and *H. salinarum* (five genes upstream and downstream the rhodopsin-encoding gene are included in the figure). SR rhodopsins (red arrows) and transducers Htr (blue arrows) are shown, as well as the respective gene orientation. Dark gray vertical bars represent transmembrane domains (TMs) as they appear in the translated protein (the prediction was performed using Phobius [48] at http:/phobius.sbc.su.se/). Several sensor-kinases (homologous to RcsC of *E. coli*) are found close to both *Sr*SR and *Sr*SRI. These proteins are likely to be involved in the phototransfer signaling pathway, thus constituting what might be defined sensory-genomic islands. (**c**) Sequence-alignments (Clustal Omega [49] at www.ebi.ac.uk) of *S. ruber* transducers *Sr*Htr and *Sr*HtrI and comparison with *H. aranensis* and *H. salinarum* Htr transducers. Only conserved, identical amino acids are shown

different organisms but also molecular tinkering, i.e. the adaptation of other genes to adjust to the novel function gained by HGT. These two rhodopsins, *Sr*SR and *Sr*SRI (SRU_2511, SRU_2579), are similar to the sensory rhodopsin I (*Hs*SRI), which mediates positive phototaxis adjusted to red light in *H. salinarum*. The genome of *S. ruber* lacks a homolog of *Hs*SRII, the other sensory rhodopsin encoded by *H. salinarum* that is photophobic to blue light. However, its sensory rhodopsins have transducers (the proteins involved in directing flagellar motion, as happens in *Hs*SRI and *Hs*SRII) that are more similar by sequence to the chemoreceptors of bacteria [50]. It is important to remark that archaeal and bacterial flagella have radically different structure and function, and could not be controlled by a similar transducing pathway. Unfortunately, there has been very little work carried out to clarify the phototactic behavior of *S. ruber*, considering that the sensory rhodopsins that it encodes have low identity between them (52%) as well as between their transducers (62%). In addition, it is not known whether transducers replicate the function of the *Hs*SRI transducer. If they replicate its induced behavior, i.e. attraction by long wavelength light, it remains unclear why there are two relatively similar receptors with two relatively different transducers within two different genomic contexts (Figs. 1 and 2).

It can be concluded that the basic molecular design of rhodopsins with the seven transmembrane helices and the retinal chromophore has provided membranes with bioenergetics, behavioral, and ion homeostatic swiss-knife tool, which has diversified to fulfill an enormously diverse set of biological roles. The exploration of their biodiversity can lead to many insights about the critical roles played by membrane proteins in cell biology. It is also essential for the escalation of the number of applications of microbial rhodopsins in optogenetics or other developing biotechnological fields.

3 Functions

3.1 Proton Pumps

Light-driven proton pumps take a distinguished place among all membrane proteins since the discovery of the first microbial rhodopsin, bacteriorhodopsin from archaea *Halobacterium salinarum* (*Hs*BR) in 1971 [1]. *Hs*BR became the first membrane protein whose three-dimensional structure was deciphered. The molecular mechanism of the *Hs*BR outward proton pumping has been scrutinized in several research publications, and here we would refer to dedicated reviews on bacteriorhodopsin functional and structural features [15, 22, 56]. Currently, as was mentioned, microbial rhodopsin superfamily includes more than 10,000 genes across all domains of life, with a significant part of light-driven proton pumps among them. Following the phylogeny of microbial rhodopsins, light-driven proton pumps have been separated into two major groups, associated with archaeal and bacterial families of microbial rhodopsins [57, 58]. Recently, one of the microbial rhodopsin subfamilies called xenorhodopsins (XeRs) was shown to exhibit inward proton pumping ability, thus broadening the list of microbial rhodopsin functional roles in nature [31, 32]. Below we will describe major functional groups of rhodopsins to provide a basis for understanding the diversity of functions and their molecular mechanisms.

3.1.1 Archaeal Outward Proton Pumps (Bacteriorhodopsins)

The best-studied microbial rhodopsin *Hs*BR is a classical representative of bacteriorhodopsins (BRs)—archaeal light-driven outward proton pumps. These proteins set standards for microbial rhodopsin photochemistry, such as a single retinal peak in absorption spectrum ($\lambda_{max} = 567$ nm for *Hs*BR) and a millisecond-timescale photocycle [59]. After illumination, *Hs*BR goes through seven intermediates, namely I, J, K, L, M, N, and O states, where the M-state exhibits a large blue-light absorption shift ($\lambda_{max} = 410$ nm), associated with the deprotonated state of the retinal Schiff base (RSB). At least four photointermediates (K-, L-, M-, and N-states) were found to be associated with structural rearrangements observed in X-ray crystallographic experiments. The photocycle of *Hs*BR decays within 20 ms and shows little response to pH change and buffer composition; however, it is altered significantly depending on the temperature and lipid environment [59]. In the dark, bacteriorhodopsin exists in a stable dark-adapted state ($\lambda_{max} = 560$ nm) that partially undergoes a transition into light-adapted state ($\lambda_{max} = 567$ nm) after light exposure. The light-adapted state is associated with all-*trans* 15-*syn* configuration of the retinal, whereas the dark-adapted state is a 1:1 mixture of the all-*trans* 15-*syn* and 13-*cis* 15-*syn* chromophore states [60]. The crystal structure of *Hs*BR revealed an asymmetric hydration pattern, associated with vectorality of proton transfer mediated by hydrogen bonds. The role of highly coordinated water molecules in the near vicinity of the RSB is crucial for proton transport, as shown by Fourier-transform infrared (FTIR) spectroscopy [61, 62]. Despite the well-studied biochemistry, *Hs*BR found very limited application in neuroscience and optogenetics, mainly due to the poor targeting to plasma membranes of heterologous hosts. However, archaeorhodopsin-3 (Arch), another light-driven proton pump that shares 60% homology with *Hs*BR, was proven to be one of the most efficient neural silencers before the discovery of anion-conducting channelrhodopsins [30, 63]. Despite the fact that the molecular mechanism of Arch is similar to *Hs*BR, Arch demonstrates a superior genetic targeting in the plasma membrane of mammalian cells, while maintaining high structural stability and robust functionality [63].

3.1.2 Bacterial Outward Proton Pumps (Proteorhodopsins)

In 1999, a gene of a microbial rhodopsin with proton pumping ability was found in marine bacteria genome. The protein was named proteorhodopsin (PR), the first bacterial light-driven proton pump [26]. Since that time, it was established that proteorhodopsins (PRs) are the most abundant and widely distributed subfamily of microbial rhodopsins, now found in many bacterial taxa and some other nonbacterial organisms [64–66]. PRs are generally classified into two groups: green-light-absorbing (GPR, $\lambda_{max} \sim 525$ nm) and blue-light-absorbing (BPR, $\lambda_{max} \sim 490$ nm) proteins according to their spectral characteristics correlated with their habitat in the world ocean. Unlike BRs that are capable of pumping protons in a wide pH range, PRs have unusually high pK_a of the counterion of the RSB (GPR D97 $pK_a = 7.1$) that results in the blockage of the proton transfer under neutral and acidic pH

[67]. This feature of PR photochemistry is associated with a highly conserved histidine residue on helix B, which is both interconnected with the aspartate counterion of the RSB and exposed to the solvent [66, 68]. For instance, in GPR the histidine (H75) imidazole group interacts directly with the carboxyl group of the primary proton acceptor from the RSB (D97), which causes the bimodal behavior of the protein [69]. At basic pH the PR photocycle enters a proton translocation mode with five photointermediates: K, M_1, M_2, N, and O-states with a 150 ms total length [70]. At neutral and acidic pH the PR photocycle consists of K, L, and N-states with the total length of 25 ms and lacks an apparent M-like state that can be correlated with the active state of the protein [71]. Despite their abundance in marine organisms and importance for evolutionary studies, PRs have found very limited application in neuroscience, partially due to their pH sensitivity.

3.1.3 Inward Proton Pumps (Xenorhodopsins, Schizorhodopsins)

In 2011 the D217E mutant of *Anabaena* sensory rhodopsin (ASR) was reported to exhibit the inward proton pumping [72]. After that a novel clade of microbial rhodopsin, dubbed xenorhodopsins (XeRs), was discovered [73]. XeRs are phylogenetically related to bacterial sensory rhodopsins and have an unusual combination of functionally relevant residues [74]. Later, XeRs were shown to function as natural light-driven inward proton pumps [31, 32]. The photocycle of XeRs has four different intermediate states: K, L, M_1, and M_2, where M_1 and M_2 states are the likely candidates to be associated with structural rearrangements during the proton transfer. Besides that, a xenorhodopsin from *Nanosolina* (*Ns*XeR) is capable of triggering action potentials in rat hippocampal neurons at 532 nm with the frequency of at least 40 Hz [32]. The ability of XeRs to depolarize the cell membrane without altering any other ion concentrations presents an important alternative approach to neuronal applications enabling complete blockage of ion transport (i.e. calcium) inside or outside the cell.

In 2019, a new clade of schizorhodopsins (SzRs) were found in Asgard archaea, the archaeal group closest to eukaryotes [75]. Further characterization revealed that schizorhodopsins function as inward proton pump, despite low phylogenetic similarity between SzRs and XeRs [76]. However, both groups share similar trimeric architecture and H^+ transport mechanism, which suggests that SzR and XeRs evolved convergently to similar function. Unlike XeRs, photocycle of SzRs have only one apparent M-state and comprise three intermediate states (K, L, M). Recently, the characterization of the subfamily of schizorhodopsins, Antarctic rhodopsins (AntRs), was reported [77].

3.2 Cation-Conducting Channelrhodopsins (CCRs)

Rhodopsins with unknown function were initially discovered in the genome of chlorophyte algae *Chlamydomonas reinhardtii* [27, 78–80]. For a long time scientists were investigating the phototaxis of green algae, and it turned out that two rhodopsin genes are responsible for its regulation in *Chlamydomonas reinhardtii*. Now these two rhodopsins are known as ChR1 and ChR2.

3.2.1 Chlorophyte Cation-Conducting Channelrhodopsins

Heterologous expression in *Xenopus laevis* oocytes of both ChR1 and ChR2 genes allowed to understand their function. ChR1 was reported to be the first known rhodopsin that functions as a channel selectively permeable for protons [27]. ChR1 action spectrum peaks at 500 nm. Illumination with light leads to the opening of a pore for protons to go down the electrochemical gradient.

Later ChR2 was reported to be a light-gated cation channel [78]. In the same work it was shown that ChR2 is able to depolarize the cells to the level sufficient to fire neurons, which potentially meant a new era in neuroscience. At the same time the authors of that work patented the protein as an optogenetic tool describing in detail an optogenetic protocol to study neurons. Both ChR1 and ChR2 comprise seven transmembrane helices and a large soluble intracellular part. The transmembrane rhodopsin-like part is fully responsible for channel activity. At the moment, the function of water-soluble cytoplasmic parts is not known. However, it was recently reported that multiple reversible phosphorylation of ChR1 could be important for photoreception in *Chlamydomonas reinhardtii*, which is disturbed when the C-terminus is truncated [81].

ChR2 is rapidly opened upon illumination (~0.2 ms) and then it is desensitized to a steady current which is approximately three times smaller than the peak current. Furthermore, ChR2 was proven to be capable of depolarizing the cells. Later, it was reported that illumination of hippocampal neurons expressing ChR2 indeed induces depolarization capable of firing action potentials [82]. This discovery was a revolution for neuroscience which now had a new powerful tool for investigation—optogenetics.

Search for new channelrhodopsins became important for the last 15 years. Since ChR2 has absorption maximum at blue light (480 nm) and it is not fast enough for some neuroscientific applications, search for natural opsins and engineering of ChRs with the desired properties helped to solve some of the problems. Currently, optogenetic toolbox includes a broad range of channelrhodopsins with different action spectra, inactivation time, and selectivity for numerous optogenetic applications (Table 1).

In 2008, the channelrhodopsin VChR1 from *Volvox carteri* which has red-shifted spectrum (545 nm) and functions also as a cation-selective channel [94] was described. Later a chimeric channelrhodopsin C1V1 (ChR1 and VChR1 chimaera) was characterized [99] with the absorption maximum at 540 nm. Also, two more

Table 1
Description of the main optogenetic tools

Opsin	Host organism	Mode of control	Wavelength sensitivity	Modulatory capabilities	References
Arch/ArchT	Halorub sodomense	Hyperpolarizing	575 nm (maximum activation)	Red-light activated proton pump with robust expression in heterologous host cells. Can be used for neural silencing in various cell types both in vitro and in vivo. Can cause large pHi swings, photocurrent amplitude up to 0.3–0.4 nA.	[63]
LR/Mac	Leptosphaeria maculans	Hyperpolarizing	525 nm (maximum activation)	Blue-light activated proton pump, can be used for neural silencing in vitro and in vivo. Can cause pH_i swings, photocurrent amplitudes up to 0.2–0.3 nA, suitable for multicolor silencing experiments	[63]
NpHR, eNpHR, NpHR 3.0	Natronomonas pharaonis	Hyperpolarizing	580 nm (maximum activation)	Light-activated chloride pump. Can be used to mediate sustained inhibition of neurons over minutes with high temporal precision. Capable of inhibiting single action potentials within high frequency spike trains both in vitro and in vivo.	[83]
ChR2, ChR2 (H134R), ChR2 chimaeras	Chlamydomonas reinhardtii	Depolarising	470 nm (maximum activation), can be varied in chimaeras	Rapid and precise neural activation tool sensitive to blue light. Can be used to evoke single spikes and defined trains of action potentials over a range of frequences. H134R mutant has higher photocurrent amplitude, but slower closing kinetics. Various mutants modulating activation spectrum and kinetics are available. Has photocurrent amplitudes up to 2 nA	[82–84]

Step function opsin (SFO) genes: ChR2 (C128A/S/T)	Chlamydomonas reinhardtii	Depolarising	470 nm (switching on), 545 nm (switching off)	Point mutants of ChR2 with slow or optically switchable kinetics. C128 mutations show prolonged activation, long open state lifetime and high light sensitivity. SFO variants can be switched on and off with blue and green light pulses respectively.	[85]
Bacteriorhodopsin-like CCRs (GtCCR4, ChrMine)	Guillardia Theta, metagenomic origin	Depolarising	530 nm (GtCCR4), 585 nm (ChrMine)	Highly effective red-light depolarization tools (ChRMine) that can elicit photocurrents up to 4–6 nA in HEK cells. Can be used for neural activation in vivo. Has almost tenfold higher Na^+/H^+ permeability ration in comparison to ChR2. At the moment one of the most efficient red-shifted activation tools.	[86, 87]
Engineered ACRs (iC1C2, iC++)	Chlamydomonas reinhardtii	Hyperpolarizing	480 nm (maximum activation)	Blue-light activated chloride channel with robust expression in heterologous host cells. Can be used for highly efficient neural inhibition with photocurrent amplitudes up to 3–5 nA.	[88, 89]
Natural ACRs (GtACR1, ZipACR, FLASH)	Guillardia Theta	Hyperpolarizing	515 nm (GtACR1)	Light-activated chloride channels with fast closing kinetics and high photocurrent amplitude efficiency (up to 8 nA). At the moment, one of the most powerful neural inhibition tools both in vitro and in vivo with robust expressions levels in various neuron cell cultures.	[87, 89]

(continued)

Table 1
(continued)

Opsin	Host organism	Mode of control	Wavelength sensitivity	Modulatory capabilities	References
MerMAIDs (MerMAID1, MerMAID6)	Metagenomic origin	Hyperpolarizing	500 nm (MerMAID1)	Intensely desensitized light-activated chloride channels with rapid closing kinetics. Could be used for effective neural inhibition in vitro. Due to the desensitization effect can block single action potentials with high precision in combination with subsequent imaging.	[90]
Xenorhodopsins (NsXeR, PoXeR)	Nanosalina	Depolarising	565 nm (NsXeR)	Red-light activated inward proton pump with robust expression in neural cell cultures. Can be used for a fast neural activation with firing frequency up to 40 Hz. XeRs do not change any intracellular sodium and calcium concentration and can be effectively used in experiment to prevent calcium permeation inside the cell.	[31, 32, 91]
eKR2, eKR2 R109Q	Krokinobacter eikastus	Hyperpolarizing	525 nm (maximum activation), could be shifted to 565 nm with mutations	Light-activated sodium pump that enables precise inhibition of action potential firing. eKR2 does not influence distribution of H^+ and Cl^- ions, that are the most used one for inhibition purposes. eKR2 R109Q mutant converts the protein into a K^+/Na^+ channel that could be used for advanced neural silencing with potassium ions.	[92, 93]

Red-shifted channelrhodopsins (Chrimson, VChR1)	Chlamydomonas noctigama	Depolarising	585 nm (Chrimson), 605 nm (ChrimsonSA)	Red-shifted cation channelrhodopsins. Similar to ChR2, Chrimson can be used to drive reliable action potential over a range of frequences. With 600 nm light, ChrimsonSA can be efficiently used for multicolor activation in combination with ChR2.	[94–96]
Ultrafast channelrhodopsins (vfChrimson, Chronos)	Chlamydomonas noctigama	Depolarising	585 nm (vfChrimson), 500 nm (Chronos)	Point mutants of Chrimson with optimized fast closing kinetics. vfChrimson can be used for a ultrafast neural activation with firing frequency up to 300 Hz (tau off = 2.8 ms). Promising candidates for specific applications such as hearing restoration.	[91, 95]
Enzymorhodopsins (HKR1, RhoPDE, RhGC)	Chlamydomonas reinhardtii	Secondary messengers control	490 nm (HKR1), 530 nm (CaRhGC)	Light-activated enzymes that induce secondary messenger (such as cAMP and cGMP) pathways with blue-green light. Can be expressed in neural cells for optical manipulation over cyclic nucleotide dynamics.	[97, 98]

chimaeras ChEF and ChIEF (ChR1/ChR2 chimaeras) were earlier shown to have an increased steady current [100].

Another approach to obtain ChRs with new properties is mutagenesis. ChR2 H134R has larger photocurrents than those of ChR2, but slower kinetics, which slightly decrease its temporal resolution [101]. Mutations at positions C128 and D156 (DC-gate, C128T/A/S and D156A) make closing kinetics significantly slower (in particular for D156A ~6.9 min). The C128T/A/S variant of ChR2 is called a "step-function opsin" and it can be used to induce prolonged depolarization [102]. A combination of two mutations, C128S and D156A, is called "stabilized step-function opsin." It can be activated with a short 470 nm flash and deactivated with a 590 nm flash (the closing time constant ~29.3 min) [99]. Other mutations helped to obtain fast variants of ChR2: ChR2 E123T is called ChETA and has a lower closing time (~4.4 ms) [85]. Some mutations can alter the selectivity of a channel: the ChR2 L132C mutant is called CatCh and it has six times higher permeability for calcium and 70 times greater light sensitivity compared to ChR2 [103], and the mutation E90K makes ChR2 even anion chloride-selective (ChloC [104]).

In 2013 a red-shifted chimera channelrhodopsin was reported [105]. The protein consists of ChIEF N-terminus, A-E, and G helices of VChR1, and F helix of VChR2 (another channelrhodopsin from *Volvox carteri*) with a mutation L171I. The chimera has a spectral maximum of around 600 nm and larger photocurrents compared to the already known red-shifted channelrhodopsin variants.

Another approach to obtain new optogenetic tools is finding natural opsins with desired properties. In 2014 the survey on the physiological characterization of opsins from over 100 species was done [95]. As a result, from a long list of different channelrhodopsins, the two most promising proteins were selected and named Chrimson (from *Chlamydomonas noctigama*) and Chronos (from *Stigeoclonium helveticum*). Chronos' optimal absorption wavelength is approximately 500 nm, while Chrimson is an extremely red-shifted channelrhodopsin with the absorption maximum at about 600 nm. An advantage of Chronos is its closing kinetics, which is very fast, 3.2 ms (for Chrimson it is ~25 ms). Moreover, these two rhodopsins were shown to allow independent stimulation of different neuronal populations with lights of a different wavelength. Chrimson and Chronos can reliably trigger action potentials utilizing 625 nm and 470 nm light, respectively. At the moment, Chrimson is widely used in optogenetics. Several other mutations of Chrimson were reported, which can considerably accelerate the closing kinetics of the protein [91]: combining the mutations together, scientists obtained the so-called fChrimson (Chrimson Y261F/S267M, ~5.7 ms) and vfChrimson (Chrimson K176R/Y261F/S267M, ~2.7 ms).

3.2.2 Cryptophyte Cation-Conducting Channelrhodopsins

In 2016, 53 proteorhodopsin-like genes were found in a complete nuclear genome of cryptophyte algae *Guillardia theta* [106]. Despite the fact, that all the functional residues at TM3 and TM7 helices coincide with those of archaeal proton pumps, nine proteins were demonstrated to function as cation conducting channelrhodopsins, and named bacteriorhodopsin-like CCRs (BCCRs) due to the same DTD motif as in bacteriorhodopsin according to the residues in functionally relevant positions in *Hs*BR (D85, T89, and D96). BCCRs were found to pass almost exclusively sodium under physiologic conditions with a 37-fold higher Na^+/H^+ selectivity ratio in comparison to those of chlorophyte CCRs. Later, the ChRMine protein from the BCCR clade was shown to have a robust expression in neurons and the ability to trigger action potentials [86]. In cultivated neurons, ChRMine induced ~4 nA excitatory currents under illumination with 585 nm light and demonstrated reliable spiking up to 40 Hz with the red-shifted light under low irradiance values. At the moment, the ChRMine has been shown to be one of the most effective red-light activation tools among channelrhodopsins that have no apparent template for structure-guided mutagenesis. Therefore, deciphering the BCCR structure should be considered as one of the important tasks to be solved in the future. In 2020, 19 more BCCR members were reported, some of which were found to have unusual desensitization and largely blue-shifed M-state (330 nm) [107].

3.2.3 Viral Channelrhodopsins Group 1 (VR1, VirChR1s)

In 2012 it was found that the complete genome of Organic Lake Phycodnavirus contains genes encoding two monophyletically split families of putative proteorhodopsins that were named viral rhodopsin group 1 and 2 [38]. Later, both families were significantly expanded with a number of genes that belong to other giant viruses by means of functional metagenomics [108, 109]. In 2019, viral rhodopsins from group 1 were reported to have a proton-pumping function with 400 ms photocycle length [40]. In 2020, a new study presented data demonstrating that rhodopsins from VR1 group (including VirR1) are cation-conducting channelrhodopsins [110]. Similar to proteorhodopsins, VR1 rhodopsins (or viral channelrhodopsins, VirChR1s) showed strong retention in cytosol when expressed in mammalian cells, which limits the possible applications of VirChR1s without further optimizations. Besides that, VirChR1s were found to be inhibited by moderate calcium concentrations (2 mM Ca^{2+}) that were not previously observed in other CCRs [110].

3.3 Anion-Conducting Channelrhodopsins (ACRs)

After the tremendous success of cation-conducting channelrhodopsins (CCRs) applications in neuroscience as neural cells activators, a major question arose as to whether channelrhodopsins could be converted into anion-conducting neuron silencers. During the recent years it was demonstrated that CCRs can be effectively transformed into ACRs, and also several clades of ACR were found in nature. Both these facts suggest that ACRs like CCRs could be involved in the vision of marine microbes, which requires further research.

3.3.1 Engineered Anion-Conducting Channelrhodopsins (eACRs)

In 2012, the 2.3 Å crystal structure of the cation channelrhodopsin chimera C1C2 revealed an internal light-activated pore of the channel and opened a possibility for enhancing CCRs functional properties by structure-guided site-directed mutagenesis [111]. The first attempts at altering CCRs selectivity mainly targeted multiple glutamate residues at helix 2 (TM2) that are involved in the formation of constriction sites and line ChR2 water-filled pores. In fact, a single replacement of a E90 glutamate with lysine in C1C2 caused an alteration of the channel selectivity to a chloride conductance, however, with some remaining cation conductance [104]. Further efforts directed to the development of more effective eACRs led to the building of ninefold mutated C1C2 variant, called iC1C2, that presented a higher photocurrent without a major pore structure alteration [112]. A big next step for eACRs was made with the creation and structural characterization of enhanced iC1C2 variant, iC++ that demonstrated a 15-fold increase of the photocurrent amplitude comparable with nACRs while maintaining strong expression in vivo [88, 89]. At the moment, dACRs present an important example of functional plasticity of channelrhodopsins and enable even more possibilities for structure-guided mutagenesis to enable a full range of kinetics and absorption for anion channelrhodopsins.

3.3.2 Natural Anion-Conducting Channelrhodopsins (nACRs)

Before 2015, only the cation channelrhodopsins from chlorophyte algae were known and all that have been tested were exclusively cation channels. However, a similar photoreception mechanism was found to be employed by phylogenetically distinct cryptophyte algae [113]. Sequencing of a complete nuclear genome of *Guillardia theta* allowed the identification of the first natural anion-conducting channelrhodopsins *Gt*ACR1 and *Gt*ACR2 [30]. In contrast with the engineered ACRs, nACRs exhibit a greater level of both peak and stationary photocurrents and show practically no permeability for cations conducted by CCRs. Later, the nACR subfamily expanded to 20 functionally confirmed proteins that presented a large variation of photocurrent amplitude, kinetics, and activation spectra [87]. Interestingly, nACRs conserve a glutamate residue at TM2 (E90 in ChR2) that was shown to play a critical role in ChR2 conductance and selectivity [104, 114]. Determination of 2.9 Å resolution structure of *Gt*ACR1 revealed a similar organization of water-accessible pores and constriction sites to those in CCRs, however with altered pore electrostatics [115, 116]. A *Gt*ACR1 photocycle has five distinguishable states: K, L, M, N, and O-states with L- and M-conducting states that generally resemble photocycle kinetics of bacteriorhodopsin [117]. However, *Gt*ACR1 demonstrates bimodal opening and closing of the channel regulated by two different mechanisms that can be described as fast opening-slow closing and slow opening-fast closing reminiscent of those in ChR2 [118, 119]. Besides that, a mechanism of closing kinetics of *Gt*ACR1 was thoroughly studied with mutations [115].

3.3.3 Intensely Desensitized Anion-Conducting Channelrhodopsins (MerMAIDs)

A recent outburst of the available world ocean metagenomic data has brought a whole new impetus to the search for potential channelrhodopsins with new functions. In 2019, a new branch of channelrhodopsins, named MerMAIDs was identified by means of functional metagenomics in contigs assembled from the *Tara Oceans* metagenomes [120, 121]. MerMAIDs have been shown to have a high peak photocurrent amplitude followed by almost complete desensitization under continuous bright light exposure. Similar to cryptophyte ACRs, MerMAIDs demonstrate anion-only channel selectivity and biphasic closing kinetics [90]. The photocycle of MerMAID1 includes three consecutive states: K, L, and M-states, where L and M-states can be associated with short-living conducting and long-living desensitized states. Besides that, MerMAID1 was found to be able to recover stationary photocurrent with a single C84T mutation (analogous to C128T mutation in ChR2), which additionally emphasizes the importance of TM3 cysteine residue for channelrhodopsin kinetics [90, 122].

3.3.4 Prasinophyte and Viral Anion-Conducting Channelrhodopsins (PyACRs and vPyACRs)

Further metagenomic search of possible channelrhodopsins delivered more channelrhodopsins subfamilies, with a notable example of a distict clade of ACRs found across prasinophytes and multiple strains of giant viruses [123]. Similar to VirChR1s, v*Py*ACRs showed stong retention in cytosol, when expressed in mammalian cells, that allowed only limited characterization of these proteins. Despite the fact that all prasinophytes lack apparent eyespot structures, ACRs are likely to play a part of the visual system although some of ACR genes were carried by giant viruses.

3.3.5 Nonalgal Anion-Conducting Channelrhodopsins with Red-Shifted Absorption (RubyACRs)

In 2020, a new clade of anion-conducting channelrhodopsins were found in genome of heterotrophic protists labyrinthulea and haptophyte algae [124]. Unlike other reported natural ACRs, RubyACRs were reported to include proteins with both blue-shifted and red-shifted absorption spectra ranging from 480 nm up to 590 nm. Besides that, the photocurrents recorded from *Al*ACR1 rhodopsin showed weakly voltage-dependent biphasic decay that suggests a different to nACRs gating mechanism.

3.4 Non-proton ion Pumps

3.4.1 Archaeal Anion Pumps (Halorhodopsins, TSA Rhodopsins)

In 1980, the first light-driven chloride ion pump named halorhodopsin (*Hs*HR) was discovered in archaeon *Halobacterium salinarum* [125]. Later, BRs and archaeal chloride pumps (halorhodopsins, HRs) appeared to have mechanistic similarities, since *Hs*BR can be converted into a chloride pump by single mutation D85T [126]. HRs are also referred to as TSA rhodopsins (T111, S115, and A122 in *Hs*HR), according to the residues in functionally relevant positions in *Hs*BR (D85, T89, and D96, respectively). The high-resolution crystal structures of *Hs*BR and *Hs*HR shed light on the molecular mechanism of ion transport by archaeal pumps [127, 128]. In the *Hs*HR structure, position of the RSB counterion (D85 in *Hs*BR) was found to be occupied by a chloride ion coordinated with the threonine 111 side chain and multiple hydrogen bonds with internal water molecules. However, despite the high similarity between *Hs*BR and *Hs*HR, the reverse substitution of T111D did not convert *Hs*HR into an H^+ pump, indicating the asymmetric direction of functional conversion [129]. The halorhodopsin photocycle includes four distinct photointermediates: K, L, N, and O, where the L state was shown to be a critical step for ion transfer [130, 131]. *Hs*HR was shown to pump all halogen ions, including bromide and iodide ions, and also the nitrate ion; however, under physiological conditions, *Hs*HR pumps exclusively chloride ions [132]. Recently, new genes attributed to TSA rhodopsins were found in cyanobacteria, and surprisingly a sulphate (SO_4^{2-}) transporter from cyanbacterium *Synechocystis* (*Sy*HR) was recently reported [133, 134]. Due to their simplicity, HRs and specifically the halorhodopsin from *Natronobacterium pharaonis* (*Np*HR) have been used for a long time for neuron silencing in optogenetic experiments. The plasma membrane expression of *Np*HR in human cells was iteratively optimized using the Kir2.1 trafficking signal [83]. The optimized protein version was labeled eNpHR3.0. It was shown that eHpHR3.0 is capable of silencing neurons when illuminated with 630/660/680 nm lasers.

3.4.2 Bacterial Anion Pumps (NTQ Rhodopsins)

In 2014, a new clade of microbial rhodopsins from marine bacteria with ion-pumping activity was reported, named NTQ rhodopsins, according to the Asn, Thr, and Gln residues in functional amino acid positions in *Hs*BR (D85, T89, and D96, respectively) [135, 136]. Despite the fact that NTQ rhodopsins are phylogenetically closer to sodium pumps (NDQ rhodopsins) than to archaeal anion pumps, their photochemistry is closer to that of halorhodopsins and not of sodium pumps. The photocycle of a bacterial anion pump from *Fulvimarina* (FR) has four photointermediate states (K, L, N, and O) and lacks an apparent blue-shifted M-state, similar to *Hs*HR [135]. The crystal structure of the NTQ rhodopsin from *Nonlabens marinus* S1-08T (ClR) at the resolution of 1.5 Å revealed a sodium pump-like architecture, which additionally supports the hypothesis of evolutionary relation of NTQ and NDQ rhodopsins [137]. The ion-pumping pathway of ClR was found to have two chloride binding sites; the first side is found in the nearest vicinity of the RSB and the second is at a cytoplasmic loop of the protein. Despite the fact that the motif replacement mutation D116T of KR2 does not convert this sodium pump into a chloride pump, the triple mutant D116T/F72G/D102N was shown to complete a successful ion pumping conversion [138]. The reverse conversion of the chloride pump into a sodium pump has not yet been achieved; however, the N110D/Q121E/S225F triple mutant of FR has been shown to function as an outward proton pump. This finding indicates that the conversion asymmetry in ion pumping may be a universal feature of microbial rhodopsins [139]. At the moment, no optogenetic applications for bacterial anion pumps have been reported.

3.4.3 Cation Pumps (NaRs)

Before 2013, rhodopsins with cation-pumping ability were generally considered to be nonexistent due to the energetic restrictions of the positively charged RSB. However, in 2013, it was reported that sodium-pumping rhodopsin naturally exists in *Krokinobacter eikastus* [28, 136]. It was shown that the microbial rhodopsin called KR2 is capable of pumping sodium and lithium ions out of the cells, but converts into an outward proton pump in their absence. In fact, it was demonstrated later that KR2 uptakes both sodium ions and protons in a competitive manner, and the rate constant of proton uptake is more than 10^3 times larger than that of sodium [140]. Therefore, KR2 pumps almost exclusively sodium ions under physiological conditions, when the concentration of sodium is much higher than that of protons (pH 8.0, 150 mM NaCl).

In this sodium-pumping mode, KR2 photocycle has four intermediate states: K, L, M, and O [28]. In contrast to other active transporters, sodium is not bound inside the protein in the ground state [28, 141]. Instead, the ion binds transiently in the core of the protein in the O-state [28]. The same observation was shown for another sodium-pumping rhodopsin from *Gillisia limnaea* (GLR) [142].

Several studies on the KR2 mutants revealed additional details of light-driven sodium pumping and ion selectivity of KR2. It is known that single mutations G263F(W) and S254A enable KR2 to pump potassium ions [143–145]. A double mutation N51P/G263W(F) enhances potassium pumping efficiency even more [143], while the N61L/G263F variant of KR2 is also capable of pumping cesium ions [146]. Strictly Na^+-pumping rhodopsin from *Dokdonia* sp. PRO95 regains the proton pumping ability when its cysteine 253 is mutated to serine, as in KR2 [147].

Recently it was shown that R109Q mutant of KR2 is passively permeable for potassium ions [92]. However, passive transport was observed only under alkaline conditions, when RSB counterions of KR2 were deprotonated. Also, the active transport of sodium in KR2-R109Q competes with passive permeability for potassium. The combination of R109Q and S70A mutations led to an increase in passive permeability and a decrease of active pumping. It was also reported that KR2 is able to silence neurons with limited efficiency [143]. Further optimization of KR2 led to engineering eKR2, which is much better localized in the plasmatic membrane of mammalian cells and can be used in optogenetics [93].

3.5 Enzymerhodopsins

Enzymerhodopsins are a major family of rhodopsin-based enzymes that are capable of triggering cellular cascades with light. Enzymerhodopsins have two-component signaling systems-like modular organization, where an N-terminal microbial rhodopsin domain is followed by a C-terminal effector domain. However, unlike archaeal sensory rhodopsins that have a cognate transducer protein, enzymerhodopsin domains are fused into a single amino acid sequence [97]. Based on their functional activity, enzymerhodopsins are classified into three groups: histidine kinase rhodopsins (HKR), rhodopsin phosphodiesterase (RhoPDE), and rhodopsin guanylyl cyclase (RhGC). All the three groups comprise eight transmembrane helices, with both termini located within the cytoplasm and share a homodimeric architecture which separates enzymerhodopsins from all other type 1 rhodopsins.

3.5.1 Histidine Kinase Rhodopsins (HKR)

Initially discovered in freshwater algae *Chlamydomonas reinhardtii*, histidine kinase rhodopsin 1 (HKR1, or *Chlamydomonas* opsin-related protein 5, COP5) is the first natural enzymorhodopsin to be reported [148]. Altogether, eight HKR sequences were reported only in *C. reinhardtii* genome with a variety of putative HKR genes in other organisms; however, transcription of only three proteins (HKR1, HKR2, and HKR4) was verified [149]. The rhodopsin domain of HKRs is connected with its effector domain by a short presumably alpha-helical linker (about 20 amino acids) and has no homology to linker domains in two-component systems [150]. The histidine kinase domain includes histidine phosphoryl transfer (DHp) and catalytical/ATP-binding (CA) domains that are homologous to the canonical histidine kinases from various organisms. Furthermore, overexpression of HKR2 protein in *V. carteri* leads to significantly increased cGMP levels, demonstrating the guanylyl cyclase activity of HKRs in vivo [151]. At the moment, photochemistry of the HKR family relies on the experimental data obtained with the HKR1 isolated rhodopsin domain. Analogous with other microbial rhodopsins, the rhodopsin fragment of HKR1 was found to be a UVA receptor (λ_{max} = 380 nm) that can be converted by UV light into a blue-light absorbing state (Rh-Bl) (λ_{max} = 490 nm) and reversed by blue light. Upon light activation, HKR1 undergoes a transition from the ground state ($\lambda_{max} = 380$ nm) to the Rh-Bl state ($\lambda_{max} = 490$ nm) followed by the red-absorbing state ($\lambda_{max} = 560$ nm) that decays in ~50 ms. Despite the fact that understanding of the molecular mechanism of the HKR function requires further investigations, the cyclase activity of the effector domain has been shown to be both light-inhibited and ATP-dependent that makes HKRs conspicuous among other histidine kinase families [152].

3.5.2 Rhodopsin Phosphodiesterase (RhoPDE)

Despite the fact that regulation of secondary messengers like cGMP is tied mainly to G protein signaling cascade and type 2 rhodopsins, recently, several type 1 rhodopsins with phosphodiesterase activity were discovered [153]. In 2017, a new rhodopsin from *Salpingoeca rosetta* (*Sr*RhoPDE) was shown to regulate cGMP hydrolysis by light in *Xenopus laevis* oocytes in a light-dependent manner. *Sr*RhoPDE has a characteristic rhodopsin spectrum susceptible to blue light ($\lambda_{max} = 492$ nm) and a long-living M-state that decays within ~7 s. The M-state photointermediate was shown to be associated with light-induced enzyme activation, and therefore could be involved in enzyme activation mechanism. The phosphodiesterase domain (PDE) of the RhoPDE crystal structure reveals 16 α-helices architecture that is known from PDEs from other organisms [154]. A putative activation mechanism was proposed by analogy with that of the human phosphodiesterase PDE2A [155]. It includes a conformational change triggered by the light-induced movement of helix 7 that propagates via a coiled-coil linker and alters the position of the allosteric center of PDE domain [155]. A multidisciplinary study later revealed the photoactivation mechanism of PDE in great detail; surprisingly, the rhodopsin domain harbored an additional eighth N-terminal α-helix [156].

3.5.3 Rhodopsin Guanylyl-Cyclase (RhGC)

The first rhodopsin guanylyl-cyclase was discovered in *Blastocladiella emersonii* (BeRhGC), where the protein was shown to play an important role for fungal phototaxis through a cGMP signaling pathway [29]. The guanylyl-cyclase domain of RhGCs belongs to class III cNMP cyclases and exhibits more activity in the presence of Mn^{2+} over Mg^{2+} [157]. Unlike other enzymerhodopsins, RhGCs have two alpha helices (helix 0 and helix 1) at N-terminus preceding the heptahelical bundle and 50 amino acid α-helical linker that connects enzymatic and rhodopsin domains. The isolated GC domain is constitutively active; however, full-length RhGCs are approximately fourfold more active under light illumination, presumably due to rhodopsin-induced stabilization. Besides that, RhGCs could be converted into an adenylyl cyclase with an E497K/C566D double mutation [98]. The rhodopsin domain of BeRhGC has a characteristic absorption spectrum that can be activated with green light ($\lambda_{max} = 530$ nm). The photocycle of RhGCs contains four intermediates: an early decaying K-state, followed by two blue-shifted L_1 and L_2 states and a major M-state that decays within 600 ms [158]. When expressed in various heterologous cells, RhGCs are capable of generating cGMP under green light with a high substrate selectivity and little enzymatic dark activity in vivo [159].

3.6 Viral Rhodopsins Group 2

In 2019, one of the viral rhodopsins from group 2 (VR2), VirR2 (OLPVRII) was structurally and functionally characterized [39]. The crystal structure of VirR2 obtained at the resolution of 1.9 Å has shown unusual pentameric architecture with a symmetrical bottle-like channel between the protomers, surprisingly reminiscent of those in pentameric ligand-gated channels. Molecular dynamics simulations suggest the pore to be able to facilitate a chloride-selective channeling. However, no direct electrophysiological data are yet available to support this statement. VirR2 has been demonstrated to have a 100 ms long photocycle with four photointermediates: K, M, N, and O-states that resemble bacteriorhodopsin-like behavior with the putative active state associated with the M-state. Similarly to ChR2, VirR2 was reported to have a weak proton pumping ability observed during black lipid membrane experiments [39, 160]. Despite the fact that VirR2 functionality has not yet been confirmed, a possible case of light-gated ion channel controlled by the configuration of the oligomeric state would be a valuable asset for some of the optogenetic applications that require precise ion selectivity.

3.7 Heliorhodopsins

Another unusual clade of microbial rhodopsin superfamily is heliorhodopsins (HeRs) that were recently discovered by means of functional metagenomics [33]. Heliorhodopsins were found to be widely distributed among all kingdoms of life, forming about ten phylogenetically different subgroups that can potentially indicate an ecologically relevant function of HeRs. In fact, a broad phylogenetic distribution of HeRs could potentially represent not only a single function, yet more studies are needed to fully understand the potential of the HeR clade. Surprisingly, HeRs are embedded in the membrane with their N-termini facing the cytoplasm, which is opposite to the orientation of all previously known rhodopsins. Despite the fact that crystal structures solved at high 1.5 Å resolution [36] of two representative HeR proteins (48C12 and *Ta*HeR) have been reported [34–36] and the authors [36] hypothesize that the protein may have enzymatic activity providing chemical reactions with proton transfer and may be involved, for instance, in the carbon cycle, their function remains unknown. Heliorhodopsins have a second-scale photocycle that includes four photointermediates: K, L, M, and O states, where the M-state is a likely candidate to represent the active state of the protein [33, 36, 161]. Interestingly, the structure of HeR-48C12 at 1.5 Å allowed determination of a surprising triangle-like electron density in the near vicinity of the RSB region that was assigned to an acetate molecule [36]. A similar finding was also evidenced in recent work on 48C12 mutants, where authors show anion binding near the RSB of the heliorhodopsin mutants [162]. Finally, HeR from *Thermoplasmatales archaeon* was shown to bind zinc ions [163].

Despite the fact that the role and function of heliorhodopsins remain to be elusive, it has been hypothesized that heliorhodopsins could fixate carbonate or nitrate in their hosts. Thus they may be involved in global ecological processes [36]. Furthermore, microbial rhodopsins with potential carbon or nitrogen fixation mechanisms could be of particular interest in artificial cell design and green bioenergetics.

4 Structures of New Rhodopsins

Due to the constantly growing number of the identified rhodopsin genes and the development of the X-ray synchrotron sources and XFELs, a considerable amount of structural information on rhodopsins has been reported during the past decade. Here we summarize the current understanding of high-resolution X-ray crystallography structures of different classes of rhodopsins and depict their function–structure relationships.

With this in view, we will describe the proteins in terms of the "cavities and gates" concept introduced in [164]. It means that rhodopsins perform their functions by operating their inner (or, in some cases, outer) cavities and restriction sites (gates) between them. Particularly, in the case of light-driven transporters, cavities are organized in a chain, which, as expected, lines the substrate pathway. All conformational changes in cavities and gates are originally triggered by the retinal isomerization.

4.1 Proteorhodopsins

Despite a great number of identified sequences of PRs, the number of their high-resolution structures is limited. Among them are crystal structures of a proteorhodopsin from bacterium *Exiguobacterium sibiricum* (ESR) [66], and two proteins isolated from the Mediterranean Sea at a depth of 12 m and the Pacific Ocean near Hawaii at a depth of 75 m (*Med12*BPR and *HOT75*BPR, respectively) [68], all reported in 2013. However, there are also structures of some other bacterial rhodopsins, which possess outward proton-pumping activity, albeit distinct from PRs, such as a xanthorhodopsin from eubacterium *Salinbacter ruber* (XR) [165], and rhodopsins from cyanobacterium *Gloeobacter violaceus* PCC 7421 (GR) [166] and from an extremely thermophilic bacterium *Thermus thermophiles* (TR) [167].

It took 13 years to obtain the first high-resolution crystal structure of a proteorhodopsin. During that time, structural information on the oligomerization of PRs was obtained using atomic

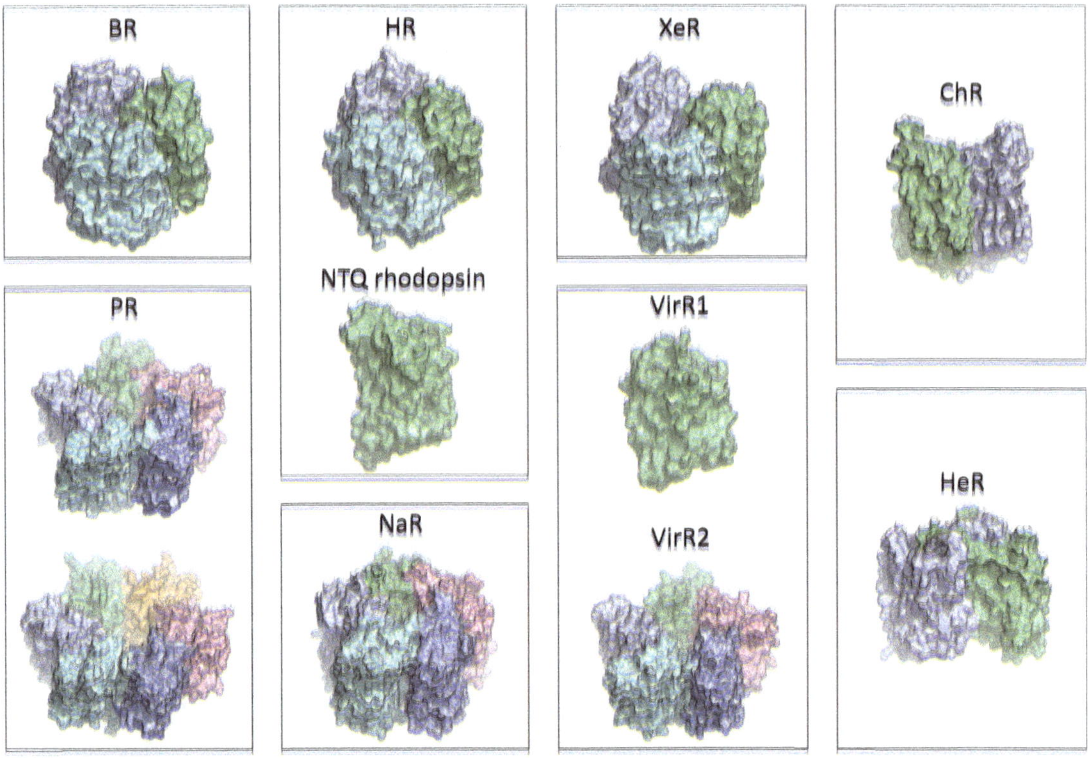

Fig. 3 Oligomeric assembly of microbial rhodopsins. Surface representation of oligomers of different classes of microbial rhodopsins: BR represents bacteriorhodopsins (PDB ID: 1C3W); HR and NTQ rhodopsin represent chloride-pumping rhodopsins from archaea and bacteria, respectively (PDB IDs: 1E12 and 5ZTK); XeR represents inward proton pumps xenorhodopsins (PDB ID: 6EYU); ChR represents CCRs and ACRs (PDB ID: 6EID); PR represents proteorhodopsins (PDB IDs: 4KLY and 4JQ6); NaR represents sodium-pumping rhodopsins (PDB ID: 6REW); VirR1 and VirR2 represent viral rhodopsins of group 1 and 2, respectively (PDB IDs: 6JO0 and 6SQG); HeR represents heliorhodopsins (PDB ID: 6SU3)

force microscopy (AFM) [168], and also solid-state and solution nuclear magnetic resonance (NMR) [169–171]. Particularly, it was shown that PRs are organized into hexamers or/and pentamers in lipid membranes (Fig. 3), and even the first seven α-helical model was reported (PDB ID: 2L6X [170]).

Atomistic insights into PRs, which appeared in 2013 first for ESR [66] and later for *Med12*BPR and *HOT75*BPR [68], revealed unique features of the proteins [172]. One of them is a characteristic distortion of the α-helical structure of the helix F in its middle part stabilized by side chains of W154 and N224 in the case of ESR, W141, and N212 in the case of *Med12*BPR and W159 and N231 in the case of *HOT75*BPR (Fig. 4). This was not found in BRs and XRs. Another feature is the presence of histidine residue (H57, H57 and H75 in ESR, *Med12*BPR and *HOT75*BPR, respectively) tightly interacting through a short 2.4 Å hydrogen bond with the primary proton acceptor from the RSB (D85, D79 and D97 in

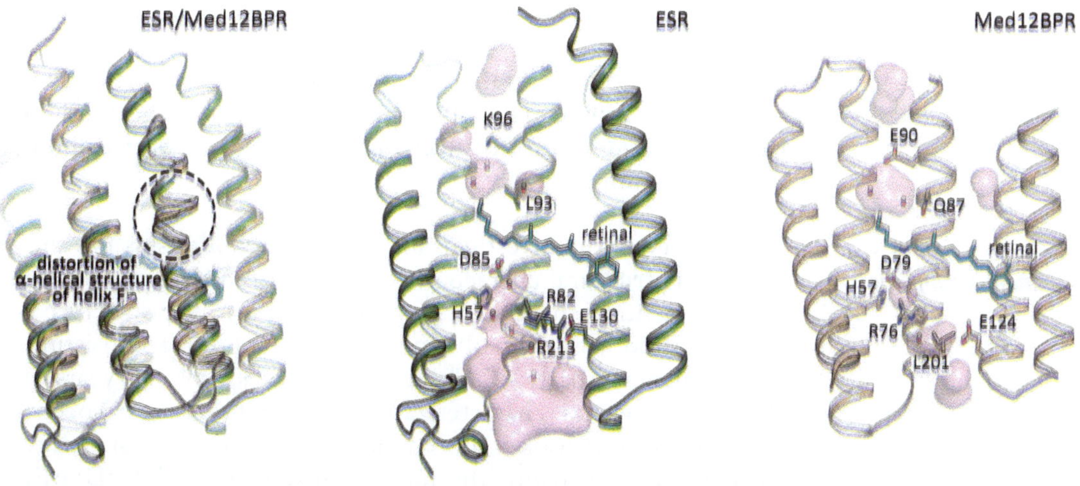

Fig. 4 Characteristic features of proteorhodopsin structures. Structural alignment of ESR (PDB ID: 4HYJ) and Med12BPD (PDB ID: 4JQ6) models and characteristic distortion of the α-helical structure of the helix F in PRs (left panel); inner cavities and key residues of ESR (middle panel); inner cavities and key residues of Med12BPR (right panel). Cavities are calculated using HOLLOW [173] and shown with pink surface

ESR, *Med12*BPR and *HOT75*BPR, respectively), increasing pK_a of the proton acceptor (Fig. 4). In the case of pentameric and hexameric models of PRs, this residue is an important part of oligomerization interface [68]. Finally, in PRs a relatively big polar cavity is present in the vicinity between the RSB and proton donor in the cytoplasmic inner part of the protein. In the case of *Hs*BR a similar cavity is formed transiently during the photocycle (in L-, M-, and N-states), while in PRs it is found already in the ground state albeit there is no hydrogen bonding chain from the RSB to the proton donor.

There are two distinct elements in the structure of ESR compared to other known models of PRs. First, the orientation of residues in the extracellular side (H57 and R82) creates a big cavity, which protrudes from the D85-H57 pair directly to the extracellular space, similar to that in C1C2 channelrhodopsin [111] (Fig. 4). Therefore, the proton shared in D85-H57 pair can be released directly to the solvent. Second, unlike in other light-driven proton pumps, ESR has a positively charged lysine residue K96 at the proton donor position at the cytoplasmic side, which is involved in the reprotonation of the RSB during the photocycle, while most commonly this position is occupied by aspartate (BRs) or glutamate (PRs) residues (Fig. 4). This could be the reason for the reversed order of protein uptake and release by ESR compared to other proton pumps [174]. It should be mentioned that there is also another group of eubacterial light-driven proton pumps with positively charged histidine residue at the cytoplasmic side in the region of the K96 residue of ESR [175, 176]. However, in that case the histidine located in the helix B is not a direct proton donor to the

RSB [175]. The structural data on these rhodopsins is not available at the moment.

There is no structural information on the intermediate states of the PRs photocycle. This dramatically limits our understanding of the proton translocation mechanism of bacterial proton pumps, which may differ notably from that of BRs.

4.2 Channelrhodopsins

The history of the investigations of channelrhodopsins has shown that their structural characterization was also a challenge. We also have to mention that the first approximation to the structure of a channel rhodopsin was obtained with the chimera consisting of complementary parts of ChR1 and ChR2 [111], 10 years after the discovery and characterization of the proteins, respectively [27, 78, 177]. It took 5 more years to obtain the first structure of the native cation channelrhodopsin, ChR2 [164] (PDB ID: 6EID) (at 2.4 Å resolution). Luckily, crystal structures of the native anion channelrhodopsins (*Gt*ACR1) appeared only 3 years after their discovery in 2015 [115, 116] (PDB IDs: 6CSM, 6EDQ), together with the model of an engineered light-gated anion channel, though at a modest resolution of 2.9 Å [89] (PDB ID: 6CSN). Finally, very soon after that, the structure of a red light-activated cation channelrhodopsin Chrimson was reported [96].

All the available structural information on channelrhodopsins is accurately summarized and analyzed in [178] and [89]. Hence, here we will briefly describe key elements of their organization. The structures demonstrate common features of the light-driven passive transporters. First, these proteins form similar dimers (Figs. 3 and 5). The protomers interact mostly through transmembrane (TM) helices 2, 3, and 4 and are cross-linked by the disulfide bonds between cysteine residues of the N-terminus of the proteins. Second, channelrhodopsins contain large inner cavities and pores, separated from each other by the constriction sites (gates) (Fig. 6). This allows a fast opening of the inner channel by the conformational rearrangements in the gates. Thus, a wide pathway for ions, lined with the chain of cavities, is prepared already in the dark state.

Although all structurally characterized channelrhodopsins can be described in terms of "cavities and gates (CaG)" concept, the organization of the cavities and gates in each particular case has particular qualities. The extracellular inner part of the proteins contain big polar cavities, filled with water molecules and in most cases separated by extracellular gates (EGs), formed around the arginine residue of TM helix 3. There is no regularity in the organization of EGs; however, they are usually linked directly to the RSB via a water-mediated hydrogen bonding network. Thus, the conformation of EGs can be easily controlled by the retinal configuration.

Another constriction site (central gate, CG) was found in the middle part of the proteins. CGs are presumably the main

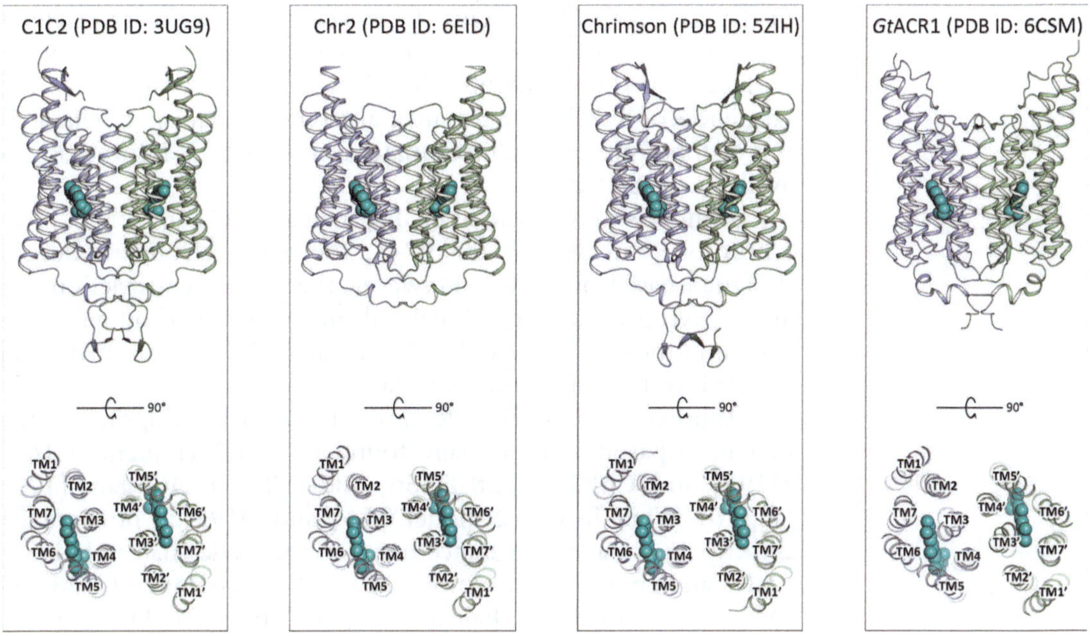

Fig. 5 Dimers of different types of channelrhodopsins. Retinal cofactor atoms are shown with cyan spheres

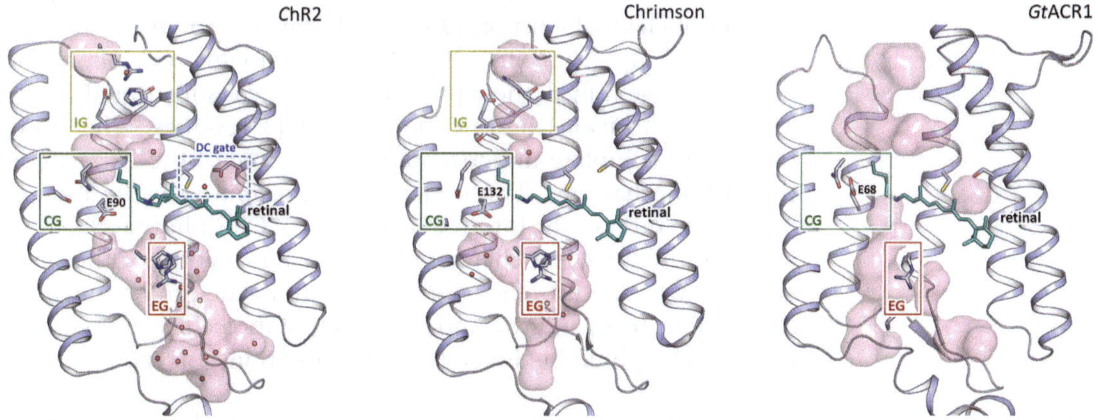

Fig. 6 Cavities and gates of different types of channelrhodopsins. Intracellular (IG), central (CG), and extracellular (EG) gates are contoured with yellow, green, and red rectangles. Retinal cofactor is colored cyan. DC gate (C128 and D156 residues) of ChR2 is contoured with dashed blue rectangle. Inner cavities are calculated using HOLLOW [173] and shown with pink surface

constriction sites (in case of *Gt*ACR1 the only one) of ChRs. The core of CGs of both natural cation and anion channelrhodopsins with known structures is formed by glutamate residue (E90 in ChR2, E132 in Chrimson and E68 in *Gt*ACR1), which is located in the middle part of TM helix 2. Analogous glutamate is absent only in the engineered Cl^- channel iC++; however in that case molecular dynamics simulations showed an unstable conformation of the CG [89]. Most likely, the existence of a tight intracellular gate (IG) in the case of iC++ contributes to the prevention of water permeability [89]. Notably, the glutamate is also absent in the sequences of BCCRs suggesting a different mechanism of channeling of those proteins. Unfortunately, no structural information is available on BCCRs.

Tight IGs were found in CCRs, such as C1C2 chimera, ChR2, Chrimson, and also in the iC++, the anion channelrhodopsin designed based on C1C2. However, there is no constriction site at the cytoplasmic part of natural anion channelrhodopsins, as shown in the case of *Gt*ACR1. Importantly, IGs are not hydrogen bonded to the RSB in the dark state.

Another characteristic feature of channelrhodopsins is the cysteine-aspartate pair initially found in C1C2 chimera (C167-D195) and ChR2 (C128-D156) and called a DC gate [122] (Fig. 6). While in C1C2 model C167 and D195 do not interact directly due to a long distance between their side chains [111], in ChR2 an important water molecule was found between C128 and D156, which mediates their interactions [164]. The DC gate was suggested to be involved in RSB reprotonation. Thus, it presumably is part of the proton pathway from the cytoplasm to the RSB. This could also explain the effects of the DC gate mutation on the kinetics of channel closure. On the other hand, no DC gates were later found in Chrimson and *Gt*ACR1. Although cysteine in TM helix 3 is present in all these proteins, aspartate is substituted by the cysteine (Chrimson) and serine (*Gt*ACR1) residues. Therefore, there is no complete understanding of the role of the DC gate in the ChR mechanisms. This understanding is limited by the absence of the structures of the intermediate states of ChR photocycle, most importantly, the open channel structure. Elucidation of the open channel structure is obligatory for the rational design of enhanced optogenetic tools and the understanding of molecular mechanisms of channeling in light-driven passive cation transporters.

4.3 Ion pumps

4.3.1 Sodium Pumps

One of the major recent discoveries is the Na^+-pumping KR2 rhodopsin from *Krokinobacter eikastus*, the first non-proton cation pump, which was unexpected. As was already mentioned, this is in complete contrast with the existing paradigm that the retinal-based non-proton cation pump cannot exist in nature. The paradigm was based on our knowledge of *Hs*BR, the most studied membrane protein.

Accepting the fact that Na^+ pumps do exist means that the molecular mechanism of ion translocation of KR2 is very different from that of the known light-driven ion pumps, which are, to a large extent, similar to *Hs*BR. Indeed, non-proton cation pumps cannot benefit from the Grotthuss mechanism, which is the basis of H^+-translocation in BRs [62]. The importance of elucidation of the mechanism of KR2 is far beyond sodium pumping. It certainly may

Table 2
Description of the KR2 structures, deposited to the Protein Data Bank. The most relevant model (PDB ID: 6YC3) is shown with bold font

PDB ID	Oligomeric form	Resolution, Å	Conformation	Remarks
3X3B	Monomeric	2.3	Compact	pH 4.0
3X3C	Monomeric	2.3	Compact	pH 8.9, soaked from pH 4.0
5JRF	Monomeric	2.5	Compact	Iodide-bound, pH 4.3
4XTL	Pentameric	1.45	Compact	pH 4.3
4XTN	Pentameric	2.2	Expanded/Compact	pH 4.9
4XTO	Pentameric	2.8	Compact	pH 5.6, dehydrated
6REW	Pentameric	2.2	Expanded	pH 8.0
6RF0	Pentameric	3.0	Compact	pH 8.0, dehydrated
6RF1	Pentameric	2.8	Expanded	pH 8.0, recovered after dehydration
6REX	Pentameric	2.7	Expanded	pH 6.0, soaked from pH 8.0
6REZ	Pentameric	2.6	Expanded/Compact	pH 5.0, soaked from pH 8.0
6RF5	Monomeric	2.3	Compact	pH 6.0, soaked from pH 4.3
6RF6	Monomeric	1.8	Compact	pH 8.0, soaked from pH 4.3
6RF7	Monomeric	2.6	Compact	pH 8.9, soaked from pH 4.3
6YC3	**Pentameric**	**2.0**	**Expanded**	**pH 8.0**
6TK6	Monomeric	1.6	Compact	pH 8.0, soaked from pH 4.4
6TK7	Monomeric	1.6	Compact	pH 4.4

help to find a rational approach to the design of other non-proton cation pumps for optogenetic applications [92, 93, 143, 144].

As we have already mentioned, there were several attempts to solve this great puzzle. Extensive research on KR2 provided the first important information; however, the ideas on the mechanisms were not conclusive [141, 143–145, 179, 180]. The major problem is that most of the existing high-resolution structures are limited only to the resting state of KR2. Moreover, the structures did not reveal the transported substrate, a sodium ion, inside the protein. This fact added more mystery to the field. Here we will describe in brief what is currently known about the structure of KR2 and what is hypothesized on the molecular mechanism of its function. It should also be noted that high-resolution structures of the light-driven sodium pumps are limited to only those of KR2. At the same time, the number of reported structures of KR2 and its functional mutants is large and may be confusing [143–145] (Table 2). Below we summarize the most recent results and describe the current understanding of light-driven sodium pump structural organization. Finally, we would like to mention that very recently, the structure of the O intermediate pentameric state of KR2 became known [181]. It revealed the sodium ion transiently bound in the core of the KR2 protomer. Soon after that structures corresponding to the K, L, M, O_1, and O_2 states of monomeric form of KR2 were reported [182]. Here we also briefly describe these important findings.

The first batch of KR2 structures was reported in 2015: three monomeric and two pentameric models were presented [143, 144]. All monomeric forms (PDB ID: 4XTL [144], 3X3B [143], 3X3C [143]) were obtained using crystals initially grown at acidic pH values. Although, in one case, the crystals were soaked in the buffer solution with neutral pH, the oligomeric state in the crystals remained monomeric, which is, as was later shown, not biologically relevant [183]. Indeed, as it was shown in 2018 by high-speed AFM, KR2 forms pentamers when reconstituted into the lipid membrane at both acidic and neutral pH values [183], while the monomeric fraction occurs only at acidic pH and only in detergent micelles and LCP-grown crystals at acidic pH [143, 144, 184]. Therefore, KR2 is supposed to be a pentamer in vivo (Fig. 3). Unfortunately, both pentameric structures (PDB ID: 4XTN [144] and 4XTO [144]) were initially obtained at pH lower than 6, where the protein preferably pumps protons [140]. Interestingly, four out of five presented models were different from each other, particularly in the organization of the RSB region, and also the position of the helix C varied in the structures.

The controversy was resolved in 2019 when the pentameric structure of KR2 was obtained using the crystals initially grown under physiological conditions [145]. It was shown that all the previously reported models were likely unphysiological, and the mechanisms of structural switches in KR2 upon pH decrease were also demonstrated. It was also noticed that dehydration of KR2 crystals might result in the conformational differences similar to those observed in the previously determined structures, which was probably the case of 4XTO model.

Nevertheless, despite several remarkable differences in the protein structures in monomeric and pentameric forms, several features of the KR2 also have common elements in all the existing models. First, there is a big water-filled cavity at the cytoplasmic part of the protein, which protrudes from the protein surface almost to the RSB, being separated from it only by the Q123 residue which is part of the NDQ motif, specific for all known sodium pumps (Fig. 7). Thus, Q123, together with S61, comprise a cytoplasmic gate, which presumably opens (or becomes permeable) upon sodium uptake (Fig. 7). Second, there is a short N-terminal α-helix, capping the inside of the protein at the extracellular side and thus creating a putative ion-release cavity near E11, E160, and R243 residues. The mutational analysis showed that the

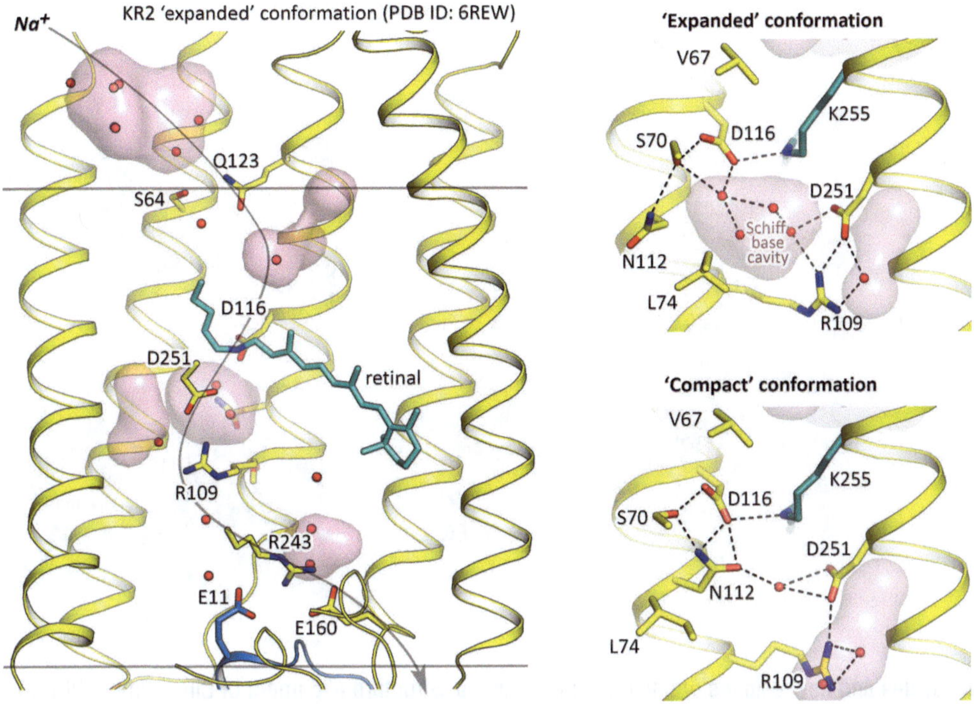

Fig. 7 Architecture of KR2 and differences between the "expanded" and "compact" conformations. Cavities are calculated using HOLLOW [173] and shown in pink surface. Gray arrow indicate putative pathway of Na^+ inside the protein protomer. Retinal cofactor is colored cyan. Dashed black lines indicate hydrogen bonds in the RSB region. Gray lines indicate hydrophobic/hydrophilic membrane core boundaries

helix is crucial for sodium pumping, although the E11A, E160A, and R243A mutations did not affect the functional properties of the protein [143, 144].

Currently, the most physiologically relevant high-resolution structure of KR2 (the pentameric form at pH 8.0 at 2.0 Å resolution, PDB ID: 6YC3 [181]) shows that the rhodopsin has a third key element: a big polar cavity near the RSB (Schiff base cavity), which is filled with four water molecules in the ground state (Fig. 7). The formation of the cavity is only possible if the N112 side chain flips to the oligomerization interface, creating space near the RSB. This conformation of the protein with a big cavity near the RSB was called "expanded" (Fig. 7). Importantly, the "expanded" conformation is only observed in the pentameric form of KR2 (Table 2), as it requires a notable shift of the extracellular half of helix C, which is presumably restrained by the crystal contacts in the crystals of monomeric protein. On the contrary, when N112 flips to the inner part of the protein and is hydrogen bonded to S70 and D116, the conformation was called "compact" (Fig. 7).

The role of the "expanded" and "compact" conformations remains unclear. It is believed that the "expanded" conformation

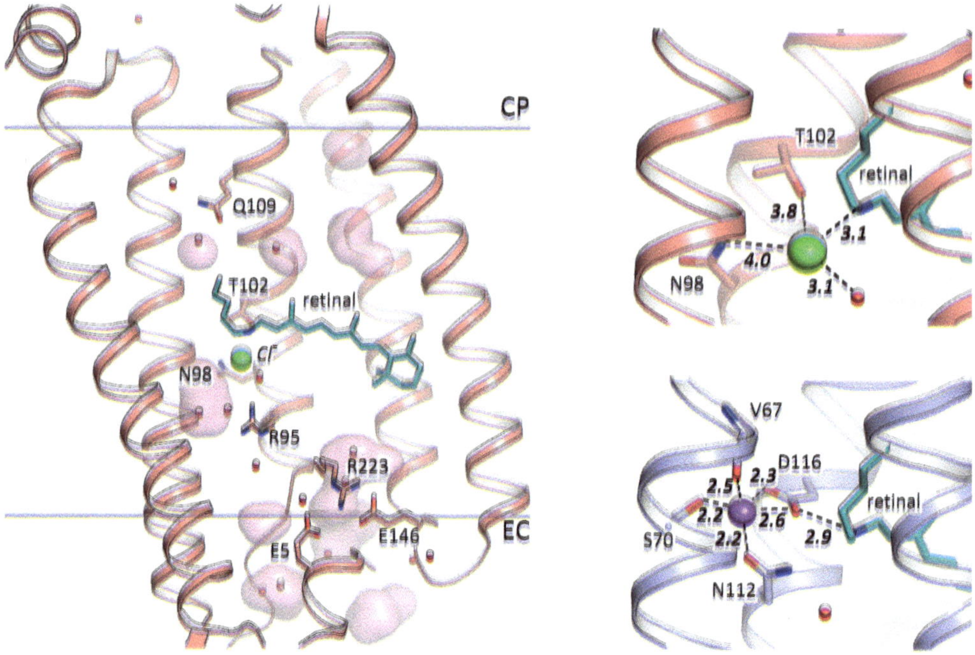

Fig. 8 Cavities and key residues of bacterial anion pumps. Structure alignment of ClR (salmon, PDB ID: 5B2N) (left panel); Cl^- (green sphere) and Na^+ (violet sphere) binding sites in the ground state of ClR (top) and the O-state of KR2 (bottom), respectively (right panels). Cavities are calculated using HOLLOW [173] and shown with pink surface

may be important to create space for sodium passage near the RSB [145]. The "compact" conformation appears in the pentameric form upon pH decrease and correlates well with the color shift associated with D116 protonation. Thus, it may be that the role of the "compact" conformation is in the stabilization of the protonated D116. Hence, the "compact" conformation in the ground state of KR2 most likely represents the nonfunctional protein, as D116 is the key residue, which acts as a primary proton acceptor from the RSB and must be deprotonated in the dark state. Although the pentameric assembly is crucial for the KR2 functioning, we have to stress that the protein in monomeric form may pump sodium at high salt concentrations to some extent.

Recently, the structure of the O intermediate state of the KR2 photocycle was reported at 2.1 Å resolution (PDB ID: 6XYT [181]). The high resolution of the data together with 100% occupancy of the intermediate in the crystals helped to reveal transient Na^+ binding site in the core of the KR2 protomer (Fig. 8). Surprisingly, the conformation of KR2 in the O-state is similar to the "compact" one. The flip of the N112 side chain to the inside of the protein allows tight coordination of Na^+ by the S70, N112, and D116 side chains. Thus, the ion is coordinated by five oxygens with the mean Na-O distance of 2.3 Å, characteristic for this type of ions

[185]. Therefore, the "compact" conformation may be crucial for the transient sodium binding in the O-state. Consequently, the switch from the "compact" to "expanded" conformation upon O-to-ground transition in KR2 guides Na^+ release to the extracellular bulk.

Also in 2020, structural changes of the monomeric form of KR2 in course of the photocycle were reported by using time-resolved serial femtosecond crystallography (TR-SFX) at SwissFEL [182]. The structures corresponding to the early stages of KR2 photocycle showed the shift of the C20 atom of retinal cofactor towards V117 residue in picoseconds (PDB ID: 6TK5) resulting in the shift of the central region of helix C occurring in nanoseconds (PDB ID: 6TK4). Consequently, in microseconds, with the formation of the L/M state, the V117 side chain is flipped from the retinal vicinity (PDB ID: 6TK3). The displacement of the helix C created the pore for Na^+ uptake; however, the pore diameter (1.4 Å) is smaller than that expected for Na^+ permeation (1.9 Å). The distance between the RSB and D116 is decreased to 2.4 Å in nanoseconds and increased back to 2.8 Å in microseconds, indicating H^+ translocation to D116 with the formation of the L/M state. In 1 ms (O_1-state, PDB ID: 6TK2), the first Na^+ binding site was found in monomeric form of KR2 between N112 and D251 residues. Next, in 20 ms (O_2-state, PDB ID: 6TK1), the ion is translocated from the first site to the second one formed near E11, E160, and N106 residues. Finally, the Na^+ ion is released from the second site to the extracellular space.

It should be stressed that the mechanism, described in the previous paragraph corresponds to the monomeric form of KR2, which is likely an artifact of the crystallization under acidic conditions and does not appear in lipid membranes [183]. It should also be noted that the structures of the intermediate states in that case were built using extrapolated electron density maps with the activation level of the protein in crystals of only 14%. The low activation level, together with the moderate resolution (2.25 Å for the early stages and 2.5 Å for the O_1 and O_2 states), makes the interpretation of the crystallographic data problematic. Thus, further structural investigations of the early KR2 intermediate states of the protein in the pentameric form are still required to completely understand the principles of the non-proton cation pumps and also for a rational design of the enhanced KR2-based optogenetic tools.

4.3.2 Anion Pumps

Archaeal Anion Pumps

The first structure of the light-driven chloride pump halorhodopsin from *Halobacterium salinarum* (*Hs*HR) was obtained in 2000 [127]. Since then, HRs have become the third class of microbial rhodopsins, for which high-resolution structures not only of the ground but also of the intermediate states are available [186]. In 2015 crystal structures of several key intermediate state of the halorhodopsin photocycle from *Natronomonas pharaonis* (*Np*HR) were obtained, which demonstrated the main steps of light-driven chloride translocation by HRs [186].

Overall, HRs are organized into trimers and show a BR-like fold (Fig. 3). The *Hs*HR architecture is very similar to that of *Hs*BR, while *Np*HR has a short α-helix at the N terminus and 18 residues longer BC loop, both located at the extracellular side of the protein (Fig. 9). The chloride ion binds near the RSB in the ground state of all HRs, while in *Np*HR the second binding site is present outside the protomer near the AB loop and helix D′ of the neighbor protomer at the cytoplasmic side of the protein (Fig. 9). With the formation of the N-state, Cl^- is translocated to the cytoplasmic side, and remains bound to the RSB. At the same moment, a chain of water molecules is formed, which protrudes from the transient Cl^- binding site to the cytosol, thus preforming the ion-release channel. Upon N-to-O transition, Cl^- is released to the cytoplasmic bulk and the protein remains in the anion-free conformation. Then, a new Cl^- comes from the extracellular side and the photocycle ends.

Bacterial Anion Pumps

Recently, in 2016, several structures of the bacterial anion pump from *Nonlabens marinus* S1-08 (ClR, NM-R3, NTQ rhodopsin) were reported by two groups [133, 137]. The best resolution was 1.56 Å; however, it corresponded to the model obtained at acidic pH 4.5 [137]. The most biologically relevant structure at the resolution of 1.58 Å was obtained at pH 8.0 [133]. Interestingly, unlike in the case of KR2, where pH dramatically affected the protein conformation [145], in ClR no structural rearrangements were found. Furthermore, in all types of crystals, grown at different pH values, the protein remained monomeric (Fig. 3). This finding is surprising, as some of the residues, forming the interface in the pentameric assembly of KR2, are also present in ClR, such as H20 and Y140, analogs of H30 and Y154 of KR2, respectively. The oligomeric state of ClR in vivo remains to be elucidated.

ClR demonstrated a fold very similar to that of KR2, which is not surprising, taking into account a high sequence identity of bacterial anion and cation pumps (~35% between ClR and KR2) (Fig. 8). Particularly, both proteins have a small N-terminal α-helix and show similar BC loop organization, different from that of BRs and HRs (Figs. 8 and 9).

Later, in 2018, another structural analysis of ClR was reported [187]. The authors compared high-resolution structures of the protein at 100 K and 294 K obtained at synchrotron and XFEL sources, respectively. It was shown that the hydrogen bonding network near Cl^- binding sites is altered at room temperature, which may explain lowered protein pumping efficiency at higher temperatures.

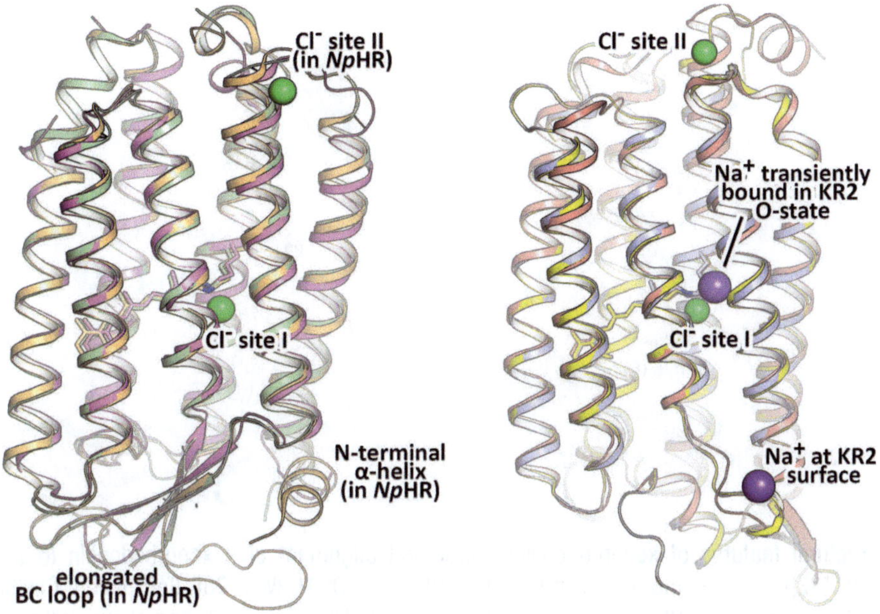

Fig. 9 Architecture of archaeal and bacterial anion pumps. Structural alignment of BR (violet, *Hs*BR, PDB ID: 1C3W), *Hs*HR (green, PDB ID: 1E12), and *Np*HR (wheat, PDB ID: 5B2N) (left panel); structural alignment of KR2 ground state (yellow, PDB ID: 6REW), KR2 O-state (lightblue, PDB ID: 6XYT [181]), and ClR (salmon, PDB ID: 5B2N) (right panel). Green spheres indicate Cl^- binding sites of HRs and ClR. Purple spheres indicate Na^+ binding sites of KR2

A recently reported model of the O-state of the pentameric form of KR2 (PDB ID: 6XYT) revealed that the transiently formed Na^+ binding site in the core of KR2 protomer is formed by N112 and D116 residues of the characteristic NDQ motif of NaRs. This is similar to Cl^- binding site I in the core of ClR, organized by N98 and T102 of the characteristic NTQ motif of bacterial anion pumps (Fig. 8). Surprisingly, the structure of the O_1-state of the monomeric form of KR2 (PDB ID: 6TK2) revealed the Na^+ binding site between N112 and D251 residues; however, in that case the overall position of the Na^+ ion is also similar to that of Cl^- in ClR.

Finally, one structure of the photoactivated state of ClR was deposited to the Protein Data Bank (PDB ID: 6ABA); however, no corresponding manuscript is available. Although some structural insights into the molecular mechanisms of the light-driven chloride pumping were described earlier for HRs, and structure-based mutational analysis of ClR revealed key residues involved in its work, a complete understanding of the bacterial anion-pumping rhodopsins requires elucidation of the high-resolution structures of the intermediate states of the photocycle.

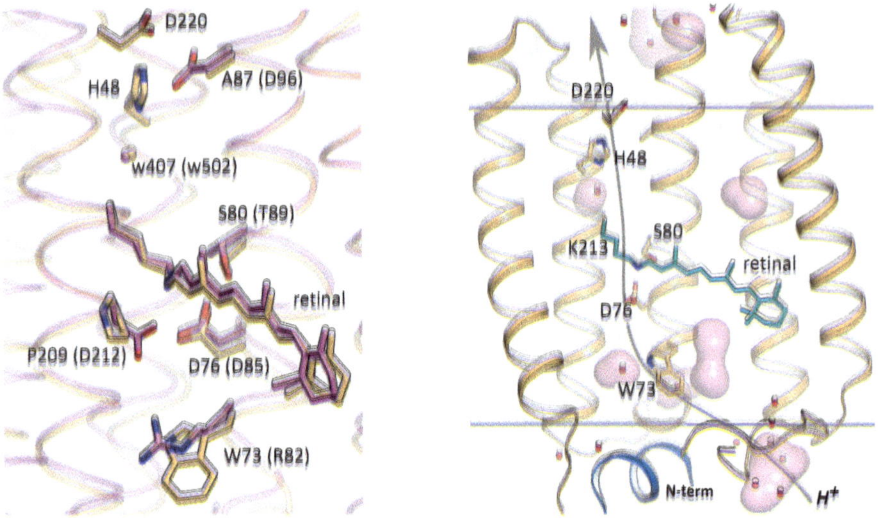

Fig. 10 Structural features of xenorhodopsins. Structural alignment of a xenorhodopsin (orange, *Ns*XeR, PDB ID: 6EYU [32]) and bacteriorhodopsin (violet, *Hs*BR, PDB ID: 1C3W [128]) (left panel). Cavities and key residues of *Ns*XeR (right panel). Cavities are calculated using HOLLOW [173] and shown with pink surface. N-terminus is colored blue

4.3.3 Inward Proton Pumps (Xenorhodopsins)

The first and the only structure of the natural light-driven inward proton pump *Ns*XeR at 2.5 Å resolution was reported in 2017 [32]. *Ns*XeR forms trimers, which are very similar to those of BRs and HRs (Fig. 3).

Dramatically different from most other rhodopsins, *Ns*XeR contains a tryptophan residue at the position of R82 in *Hs*BR (Fig. 10). The bulky positively charged arginine is one of the key determinants of unidirectional ion transport in microbial rhodopsins. Indeed, the substitution of the arginine by neutral glutamine leads to a conversion of the sodium pump KR2 into a light-gated potassium channel [92] and the proton pump *Coccomyxa* Rhodopsin (CsR) into a light-gated proton channel [188]. It was suggested that the highly conservative tryptophan in the position of the arginine might be one of the most important features of xenorhodopsins [32].

Another notable feature is the absence of the charged amino acid in the position of D96 in *Hs*BR (Fig. 10). In *Ns*XeR, H48 in helix B and D220 in helix G are located in that region, together forming a salt bridge. H48 is located closer to the RSB in the ground state of *Ns*XeR, than D96 in *Hs*BR. H48 is also hydrogen bonded to the water molecule (Wat407), similar to what is found in many microbial rhodopsins (Fig. 10). In *Hs*BR an analogous water molecule (Wat502) together with several other water molecules plays a role in the formation of a continuous hydrogen bond chain from the RSB to D96 in L and N states [14, 24, 189]. It is therefore likely that in *Ns*XeR upon retinal isomerization and following RSB reorientation to the cytoplasmic side, a similar chain is formed between RSB and the H48-D220 pair. While in *Hs*BR the deprotonation of the RSB does not proceed to D96, as it is protonated in the L-state, in XeRs it is possible that the pK_a of the H48-D220 pair makes the transfer of the proton from the RSB to the pair energetically favorable. At the same time, the absence of arginine on the extracellular side allows proton influx for RSB reprotonation from the outside of the cell.

Unfortunately, at the moment, there are no available structural data on XeR intermediates, which limits our understanding of the mechanism of the inward light-driven proton pumping. Further investigations, such as time-resolved serial femtosecond and millisecond crystallography at XFELs and synchrotron sources, may shed light on the principles of XeR work.

4.3.4 Viral Rhodopsins

Very recently crystal structures of rhodopsins from giant viruses have become available. The structures of two viral rhodopsins of group 1 (DTS motif rhodopsin from *Phaeocystis globosa* virus 12T, VirR1 and a rhodopsin from Organic Lake phycodnavirus, OLPVR1) were solved at 1.65 Å and 1.4 Å, respectively [40, 110]. The atomic structure of a viral rhodopsin of group 2 (OLPVRII, in this review we refer to is as VirR2) was solved at 1.9 Å [39].

VirR1 and OLPVR1 are very similar; therefore we will mostly refer to VirR1 when describing viral rhodopsins of group 1 since its structure appeared first. Both VirR1 and VirR2 share a similar proteorhodopsin-like fold, albeit with notable differences (Fig. 11). CD loop of VirR1 contains a short α-helix, absent in VirR2 and oriented parallel to the membrane surface and located close to the cytoplasmic hydrophilic/hydrophobic membrane core boundary. The position of helix A differs by almost 2 Å between proteins. Such a difference may originate from the oligomeric assembly of VirR2, as interprotomer contacts are formed mostly through the helices A and B. The protein forms pentamers in various environments, such as liposomes, detergent micelles and lipid matrix of LCP-grown protein crystals [39]. The pentameric assembly is also evidenced by the almost absolute conservation of the amino acid residues, responsible for strong protomer interactions. Therefore, VirR2 (and the whole group 2 of viral rhodopsins) is supposed to form pentamers in vivo (Fig. 3). The unusual feature of the pentamer is also that there is a specific pore in the center of the oligomer (Fig. 11). The pore has a narrow vestibule at the cytoplasmic side, which is capped by five positively charged residues (R29 of each protomer) and is blocked by the hydrocarbon chain in the dark state of VirR2. In between the hydrocarbon chain and the R29-ring five water molecules are coordinated by the backbone oxygens of helix A of the protein, thus forming a tight water pentagon. In the central part of the pore, the vestibule ends with

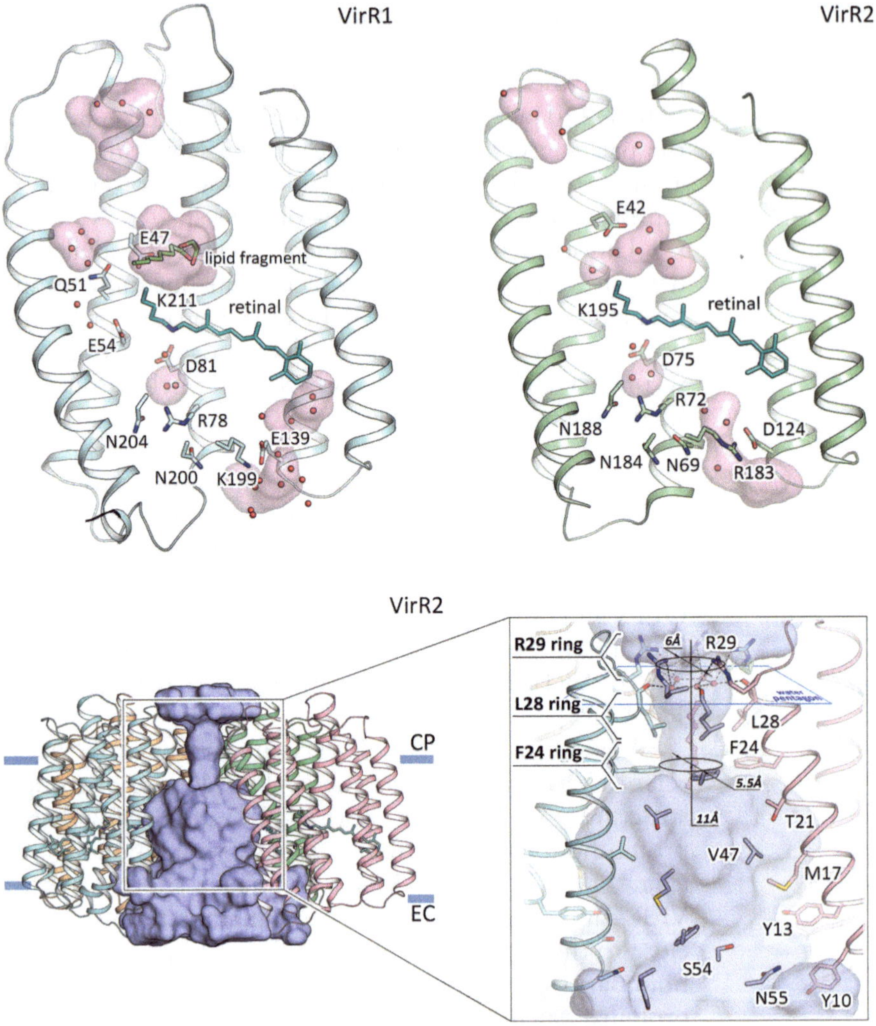

Fig. 11 Structures of viral rhodopsins. Structure of viral rhodopsin of group 1 (cyan, VirR1, PDB ID: 6JO0 [40]) and group 2 (green, VirR2, PDB ID: 6SQG [39]) (top panels). Lipid fragment penetrating VirR1 between helices F and G in the middle part of the protein is colored green. Retinal cofactor is colored teal. Inner cavities are calculated using HOLLOW [173] and shown with pink surface. Section view of the ViR2 pentamer and central pore profile (bottom panels). Central pore is shown with blue surface

the hydrophobic barrier, provided by the bulky residues (L24 and F28 in VirR2). Thus, the shape and properties of the vestibule are determined by the helix A of VirR2. The extracellular half of the pore is wide and relatively polar. Based on an unusual profile and electrostatics of the pore, it was suggested that VirR2 might act as a pentameric light-gated anion channel [39].

VirR1 and OLPVR1, on the other hand, are monomeric in crystals. However, its oligomerization assembly in lipids and in vivo was not investigated and remains unknown (Fig. 3). It was also suggested from the analysis of crystal packing of OLPVR1 that viral rhodopsins of group 1 might form dimers [110]. The conservative amino acids, forming interprotomer contacts in VirR2 (and group 2 of viral rhodopsins) are absent in VirR1 (and also in group 1 of viral rhodopsins). Therefore, it is unlikely that viral rhodopsins of group 1 are organized into similar pentamers, as group 2.

A distinct feature of VirR1 is the presence of the glutamate residue E54 (E51 in OLVPR1) in the helix B, which is similar to the glutamates of CGs of channelrhodopsins (Figs. 6 and 11). Moreover, the region near E54 is polar and filled with several water molecules, which is unusual for the known proton pumps, such as BRs and PRs (Fig. 11).

Both VirR1 and VirR2 lack a proton donor residue in helix C at the position of D96 in *Hs*BR. Instead, they possess a glutamate residue (E47 in VirR1, E44 in OLPVR1, and E42 in VirR2) in helix B, which is thus located closer to the RSB (Fig. 11). In VirR2, E42 is surrounded by numerous water molecules, located in a big polar cavity in the cytoplasmic part of the protein. The hydrogen bond network protrudes from E42 almost to the oligomerization interface and R29 ring, surrounding the central pore. The network also involves helices A and B. The network was suggested to be a key element for triggering the channel opening in VirR2, as it connects the core of the protomer with the central pore.

Surprisingly, in VirR1 and OLPVR1 the lipid molecule was found in the middle of the membrane bilayer, protruding between helices F and G and entering the proteins with the polar glycerol head (Fig. 11). The lipid is likely to be a crystallization artifact, occupying a large polar cavity at the cytoplasmic side of the protein near E47 (E44 in the case of OLPVR1), which can serve as an entrance for ions. However, the functional assays of VirR1 were only performed in *Escherichia coli* cell suspension, and thus it can be that VirR1 is not only a proton pump. The structures of VirR1 and OLPVR1, particularly the presence of E54 (E51) in the helix B and organization of large inner polar cavities, is characteristic for channelrhodopsins. Thus, these proteins might serve as models for structural studies of light-gated ion channels and deciphering the molecular mechanisms of their work. Structural investigations of the intermediate states of both VirR1 and VirR2 may also shed additional light on the function and biological role of viral rhodopsins.

4.3.5 Heliorhodopsins

Very soon after their discovery, crystal structures of two HeRs became available. One structure corresponds to the archaeal heliorhodopsin *Ta*HeR, solved at 2.4 Å [34], while the other two—to the bacterial heliorhodopsin 48C12 [35, 36], which was originally described in 2018 [33]. For 48C12, two high-resolution (both at 1.5 Å) models [36] and one low-resolution model (2.7 Å) [35] are available.

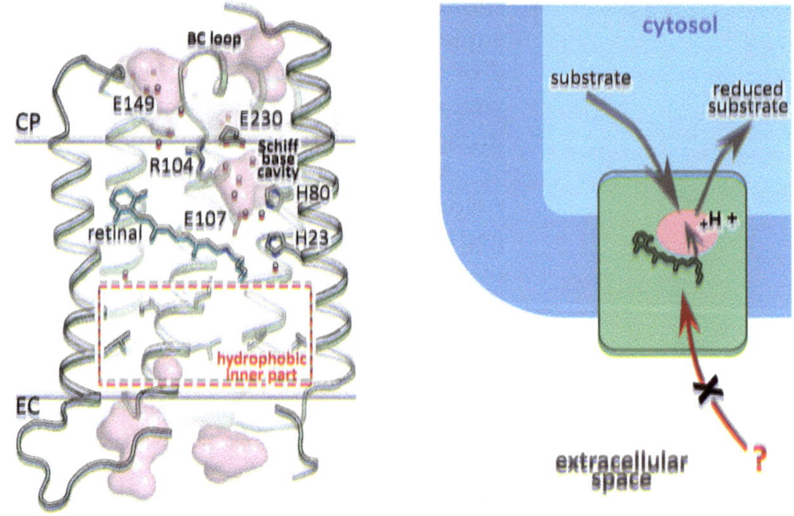

Fig. 12 Structure of HeR 48C12 and putative biological function of HeRs. Structure of 48C12 heliorhodopsin (green, PDB ID: 6SU3) (left panels). Retinal cofactor is colored teal. Inner cavities are calculated using HOLLOW [173] and shown with pink surface; putative scheme of HeRs function as an enzyme in cells (right panels). Green rounded rectangle represents HeR. Pink ellipse represents the Schiff base cavity ("active center"). Gray arrows indicate the ways of substrate and H^+ in HeR. Red arrow indicates blocked penetration of ions and substrates inside HeR from the extracellular bulk due to extremely hydrophobic and tight inner extracellular region of the protein

HeRs form dimers in lipid environments, as demonstrated by high-speed atomic force microscopy (HS-AFM) and also evidenced in LCP-grown crystals of the *Ta*HeR and 48C12 (Fig. 3) [34].

HeRs have several key features, which distinguish them from all other type 1 rhodopsins. First, both bacterial and archaeal HeRs have a short α-helix in the BC loop (Fig. 12). Second, the AB loop is relatively long and contains a β-sheet, which is oriented parallel to the membrane plane and protrudes in the direction of the second protomer of the dimer. Third, the inner extracellular part of the protein is completely hydrophobic and has no water-accessible cavities, which may be the reason for the absence of any pumping/channeling activity in HeRs (Fig. 12). Fourth, on the other side, the inner cytoplasmic part has a big polar cavity (Schiff base cavity), which is surrounded by many rechargeable residues and thus may change its electrostatic properties depending on the surrounding pH value (Fig. 12). As a demonstration, it was shown that 48C12 could accommodate anions near the RSB at acidic pH, and the crystal structure of 48C12 at pH 4.3 revealed the acetate anion bound to the E109 residue in the Schiff base cavity. At neutral pH values, the cavity is filled with seven water molecules, which are the core of the hydrogen bonding network in the cytoplasmic part of HeRs. This cluster of water molecules is supposed to be an acceptor of the proton released from the RSB upon retinal isomerization. As the proton is never released to the protein surface during the photocycle [33], it is likely that the proton is returned to the RSB from the water cluster at the end of the photocycle.

On the other hand, the Schiff base cavity was hypothesized to play a role of an "active center" of HeRs [34, 36], and therefore these proteins may function as enzymes, which catalyze redox reactions like H^+ + substrate → reduced substrate (Fig. 12). In this assumption, the substrate may be (transiently) accommodated in the Schiff base cavity, while H^+ comes from the protonated RSB during the photocycle. Importantly, if this hypothesis is correct, it would be incredible evidence of the natural adaptation of rhodopsins to perform various functions. Indeed, HeRs are perfectly designed by nature to be simultaneously (1) isolated from the extracellular space by a completely hydrophobic and tight inner region and (2) easily accessible from the cytoplasm for putative substrate binding. This is in complete contrast to other type 1 rhodopsins, which usually have more polar and loose extracellular and less polar and tight cytoplasmic parts. For illustration, in *Hs*BR cytoplasmic inner part has only D96 and T46, isolated from the bulk and the RSB by hydrophobic residues, while the extracellular side contains several water-filled cavities (in the case of channelrhodopsin C1C2 a large cavity protrudes from the extracellular bulk directly to the RSB). Therefore, to create the first HeRs in the process of evolution, it was much easier to invert microbial rhodopsins inside the membrane than to mutate both extracellular and cytoplasmic parts of the protein to exchange their composition.

5 Optogenetic Applications

Before the arrival of optogenetics in the early 2000s [78, 82, 84, 101, 190], a broad spectrum of electrode stimulation of the central nervous system (CNS) and the peripheral nervous system was applied with remarkable success for medical treatment. Just to name a few, cochlear implants for the restoration of hearing, electrode stimulation of the nucleus subthalamicus in the brain to remove the Parkinson disease phenomena and Tourette syndrome, subcutaneous pain treatment and electrical stimulation of the heart by electrical pacemakers, and many other applications. These applications are purely electrical without any chemical or biological interventions. The optogenetic approach, as it has been proven in animal experiments, does not simply replace this approach, but it has opened completely new ways in neuroscience and muscle physiology. Due to its versatile possibilities, optogenetics is used in cell biology, neuroscience, and muscle physiology. A short overview is given below.

Optogenetic applications allow the study of the activity of electrically excitable cells with extremely high spatiotemporal precision. The light activation or inhibition of these cells is electrode free and, therefore, minimally invasive [84]. Basically, two methods are used for heterologous expression, the transduction via the use of viruses as gene ferries for the respective optogenes in cell culture as well as in living animals. The working horses for the activation (depolarization) of cells are still the cation-selective channel ChR2 and the inactivating (hyperpolarizing) ion pumps HR and Arch allowing the multimodal control of the cells. A lucky case is that in almost all of the applications, the chromophore of the opsins is delivered by host cells for free, meaning no extra retinal has to be added to form the respective microbial rhodopsin. More recently discovered red light-absorbing ChRs, e.g. Chrimson [95] have remarkably improved the light activation in tissue because of deeper penetration by the reduced light scattering compared to the original blue light absorption by ChR2. Normally in the case of rodents, transgenic animals are produced, which allows the functional expression in different organs or more specific in different cell types. The virus approach is used preferentially in studies with biomedical perspectives. The most promising attempt for its application to neuronal diseases is the recovery of vision for the cure of macular degeneration [191, 192]. First clinical trials, as well as the construction of optical cochlea implants [193] and the application in the peripheral nervous system for pain relief, are underway. Optogenetic manipulation in the brain of mice allowed precise topical allocation with high cell specificity for behavior studies such as aggression [194, 195].

An absolute requirement for the application of optogenetics in humans is that the optogenes have to be tested in non-human primates (NHP). An excellent review with the references therein about the applications, e.g. analysis of neural circuits and the neural basis of behavior, such as sensation and cognition in NHP is given in [196]. Also, the possible biomedical applications of optogenetics in the treatment of deafness and Parkinson's disease are discussed.

6 Outlook

To pursue a refined knowledge of the structure–function relationships in rhodopsins is important for many reasons that go beyond their potential biotechnological applications. The capacity to predict function from sequence is a major challenge for modern biology. If it could be solved it would allow enormous advance in all the fields of human endeavor from medicine to ecology. The knowledge obtained from rhodopsins can also enlighten our understanding of the cell membrane function and the ways in which it affects several cellular mechanisms particularly bioenergetics and ionic homeostasis (for example, intracellular pH). Rhodopsins also represent a beautiful model of how microbes can adjust to a molecular model to perform a multitude of tasks by very elegant, minimal modifications of their sequence or the interaction with other cell components.

Huge abundance of microbial rhodopsins, diversity of their functions, and their enormous light-harversting power point towards their global role in life and global processes on our planet. Discovering new rhodopsins with novel functions, revealing molecular mechanisms of their function, understanding their role in climate and ecology, their applications in science, biotechnologies, and medicine are major challenges in the research area of the discussion.

Acknowledgment

This work was supported by Russian Science Foundation (RSF) Project 21-64-00018.

References

1. Oesterhelt D, Stoeckenius W (1971) Rhodopsin-like protein from the purple membrane of Halobacterium halobium. Nat New Biol 233:149–152
2. Matsuno-Yagi A, Mukohata Y (1977) Two possible roles of bacteriorhodopsin; a comparative study of strains of Halobacterium halobium differing in pigmentation. Biochem Biophys Res Commun 78:237–243
3. Hildebrand E, Dencher N (1975) Two photosystems controlling behavioral responses of Halobacterium halobium. Nature 257:46
4. Takahashi T, Tomioka H, Kamo N et al (1985) A photosystem other than PS370 also mediates the negative phototaxis of Halobacterium halobium. FEMS Microbiol Lett 28:161
5. Henderson R, Unwin PNT (1975) Three-dimensional model of purple membrane obtained by electron microscopy. Nature 257:28–32
6. Landau EM, Rosenbusch JP (1996) Lipidic cubic phases: a novel concept for the crystallization of membrane proteins. Proc Natl Acad Sci 93:14532–14535
7. Pebay-Peyroula E, Rummel G, Rosenbusch JP et al (1997) X-ray structure of bacteriorhodopsin at 2.5 angstroms from microcrystals grown in lipidic cubic phases. Science 277: 1676
8. Luecke H, Richter HT, Lanyi JK (1998) Proton transfer pathways in bacteriorhodopsin at 2.3 angstrom resolution. Science 280:1934
9. Neutze R, Pebay-Peyroula E, Edman K et al (2002) Bacteriorhodopsin: a high-resolution structural view of vectorial proton transport. Biochim Biophys Acta Biomembr 1565:144
10. Royant A, Edman K, Ursby T et al (2000) Helix deformation is coupled to vectorial proton transport in the photocycle of bacteriorhodopsin. Nature 406:645–648
11. Schobert B, Cupp-Vickery J, Hornak V et al (2002) Crystallographic structure of the K intermediate of bacteriorhodopsin: conservation of free energy after photoisomerization of the retinal. J Mol Biol 321:715
12. Luecke H, Schobert B, Richter HT et al (1999) Structural changes in bacteriorhodopsin during ion transport at 2 angstrom resolution. Science 286:255
13. Efremov R, Gordeliy VI, Heberle J et al (2006) Time-resolved microspectroscopy on a single crystal of bacteriorhodopsin reveals lattice-induced differences in the photocycle kinetics. Biophys J 91:1441–1451
14. Kouyama T, Nishikawa T, Tokuhisa T et al (2004) Crystal structure of the L intermediate of Bacteriorhodopsin: evidence for vertical translocation of a water molecule during the proton pumping cycle. J Mol Biol 335: 531–546

15. Lanyi JK (2004) Bacteriorhodopsin. Annu Rev Physiol 66:665–688
16. Edmonds BW, Luecke H (2004) Atomic resolution structures and the mechanism of ion pumping in bacteriorhodopsin. Front Biosci 9:1556–1566
17. Nango E, Royant A, Kubo M et al (2016) A three-dimensional movie of structural changes in bacteriorhodopsin. Science 354: 1552–1557
18. Borshchevskiy V, Round E, Erofeev I et al (2014) Low-dose X-ray radiation induces structural alterations in proteins. Acta Crystallogr D Biol Crystallogr 70:2675–2685
19. Borshchevskiy V, Efremov R, Moiseeva E et al (2010) Overcoming merohedral twinning in crystals of bacteriorhodopsin grown in lipidic mesophase. Acta Crystallogr D Biol Crystallogr 66:26–32
20. Borshchevskiy VI, Round ES, Popov AN et al (2011) X-ray-radiation-induced changes in bacteriorhodopsin structure. J Mol Biol 409: 813–825
21. Borshchevskiy V, Gordeliy V (2012) Crystallization of membrane proteins: merohedral twinning of crystals. In: Modern aspects of bulk crystal and thin film preparation. InTechOpen, London
22. Wickstrand C, Dods R, Royant A et al (2015) Bacteriorhodopsin: would the real structural intermediates please stand up? Biochim Biophys Acta 1850:536
23. Nogly P, Weinert T, James D et al (2018) Retinal isomerization in bacteriorhodopsin captured by a femtosecond x-ray laser. Science 361:eaat0094
24. Weinert T, Skopintsev P, James D et al (2019) Proton uptake mechanism in bacteriorhodopsin captured by serial synchrotron crystallography. Science 365:61
25. Nass Kovacs G, Colletier JP, Grünbein ML et al (2019) Three-dimensional view of ultrafast dynamics in photoexcited bacteriorhodopsin. Nat Commun 10:3177
26. Beja O, Aravind L, Koonin EV et al (2000) Bacterial rhodopsin: evidence for a new type of phototrophy in the sea. Science 289: 1902–1906
27. Nagel G (2002) Channelrhodopsin-1: a light-gated proton channel in green algae. Science 296:2395–2398
28. Inoue K, Ono H, Abe-Yoshizumi R et al (2013) A light-driven sodium ion pump in marine bacteria. Nat Commun 4:1678
29. Avelar GM, Schumacher RI, Zaini PA et al (2014) A rhodopsin-guanylyl cyclase gene fusion functions in visual perception in a fungus. Curr Biol 24:1234–1240
30. Govorunova EG, Sineshchekov OA, Janz R et al (2015) Natural light-gated anion channels: a family of microbial rhodopsins for advanced optogenetics. Science 349:647–650
31. Inoue K, Ito S, Kato Y et al (2016) A natural light-driven inward proton pump. Nat Commun 7:13415
32. Shevchenko V, Mager T, Kovalev K et al (2017) Inward H^+ pump xenorhodopsin: mechanism and alternative optogenetic approach. Sci Adv 3:e1603187
33. Pushkarev A, Inoue K, Larom S et al (2018) A distinct abundant group of microbial rhodopsins discovered using functional metagenomics. Nature 558:595–599
34. Shihoya W, Inoue K, Singh M et al (2019) Crystal structure of heliorhodopsin. Nature 574:132–136
35. Lu Y, Zhou XE, Gao X et al (2020) Crystal structure of heliorhodopsin 48C12. Cell Res 30:88–90
36. Kovalev K, Volkov D, Astashkin R et al (2020) High-resolution structural insights into the heliorhodopsin family. Proc Natl Acad Sci 117:4131–4141
37. Gómez-Consarnau L, Raven JA, Levine NM et al (2019) Microbial rhodopsins are major contributors to the solar energy captured in the sea. Sci Adv 5:eaaw8855
38. Yutin N, Koonin EV (2012) Proteorhodopsin genes in giant viruses. Biol Direct 7:34
39. Bratanov D, Kovalev K, Machtens J-P et al (2019) Unique structure and function of viral rhodopsins. Nat Commun 10:4939
40. Needham DM, Yoshizawa S, Hosaka T et al (2019) A distinct lineage of giant viruses brings a rhodopsin photosystem to unicellular marine predators. Proc Natl Acad Sci 116: 20574–20583
41. Dassarma S, Schwieterman EW (2018) Early evolution of purple retinal pigments on Earth and implications for exoplanet biosignatures. Int J Astrobiol 20:241
42. Olson DK, Yoshizawa S, Boeuf D et al (2018) Proteorhodopsin variability and distribution in the North Pacific Subtropical Gyre. ISME J 12:1047–1060
43. Becker EA, Yao AI, Seitzer PM et al (2016) A large and phylogenetically diverse class of type 1 opsins lacking a canonical retinal binding site. PLoS One 11:1–20
44. Frigaard NU, Martinez A, Mincer TJ et al (2006) Proteorhodopsin lateral gene transfer

between marine planktonic Bacteria and Archaea. Nature 439:847–850
45. Slamovits CH, Okamoto N, Burri L et al (2011) A bacterial proteorhodopsin proton pump in marine eukaryotes. Nat Commun 2: 183
46. Nakajima Y, Tsukamoto T, Kumagai Y et al (2018) Presence of a haloarchaeal halorhodopsin-like Cl- pump in marine bacteria. Microbes Environ 33:89–97
47. Antón J, Rosselló-Mora R, Rodríguez-Valera F et al (2000) Extremely halophilic Bacteria in crystallizer ponds from solar salterns. Appl Environ Microbiol 66:3052
48. Kall L, Krogh A, Sonnhammer ELL (2007) Advantages of combined transmembrane topology and signal peptide prediction--the Phobius web server. Nucleic Acids Res 35: W429–W432
49. Sievers F, Wilm A, Dineen D et al (2011) Fast, scalable generation of high-quality protein multiple sequence alignments using Clustal Omega. Mol Syst Biol 7:539
50. Mongodin EF, Nelson KE, Daugherty S et al (2005) The genome of Salinibacter ruber: convergence and gene exchange among hyperhalophilic bacteria and archaea. Proc Natl Acad Sci U S A 102:18147
51. Balashov SP, Imasheva ES, Boichenko VA et al (2005) Xanthorhodopsin: a proton pump with a light-harvesting carotenoid antenna. Science 309:2061–2064
52. Balashov SP, Imasheva ES, Wang JM et al (2008) Excitation energy-transfer and the relative orientation of retinal and carotenoid in xanthorhodopsin. Biophys J 95:2402–2414
53. Galinski EA (1995) Osmoadaptation in bacteria. Adv Microb Physiol 37:273–328
54. Oren A, Heldal M, Norland S et al (2002) Intracellular ion and organic solute concentrations of the extremely halophilic bacterium Salinibacter ruber. Extremophiles 6:491–498
55. Oren A (2008) Microbial life at high salt concentrations: phylogenetic and metabolic diversity. Saline Syst 4:2
56. Wickstrand C, Nogly P, Nango E et al (2019) Bacteriorhodopsin: structural insights revealed using X-ray lasers and synchrotron radiation. Annu Rev Biochem 88:59
57. Sharma AK, Spudich JL, Doolittle WF (2006) Microbial rhodopsins: functional versatility and genetic mobility. Trends Microbiol 14: 463
58. Pinhassi J, DeLong EF, Béjà O et al (2016) Marine bacterial and archaeal ion-pumping rhodopsins: genetic diversity, physiology, and ecology. Microbiol Mol Biol Rev 80:929
59. Chizhov I, Chernavskii DS, Engelhard M et al (1996) Spectrally silent transitions in the bacteriorhodopsin photocycle. Biophys J 71: 2329–2345
60. Hofrichter J, Henry ER, Lozier RH (1989) Photocycles of bacteriorhodopsin in light- and dark-adapted purple membrane studied by time-resolved absorption spectroscopy. Biophys J 56:693
61. Kandori H (2004) Hydration switch model for the proton transfer in the Schiff base region of bacteriorhodopsin. Biochim Biophys Acta 1658:72
62. Gerwert K, Freier E, Wolf S (2014) The role of protein-bound water molecules in microbial rhodopsins. Biochim Biophys Acta - Bioenerg 1837:606–613
63. Chow BY, Han X, Dobry AS et al (2010) High-performance genetically targetable optical neural silencing by light-driven proton pumps. Nature 463:98
64. Martinez-Garcia M, Swan BK, Poulton NJ et al (2012) High-throughput single-cell sequencing identifies photoheterotrophs and chemoautotrophs in freshwater bacterioplankton. ISME J 6:113
65. Venter JC, Remington K, Heidelberg JF et al (2004) Environmental genome shotgun sequencing of the Sargasso sea. Science 304: 66
66. Gushchin I, Chervakov P, Kuzmichev P et al (2013) Structural insights into the proton pumping by unusual proteorhodopsin from nonmarine bacteria. Proc Natl Acad Sci 110: 12631–12636
67. Dioumaev AK, Brown LS, Shih J et al (2002) Proton transfers in the photochemical reaction cycle of proteorhodopsin. Biochemistry 41:5348
68. Ran T, Ozorowski G, Gao Y et al (2013) Cross-protomer interaction with the photoactive site in oligomeric proteorhodopsin complexes. Acta Crystallogr D Biol Crystallogr 69:1965
69. Hempelmann F, Hölper S, Verhoefen MK et al (2011) His75-Asp97 cluster in green proteorhodopsin. J Am Chem Soc 133:4645
70. Váró G, Brown LS, Lakatos M et al (2003) Characterization of the photochemical reaction cycle of proteorhodopsin. Biophys J 84: 1202
71. Lakatos M, Lanyi JK, Szakács J et al (2003) The photochemical reaction cycle of proteorhodopsin at low pH. Biophys J 84:3252
72. Kawanabe A, Furutani Y, Jung KH et al (2011) An inward proton transport using anabaena sensory rhodopsin. J Microbiol 49:1

73. Ugalde JA, Podell S, Narasingarao P et al (2011) Xenorhodopsins, an enigmatic new class of microbial rhodopsins horizontally transferred between archaea and bacteria. Biol Direct 6:52
74. Vogeley L, Sineshchekov OA, Trivedi VD et al (2004) Anabaena sensory rhodopsin: a photochromic color sensor at 2.0 Å. Science 306: 1390
75. Bulzu P-A, Andrei A-Ş, Salcher MM et al (2019) Casting light on Asgardarchaeota metabolism in a sunlit microoxic niche. Nat Microbiol 4:1129–1137
76. Inoue K, Tsunoda SP, Singh M et al (2020) Schizorhodopsins: a family of rhodopsins from Asgard archaea that function as light-driven inward H^+ pumps. Sci Adv 6:eaaz2441
77. Harris A, Lazaratos M, Siemers M et al (2020) Mechanism of inward proton transport in an antarctic microbial rhodopsin. J Phys Chem B 124:4851–4872
78. Nagel G, Szellas T, Huhn W et al (2003) Channelrhodopsin-2, a directly light-gated cation-selective membrane channel. Proc Natl Acad Sci 100:13940–13945
79. Harz H, Hegemann P (1991) Rhodopsin-regulated calcium currents in Chlamydomonas. Nature 351:489
80. Ehlenbeck S, Gradmann D, Braun FJ et al (2002) Evidence for a light-induced H+ conductance in the eye of the green alga Chlamydomonas reinhardtii. Biophys J 82:740
81. Böhm M, Boness D, Fantisch E et al (2019) Channelrhodopsin-1 phosphorylation changes with phototactic behavior and responds to physiological stimuli in chlamydomonas. Plant Cell 31:886
82. Boyden ES, Zhang F, Bamberg E et al (2005) Millisecond-timescale, genetically targeted optical control of neural activity. Nat Neurosci 8:1263–1268
83. Gradinaru V, Zhang F, Ramakrishnan C et al (2010) Molecular and cellular approaches for diversifying and extending optogenetics. Cell 141:154–165
84. Zhang F, Aravanis AM, Adamantidis A et al (2007) Circuit-breakers: optical technologies for probing neural signals and systems. Nat Rev Neurosci 8:577
85. Gunaydin LA, Yizhar O, Berndt A et al (2010) Ultrafast optogenetic control. Nat Neurosci 13:387–392
86. Marshel JH, Kim YS, Machado TA et al (2019) Cortical layer-specific critical dynamics triggering perception. Science 365: eaaw5202
87. Govorunova EG, Sineshchekov OA, Rodarte EM et al (2017) The expanding family of natural anion channelrhodopsins reveals large variations in kinetics, conductance, and spectral sensitivity. Sci Rep 7:43358
88. Berndt A, Lee SY, Wietek J et al (2016) Structural foundations of optogenetics: determinants of channelrhodopsin ion selectivity. Proc Natl Acad Sci U S A 113:822
89. Kato HE, Kim YS, Paggi JM et al (2018) Structural mechanisms of selectivity and gating in anion channelrhodopsins. Nature 561: 349
90. Oppermann J, Fischer P, Silapetere A et al (2019) MerMAIDs: a family of metagenomically discovered marine anion-conducting and intensely desensitizing channelrhodopsins. Nat Commun 10:3315
91. Mager T, La Morena DLD, Senn V et al (2018) High frequency neural spiking and auditory signaling by ultrafast red-shifted optogenetics. Nat Commun 9:1750
92. Vogt A, Silapetere A, Grimm C et al (2019) Engineered passive potassium conductance in the KR2 sodium pump. Biophys J 116:1941
93. Grimm C, Silapetere A, Vogt A et al (2018) Electrical properties, substrate specificity and optogenetic potential of the engineered light-driven sodium pump eKR2. Sci Rep 8:9316
94. Zhang F, Prigge M, Beyrière F et al (2008) Red-shifted optogenetic excitation: a tool for fast neural control derived from Volvox carteri. Nat Neurosci 11:631
95. Klapoetke NC, Murata Y, Kim SS et al (2014) Independent optical excitation of distinct neural populations. Nat Methods 11: 338–346
96. Oda K, Vierock J, Oishi S et al (2018) Crystal structure of the red light-activated channelrhodopsin Chrimson. Nat Commun 9:3949
97. Mukherjee S, Hegemann P, Broser M (2019) Enzymerhodopsins: novel photoregulated catalysts for optogenetics. Curr Opin Struct Biol 57:118
98. Scheib U, Broser M, Constantin OM et al (2018) Rhodopsin-cyclases for photocontrol of cGMP/cAMP and 2.3 Å structure of the adenylyl cyclase domain. Nat Commun 9: 2046
99. Yizhar O, Fenno LE, Prigge M et al (2011) Neocortical excitation/inhibition balance in information processing and social dysfunction. Nature 477:171–178
100. Lin JY, Lin MZ, Steinbach P et al (2009) Characterization of engineered channelrhodopsin variants with improved properties and kinetics. Biophys J 96:1803–1814

101. Nagel G, Brauner M, Liewald JF et al (2005) Light activation of Channelrhodopsin-2 in excitable cells of caenorhabditis elegans triggers rapid behavioral responses. Curr Biol 15: 2279
102. Berndt A, Yizhar O, Gunaydin LA et al (2009) Bi-stable neural state switches. Nat Neurosci 12:229–234
103. Kleinlogel S, Feldbauer K, Dempski RE et al (2011) Ultra light-sensitive and fast neuronal activation with the Ca2+-permeable channelrhodopsin CatCh. Nat Neurosci 14:513–518
104. Wietek J, Wiegert JS, Adeishvili N et al (2014) Conversion of channelrhodopsin into a light-gated chloride channel. Science 344: 409–412
105. Lin JY, Knutsen PM, Muller A et al (2013) ReaChR: a red-shifted variant of channelrhodopsin enables deep transcranial optogenetic excitation. Nat Neurosci 16:1499–1508
106. Govorunova EG, Sineshchekov OA, Spudich JL (2016) Structurally distinct cation channelrhodopsins from Cryptophyte algae. Biophys J 110:2302–2304
107. Sineshchekov OA, Govorunova EG, Li H et al (2020) Conductance mechanisms of rapidly desensitizing cation channelrhodopsins from cryptophyte algae. MBio 11:e00657–e00620
108. López JL, Golemba M, Hernández E et al (2017) Microbial and viral-like rhodopsins present in coastal marine sediments from four polar and subpolar regions. FEMS Microbiol Ecol 93:9
109. Philosof A, Béjà O (2013) Bacterial, archaeal and viral-like rhodopsins from the Red Sea. Environ Microbiol Rep 5:475
110. Zabelskii D, Alekseev A, Kovalev K et al (2020) Viral rhodopsins 1 are an unique family of light-gated cation channels. Nat Commun 11:5707
111. Kato HE, Zhang F, Yizhar O et al (2012) Crystal structure of the channelrhodopsin light-gated cation channel. Nature 482:369
112. Berndt A, Lee SY, Ramakrishnan C et al (2014) Structure-guided transformation of channelrhodopsin into a light-activated chloride channel. Science 344:420–424
113. Sineshchekov OA, Govorunova EG, Jung K-H et al (2005) Rhodopsin-mediated photoreception in cryptophyte flagellates. Biophys J 89:4310–4319
114. Eisenhauer K, Kuhne J, Ritter E et al (2012) In channelrhodopsin-2 Glu-90 is crucial for ion selectivity and is deprotonated during the photocycle. J Biol Chem 287:6904
115. Kim YS, Kato HE, Yamashita K et al (2018) Crystal structure of the natural anion-conducting channelrhodopsin GtACR1. Nature 561:343–348
116. Li H, Huang C-Y, Govorunova EG et al (2019) Crystal structure of a natural light-gated anion channelrhodopsin. elife 8: e41741
117. Sineshchekov OA, Li H, Govorunova EG et al (2016) Photochemical reaction cycle transitions during anion channelrhodopsin gating. Proc Natl Acad Sci 113:E1993–E2000
118. Sineshchekov OA, Govorunova EG, Li H et al (2015) Gating mechanisms of a natural anion channelrhodopsin. Proc Natl Acad Sci 112: 14236–14241
119. Kuhne J, Vierock J, Tennigkeit SA et al (2019) Unifying photocycle model for light adaptation and temporal evolution of cation conductance in channelrhodopsin-2. Proc Natl Acad Sci U S A 116:9380
120. Bork P, Bowler C, De Vargas C et al (2015) Tara Oceans studies plankton at Planetary scale. Science 348:873
121. Gregory AC, Zayed AA, Conceição-Neto N et al (2019) Marine DNA viral macro- and microdiversity from pole to pole. Cell 177: 1109
122. Nack M, Radu I, Gossing M et al (2010) The DC gate in channelrhodopsin-2: crucial hydrogen bonding interaction between C128 and D156. Photochem Photobiol Sci 9:194
123. Rozenberg A, Oppermann J, Wietek J et al (2020) Lateral gene transfer of anion-conducting channelrhodopsins between green algae and giant viruses. Microbiology 30:4910
124. Govorunova EG, Sineshchekov OA, Li H et al (2020) RubyACRs, nonalgal anion channelrhodopsins with highly red-shifted absorption. Proc Natl Acad Sci 117:22833–22840
125. Matsuno-Yagi A, Mukohata Y (1980) ATP synthesis linked to light-dependent proton uptake in a red mutant strain of Halobacterium lacking bacteriorhodopsin. Arch Biochem Biophys 199:297–303
126. Sasaki J, Brown LS, Chon YS et al (1995) Conversion of bacteriorhodopsin into a chloride ion pump. Science 269:73
127. Kolbe M, Besir H, Essen LO et al (2000) Structure of the light-driven chloride pump halorhodopsin at 1.8 Å resolution. Science 288:1390
128. Luecke H, Schobert B, Richter HT et al (1999) Structure of bacteriorhodopsin at 1.55 Å resolution. J Mol Biol 291:899–911
129. Muroda K, Nakashima K, Shibata M et al (2012) Protein-bound water as the

determinant of asymmetric functional conversion between light-driven proton and chloride pumps. Biochemistry 51:4677

130. Gmelin W, Zeth K, Efremov R et al (2007) The crystal structure of the L1 intermediate of halorhodopsin at 1.9 Å resolution†. Photochem Photobiol 83:369
131. Lanyi JK, Vodyanoy V (1986) Flash spectroscopic studies of the kinetics of the halorhodopsin photocycle. Biochemistry 25:1465
132. Chizhov I, Engelhard M (2001) Temperature and halide dependence of the photocycle of halorhodopsin from Natronobacterium pharaonis. Biophys J 81:1600–1612
133. Hasemi T, Kikukawa T, Kamo N et al (2016) Characterization of a cyanobacterial chloride-pumping rhodopsin and its conversion into a proton pump. J Biol Chem 291:355
134. Niho A, Yoshizawa S, Tsukamoto T et al (2017) Demonstration of a light-driven SO42-transporter and its spectroscopic characteristics. J Am Chem Soc 139:4376
135. Inoue K, Koua FHM, Kato Y et al (2014) Spectroscopic study of a light-driven chloride ion pump from marine bacteria. J Phys Chem B 118:11190–11199
136. Yoshizawa S, Kumagai Y, Kim H et al (2014) Functional characterization of flavobacteria rhodopsins reveals a unique class of light-driven chloride pump in bacteria. Proc Natl Acad Sci U S A 111:6732
137. Kim K, Kwon S-K, Jun S-H et al (2016) Crystal structure and functional characterization of a light-driven chloride pump having an NTQ motif. Nat Commun 7:12677
138. Inoue K, Nomura Y, Kandori H (2016) Asymmetric functional conversion of eubacterial light-driven ion pumps. J Biol Chem 291:9883–9893
139. Kaneko A, Inoue K, Kojima K et al (2017) Conversion of microbial rhodopsins: insights into functionally essential elements and rational protein engineering. Biophys Rev 9:861
140. Kato Y, Inoue K, Kandori H (2015) Kinetic analysis of H+–Na+ selectivity in a light-driven Na+-pumping rhodopsin. J Phys Chem Lett 6:5111–5115
141. Kandori H, Inoue K, Tsunoda SP (2018) Light-driven sodium-pumping rhodopsin: a new concept of active transport. Chem Rev 118:10646–10658
142. Balashov SP, Imasheva ES, Dioumaev AK et al (2014) Light-driven Na+ pump from Gillisia limnaea: a high-affinity Na+ binding site is formed transiently in the photocycle. Biochemistry 53:7549–7561
143. Kato HE, Inoue K, Abe-Yoshizumi R et al (2015) Structural basis for Na+ transport mechanism by a light-driven Na+ pump. Nature 521:48–53
144. Gushchin I, Shevchenko V, Polovinkin V et al (2015) Crystal structure of a light-driven sodium pump. Nat Struct Mol Biol 22:390–396
145. Kovalev K, Polovinkin V, Gushchin I et al (2019) Structure and mechanisms of sodium-pumping KR2 rhodopsin. Sci Adv 5:eaav2671
146. Konno M, Kato Y, Kato HE et al (2016) Mutant of a light-driven sodium ion pump can transport cesium ions. J Phys Chem Lett 7:51–55
147. Mamedov AM, Bertsova YV, Anashkin VA et al (2018) Identification of the key determinant of the transport promiscuity in Na+-translocating rhodopsins. Biochem Biophys Res Commun 499:600–604
148. Kateriya S (2004) "Vision" in single-celled algae. News Physiol Sci 19:133–137
149. Luck M, Mathes T, Bruun S et al (2012) A photochromic histidine kinase rhodopsin (HKR1) that is bimodally switched by ultraviolet and blue light. J Biol Chem 287:40083–40090
150. Dunin-Horkawicz S, Lupas AN (2010) Comprehensive analysis of HAMP domains: implications for transmembrane signal transduction. J Mol Biol 397:1156
151. Tian Y, Gao S, von der Heyde EL et al (2018) Two-component cyclase opsins of green algae are ATP-dependent and light-inhibited guanylyl cyclases. BMC Biol 16:144
152. Kianianmomeni A, Hallmann A (2014) Algal photoreceptors: in vivo functions and potential applications. Planta 239:1
153. Yoshida K, Tsunoda SP, Brown LS et al (2017) A unique choanoflagellate enzyme rhodopsin exhibits light-dependent cyclic nucleotide phosphodiesterase activity. J Biol Chem 292:7531–7541
154. Lamarche LB, Kumar RP, Trieu MM et al (2017) Purification and characterization of RhoPDE, a retinylidene/phosphodiesterase fusion protein and potential optogenetic tool from the choanoflagellate Salpingoeca rosetta. Biochemistry 56:5812
155. Pandit J, Forman MD, Fennell KF et al (2009) Mechanism for the allosteric regulation of phosphodiesterase 2A deduced from the X-ray structure of a near full-length construct. Proc Natl Acad Sci U S A 106:18225
156. Ikuta T, Shihoya W, Sugiura M et al (2020) Structural insights into the mechanism of

rhodopsin phosphodiesterase. Nat Commun 11:5605

157. Rauch A, Leipelt M, Russwurm M et al (2008) Crystal structure of the guanylyl cyclase Cya2. Proc Natl Acad Sci U S A 105: 15720

158. Trieu MM, Devine EL, Lamarche LB et al (2017) Expression, purification, and spectral tuning of RhoGC, a retinylidene/guanylyl cyclase fusion protein and optogenetics tool from the aquatic fungus Blastocladiella emersonii. J Biol Chem 292:10379

159. Scheib U, Stehfest K, Gee CE et al (2015) The rhodopsin-guanylyl cyclase of the aquatic fungus Blastocladiella emersonii enables fast optical control of cGMP signaling. Sci Signal 8:rs8

160. Feldbauer K, Zimmermann D, Pintschovius V et al (2009) Channelrhodopsin-2 is a leaky proton pump. Proc Natl Acad Sci U S A 106: 12317

161. Shibukawa A, Kojima K, Nakajima Y et al (2019) Photochemical characterization of a new heliorhodopsin from the gram-negative eubacterium Bellilinea caldifistulae (BcHeR) and comparison with heliorhodopsin-48C12. Biochemistry 58:2934

162. Singh M, Katayama K, Béjà O et al (2019) Anion binding to mutants of the Schiff base counterion in heliorhodopsin 48C12. Phys Chem Chem Phys 21:23663

163. Hashimoto M, Katayama K, Furutani Y et al (2020) Zinc binding to heliorhodopsin. J Phys Chem Lett 11:8604–8609

164. Volkov O, Kovalev K, Polovinkin V et al (2017) Structural insights into ion conduction by channelrhodopsin 2. Science 358: eaan8862

165. Luecke H, Schobert B, Stagno J et al (2008) Crystallographic structure of xanthorhodopsin, the light-driven proton pump with a dual chromophore. Proc Natl Acad Sci U S A 105: 16561

166. Morizumi T, Ou WL, Van Eps N et al (2019) X-ray crystallographic structure and oligomerization of gloeobacter rhodopsin. Sci Rep 9:11283

167. Tsukamoto T, Mizutani K, Hasegawa T et al (2016) X-ray crystallographic structure of thermophilic rhodopsin: implications for high thermal stability and optogenetil function. J Biol Chem 291:12223

168. Klyszejko AL, Shastri S, Mari SA et al (2008) Folding and assembly of proteorhodopsin. J Mol Biol 376:35

169. Yang J, Aslimovska L, Glaubitz C (2011) Molecular dynamics of proteorhodopsin in lipid bilayers by solid-state NMR. J Am Chem Soc 133:4874

170. Reckel S, Gottstein D, Stehle J et al (2011) Solution NMR structure of proteorhodopsin. Angew Chem Int Ed Eng 50:11942

171. Shi L, Lake EMR, Ahmed MAM et al (2009) Solid-state NMR study of proteorhodopsin in the lipid environment: secondary structure and dynamics. Biochim Biophys Acta Biomembr 1788:2563

172. Gushchin I, Gordeliy V (2018) Microbial rhodopsins. Subcell Biochem 87:19

173. Ho BK, Gruswitz F (2008) HOLLOW: generating accurate representations of channel and interior surfaces in molecular structures. BMC Struct Biol 8:49

174. Petrovskaya LE, Balashov SP, Lukashev EP et al (2015) ESR — a retinal protein with unusual properties from Exiguobacterium sibiricum. Biochem Mosc 80:688–700

175. Harris A, Ljumovic M, Bondar AN et al (2015) A new group of eubacterial light-driven retinal-binding proton pumps with an unusual cytoplasmic proton donor. Biochim Biophys Acta Bioenerg 1847:1518

176. Maliar N, Okhrimenko IS, Petrovskaya LE et al (2020) Novel pH-sensitive microbial rhodopsin from Sphingomonas paucimobilis. Dokl Biochem Biophys 495:342–346

177. Lawson MA, Zacks DN, Derguini F et al (1991) Retinal analog restoration of photophobic responses in a blind Chlamydomonas reinhardtii mutant. Evidence for an archaebacterial like chromophore in a eukaryotic rhodopsin. Biophys J 60:1490

178. Rappleye M, Berndt A (2019) Structural basis for ion selectivity and engineering in channelrhodopsins. Curr Opin Struct Biol 57:176

179. Nogly P, Standfuss J (2015) Light-driven Na^+ pumps as next-generation inhibitory optogenetic tools. Nat Struct Mol Biol 22:351

180. Maliar N, Kovalev K, Baeken C et al (2020) Crystal structure of the N112A mutant of the light-driven sodium pump KR2. Crystals 10: 496

181. Kovalev K, Astashkin R, Gushchin I et al (2020) Molecular mechanism of light-driven sodium pumping. Nat Commun 11:2137

182. Skopintsev P, Ehrenberg D, Weinert T et al (2020) Femtosecond-to-millisecond structural changes in a light-driven sodium pump. Nature 583:314–318

183. Shibata M, Inoue K, Ikeda K et al (2018) Oligomeric states of microbial rhodopsins determined by high-speed atomic force microscopy and circular dichroic spectroscopy. Sci Rep 8:8262

184. Gushchin I, Shevchenko V, Polovinkin V et al (2016) Structure of the light-driven sodium pump KR2 and its implications for optogenetics. FEBS J 283:1232–1238
185. Zheng H, Chruszcz M, Lasota P et al (2008) Data mining of metal ion environments present in protein structures. J Inorg Biochem 102:1765–1776
186. Kouyama T, Kawaguchi H, Nakanishi T et al (2015) Crystal structures of the L_1, L_2, N, and O states of pharaonis halorhodopsin. Biophys J 108:2680
187. Yun JH, Li X, Park JH et al (2019) Non-cryogenic structure of a chloride pump provides crucial clues to temperature-dependent channel transport efficiency. J Biol Chem 294:794
188. Fudim R, Szczepek M, Vierock J et al (2019) Design of a light-gated proton channel based on the crystal structure of coccomyxa rhodopsin. Sci Signal 12:eaav4203
189. Nango E, Royant A, Kubo M et al (2016) A three-dimensional movie of structural changes in bacteriorhodopsin. Science 354:1552
190. Zemelman BV, Lee GA, Ng M et al (2002) Selective photostimulation of genetically ChARGed neurons. Neuron 33:15–22
191. Busskamp V, Duebel J, Balya D et al (2010) Genetic reactivation of cone photoreceptors restores visual responses in retinitis pigmentosa. Science 329:413
192. Li M, Atmaca-Sonmez P, Othman M et al (2006) CFH haplotypes without the Y402H coding variant show strong association with susceptibility to age-related macular degeneration. Nat Genet 38:1049
193. Dombrowski T, Rankovic V, Moser T (2019) Toward the optical cochlear implant. Cold Spring Harb Perspect Med 9:a033225
194. Anderson DJ (2012) Optogenetics, sex, and violence in the brain: implications for psychiatry. Biol Psychiatry 71:1081
195. Josselyn SA, Tonegawa S (2020) Memory engrams: recalling the past and imagining the future. Science 367:eaaw4325
196. El-Shamayleh Y, Horwitz GD (2019) Primate optogenetics: progress and prognosis. Proc Natl Acad Sci U S A 116:26195

Chapter 2

Molecular Biology of Microbial Rhodopsins

Martin Engelhard

Abstract

Research on type 1 rhodopsins spans now a history of 50 years. Originally, just archaeal ion pumps and sensors have been discovered. However, with modern genetic techniques and gene sequencing tools, more and more proteins were identified in all kingdoms of life. Spectroscopic and other biophysical studies revealed quite diverse functions. Ion pumps, sensors, and channels are imprinted in the same seven-helix transmembrane protein scaffold carrying a retinal prosthetic group. In this review, molecular biology methods are described, which enabled the elucidation of their function and structure leading to optogenetic applications.

Key words Purification of rhodopsins, Heterologous expression, Cell free expression, Molecular biology of optogenetics

1 Introduction

Membrane proteins have been in the focus of intensive research from a slow start in the beginning almost 150 years ago to an exponential growth since the 1980s [1]. Developments of new methods and methodology were decisive to pave the way towards unravelling structure and function of this family of proteins. Hurdles that had to be overcome, reflected the limiting experimental repertoire available to researchers in those early years. Consequently, those tissues were first investigated, which displayed properties easily to be detected. One branch of investigations focussed on excitable tissues like nerve and muscle. Another line of research concentrated on colored membrane fragments like the retina and almost 100 years later on purple membrane. The next step in the investigation of membrane proteins afforded the isolation and purification of a particular target molecule. The succeeding hurdle that had to be dealt with required a solubilization step of the membrane and appropriate chromatography. These steps contained the risk of denaturing the protein. However, the purification of photoactive pigments had the advantage that the loss of their

Valentin Gordeliy (ed.), *Rhodopsin: Methods and Protocols*, Methods in Molecular Biology, vol. 2501,
https://doi.org/10.1007/978-1-0716-2329-9_2,

absorption indicated their loss of activity. This facilitated developing mild conditions for purification.

Once, these former obstacles were solved, the question of obtaining material in manageable amounts had to be encountered. Tissues like retina provided sufficient quantities. For bacterial membrane proteins, like purple membrane, strains were tested for expression levels, again being helped by the intensity of the colored colonies. These homologous expression systems were followed by heterologous techniques, which were enabled by progress in molecular genetics. The expeditious development of Molecular Biology in all of its facets allowed the expression of membrane proteins in not only eubacterial, archaeal, and eukaryotic cells but also provided the resources to express target proteins in living tissues. In the following, these developments will be outlined and the most recent technical innovations described for rhodopsins with the emphasis on microbial rhodopsins.

2 Historical Perspectives

A young physiologist and histologist Franz Boll made the first observation of a colored protein in 1877 (for an English translation *see* [2]). He described that in light-exposed parts of a vertebrate retina, the color of a purple red substance was bleached and concluded that "the objective alteration of the rod outer segments by light is without a doubt part of the act of seeing." The physiologist Willibald Kühne followed up this pioneering work (*see* [3] for the seminal accomplishments of F. Boll and W. Kühne). A most important finding of W. Kühne was that the pigment could be isolated and characterized in bile acids [4]. From this result, he concluded that the retina contains a protein like substance. This first solubilization of a membrane protein was a decisive accomplishment as it prepared the way for purification of membrane proteins in general. Later on, these solubilization studies on rhodopsin were further investigated by various researchers including K. Tansley [5] and G. Wald [6]. From the 1930s, George Wald has tremendously furthered and stimulated the research on rhodopsin. In his 1968 Nobel lecture [7], G. Wald gives also a historical review and the current literature of this time period. Until the 1960s, he and his colleagues discovered the four vertebrate visual pigments, and the Schiff base linkage to a Lys residue could be ascertained. The mechanism of bleaching and regeneration of rhodopsin was analyzed and the electrical activity of the retina (electroretinogram) in response to a light stimulus was first measured.

An unexpected discovery was made in 1971, when W. Stoeckenius and D. Oesterhelt [8] described a rhodopsin-like protein in the purple membrane from *H. salinarium* [9]. The sole protein embedded in the purple membrane was named by the authors as bacteriorhodopsin (BR). These initial findings and the subsequent intensive research benefitted decisively from fruitful decades of rhodopsin investigations. Because BR was so easily and in large quantities available and due to its apparent similarity to rhodopsin, BR gained the interest of many laboratories, mainly from the USA, Japan, and Germany. It is interesting to note that the research on both rhodopsin and BR triggered the application and development of new biophysical methods like FTIR- [10, 11], Resonance Raman [12, 13], solid-state NMR- [14, 15], and EPR-spectroscopy [16] (for a recent review, *see* [17]). These methods gained their full strength if isotope-labeled samples (D, ^{13}C, ^{15}N) were used, either specifically labeled retinal [14], or amino acids (e.g. [4-^{13}C] Asp [15, 18]). If isotope labeled amino acids had to be incorporated into the protein, media based on a cocktail of amino acids had to be available. Fortunately, this was the case for *H. salinarum*. A further methodology, decisive for the development of the new field of optogenetics was pioneered by E. Bamberg and his coworkers. First, they studied photocurrents, which had been generated by light-activated BR on planar bilayer membranes [19]. Subsequently, the successful expression of BR in *Xenopus laevis* oocytes [20] enabled further to elucidate electrophysiological properties of pumps and channels.

Another aspect concerning the advancement of new methodologies concerned structure determination of membrane proteins [21]. R. Henderson and N. Unwin achieved, by applying a tilt series of electron microscopy (EM) maps, a three-dimensional structure of BR (7 Å resolution) [22]. In this work, they established the seven-helix structure of BR, which was later also demonstrated for rhodopsin [23]. This archetypal secondary structure has also been demonstrated for the whole family of microbial rhodopsins. The first X-ray structure of BR at 2.5 Å was published by E. Pebay-Peyroula et al. [24]. The authors made use of lipidic cubic phases [25]. This new method for the crystallization of membrane proteins turned out to be very successful, especially for the crystallization of microbial rhodopsins [26]. In 2000, Okada and colleagues published the X-ray structure of rhodopsin [27].

An important step forward was the elucidation of the primary structure of rhodopsin [23, 28] and BR [29] by protein sequencing. Sequencing of a membrane protein was (and still is) particularly challenging. Because of the hydrophobic nature of some of the proteolytic and CNBr peptides, they were insoluble in normal chromatographic buffers. Consequently, quite harsh conditions (e.g. 88% formic acid) had to be applied [30]. At the same time J. Nathans and D. Hogness were able to determine the nucleotide sequence of the cloned bovine rhodopsin DNA, which yielded its complete amino acid sequence [31]. The genomic sequence of BR was obtained by G. Khorana's group in 1981 [32]. His laboratory was also able to characterize the bacterioopsin (BO) mRNA [33].

The function of BR as a proton pump has been determined quite early by D. Oesterhelt and W. Stoeckenius [34]. On light excitation, BR undergoes a photocycle with several intermediates [35]. Following the nomenclature of the rhodopsin cycle (with the intermediates Lumi and Meta) [7], Lozier et al. named the intermediates of the BR photocycle K, L, M, N, and O. Contrary to rhodopsin, BR returns to its ground state without losing retinal. All-*trans* retinal bound to Lys212 via a protonated Schiff-base isomerises to 13-cis retinal and returns to its original state after completing the photocycle in the millisecond range. E. Racker and W. Stoeckenius utilized the function of BR as proton pump in an early Synthetic Biology experiment. They reconstituted both BR and mitochondrial ATPase in vesicles. In the presence of ADP and P_i light excitation of BR-generated ATP [36]. This work was an elegant experimental proof of the chemiosmosis theory of Peter Mitchell (Nobel lecture 1979 [37]).

BR should not remain the only pigment for long. In 1975, E. Hildebrand and N. Dencher reported on phototactic behavior of *H. salinarum* [38]. They showed that the bacteria are attracted to light >520 nm and repelled to light <520 nm. R. Bogomolni and J. Spudich identified a pigment, which they called slow rhodopsin (SR, later renamed sensory rhodopsin I, *Hs*SRI), which displayed an action spectrum with λ_{max} at about 590 nm [39]. During constant illumination of the cells at 570 nm an intermediate absorbing at 373 nm is accumulated, which can be excited by a second photon process resulting in photophobic response of the bacteria. A second sensory rhodopsin was discovered by a Japanese laboratory. T. Takahashi and colleagues described a photosystem mediating negative phototaxis [40], which they named phoborhodopsin (later renamed sensory rhodopsin II, *Hs*SRII). The sensory rhodopsins form a complex with cognate transducers [41–43], which relay the incoming signal via a cytoplasmic two-component signaling cascade homologous to the chemotactic system of enteric bacteria like *E. coli*.

A fourth archetypical microbial rhodopsin was described by Y. Mukohata [44] which he named Halorhodopsin [45]. Described first as a Na^+-pump, B. Schobert and J. Lanyi recognized this new pigment as inward-directed chloride pump [46] (for a review on HR *see* [47]).

As it turned out, microbial rhodopsins were not only identified in *H. salinarum*. Examining *Natronobacteria* (now named *Natronomonas pharaonis*) [48] from salt lakes at Wadi Natrun in Egypt, D. Bivin and W. Stoeckenius discovered two rhodopsin-like proteins, which they determined to possess sensory rhodopsin- and halorhodopsin-like properties [49]. It turned out, that these new rhodopsins (*Np*HR and *Np*SRI) had favorable features compared to those of *Hs*HR and *Hs*SRI. An important advantage was also that both rhodopsins could be heterologously expressed in *E. coli* by adding retinal at the initiation of induction [50, 51]. This furthered the opportunities for accessing site-directed mutants and isotope-labeled samples.

The search for new microbial rhodopsins made a great step forward when O. Béjà and colleagues [52] described a microbial rhodopsin, which they discovered through gene analysis of marine bacterioplankton. The amino acid sequence encoded in the genome of an uncultivated gamma-proteobacterium resembled closely those of archaeal rhodopsins. The authors were able to express this protein in *E. coli* and prove its function as light-driven proton pump. The authors named this new microbial rhodopsin Proteorhodopsin and conclude that "archaeal-like rhodopsins are broadly distributed among different taxa." This conclusion was shortly verified [53] later by shot-gun sequencing of marine bacteria from the Sargasso Sea [54].

Another tier of research that led to the discovery of light gated channels (named channelrhodopsin) dates back around 150 years when researchers started to study motile behavior of microalgae. Studying the action spectra of phototactic movement of *Chlamydomonas*, K. Foster postulated the light receptor as a rhodopsin-like protein [55, 56]. P. Hegemann while staying as postdoc in K. Foster's lab pursued this research. Using an early approach of V. Sineshchekov's group [57], P. Hegemann could show that the photoreceptor is closely linked to an ion-channel [58, 59]. It turned out, that the proteins were elusive because of their instability and heterogeneity. However, the determination of the DNA sequence of the proteins [60, 61] enabled their expression in oocytes. This work was done in collaboration with G. Nagel and E. Bamberg. Because of the nature of the two rhodopsins as two directly light-gated ion channels, the authors coined their names Channelrhodopsin-1 (ChR1) and Channelrhodopsin-2 (ChR2) [60, 62]. This result can be seen as the root of the new field of optogenetics.

The discovery of new microbial rhodopsins has seen an exponential increase since the earlier work of O. Béjà [52] and C. Venter [54]. This family of rhodopsins has now been identified in all kingdoms of life. The amazing variety of different functions has attracted a multitude of laboratories. The archetypical function as proton- and chloride pump, sensors, and channels has been enlarged to include now cation- and anion channels, Na^+-pump, inward directed H^+-pump, enzyme, gene expression regulator and Cl^-/SO_4^- transporter (for a review *see* [63], refer also to V. Gordeliy et al., this issue).

3 Isolation and Purification of Microbial Rhodopsins

Heterologous expression: The elucidation of the function and properties of proteins can be and is done at different levels. In general, on the molecular level high-end biophysical methods e.g. NMR-, FTIR-spectroscopy, EM, or X-ray crystallography are applied. It is of utmost importance to put major effort to obtain highly purified proteins. The most sophisticated methods fail if the samples are inhomogeneous. Another level would be to study membrane proteins in a more natural environment like vesicles or heterologous host cells. In this instance, e.g. EM or techniques-based fluorescence spectroscopy can be applied. Using vesicles or cells transport measurements can be undertaken or change of membrane potential can be determined. The physiological properties and their importance for the integrity of the native cell can be studied. The primary aim of these methods and techniques is to investigate the structure, mechanism of action, and physiological function of the target protein. A different objective of research is related to the utilization of proteins as vehicles for different biochemical, bioengineering, or biomedical applications. Cases in point are the use of BR as information processing media [64] or more recently manipulation of neurons by channelrhodopsin (see e.g. [65]).

The isolation of membrane proteins from their native organism is only in a very few cases possible, like rhodopsin from bovine retina or purple membrane from *H. salinarum*. With the advent of Molecular Biology technologies, heterologous expression systems became accessible. The first archaeal rhodopsin, the sensor *Np*SRII, was expressed with good yields in *E. coli* in 1997 [50]. On the other hand, it was generally not possible to use bacterial expression systems for rhodopsin. Eukaryotic proteins very often undergo posttranslational modifications like glycosylation, which do not occur in bacteria. Consequently, bovine rhodopsin was first expressed in monkey kidney COS-1 cells [66] and in human embryonic kidney cells [67]. Up to now, various systems have been developed and applied. For a recent review discussing four of the most commonly used expression systems for eukaryotic membrane proteins refer to [68]. This summary focusses on expression systems from bacterial, yeast, insect, and mammalian cells. It should be noted that the heterologous expression systems allows generating site-specific mutations.

Table 1 lists references to expression systems of selected rhodopsins from various organisms. It is obvious that favorable expression in *E. coli* is generally feasable.

It is interesting to point to an early experiment expressing BR in yeast. G. Büldt and coworkers were able to direct BR into the inner mitochondrial membrane (IM) of *Schizosaccharomyces pombe* [93]. In order to direct BR specifically to the IM, the bacterioopsin

Table 1
Expression systems employed for various type 1 rhodopsins

Microbial rhodopsins		
Proton pumps	Outward	**BR** (*H. salinarum* [8], *E. coli* [9, 69, 70], COS-1/sf9 cells [71]); **PR** (*E. coli* [52])
	Inward	**ASR** (*E. coli* [72]), **Xenorhodopsin** (*E. coli*, human embryonic kidney-, neuroblastoma-, rat hippocampal neuronal cells [73])
Anion pumps	Archaeal	***Np*HR** (*E. coli* [51], *X. laevis* oocytes [74])
	Bacterial	**ClR** (*E. coli* [75, 76])
Cation pumps		**KR2** (*E. coli* [77])
Channelrhodopsins	Cation	**ChR1** (*X. laevis* oocytes [60]), **ChR2** (*X. laevis* oocytes [62], *Leishmania tarentolae* [78], COS cells [79]), **ChRmine** (hippocampal neurons [80])
	Anion	***Gt*ACR1** (*P. pastoris* [81], Sf9 cells [82]), **MerMAID1** (*P. pastoris* [83])
Sensor rhodopsins		**HsSRI** (*E. coli* [84]), ***Sr*SRI** (*E. coli* [85]), **NpSRII** (*E. coli* [50, 51])
Viral rhodopsins		**OLPVRII** (*E. coli* [86]), **VirR$_{DTS}$** (*E. coli* [87])
Heliorhodopsin		**HeR I**: (*E. coli* [88])
Enzyme rhodopsins	Phosphodiesterase	**RhoPDE/Rh-PDE**: (HEK 293 [89, 90])
	Guanyl cyclase	**RhGC**: (*P. pastoris* [91])
	His kinase rhodopsin	**HKR1**: (*P. pastoris* [92])

ASR: *Anabena* sensory rhodopsin; BR: bacteriorhodopsin; ChR1/ChR2: channelrhodopsin-1/-2; ChRmine: channelrhodopsin from *Tiarina fusus*; ClR: chloride-pumping rhodopsin; GtACR1: anion-conducting ChRs from *Guillardia theta*; HeR: heliorhodopsin; HRK1: histidine kinase rhodopsin *Hs*SRI: *H. salinarum* sensory rhodopsin I; KR2: *Krokinobacter eikastus* sodium pump; MerMAID: distinct anion-conducting ChRs; *Np*HR: N. *pharaonis* halorhodopsin; *Np*SRII: *N. pharaonis* sensory rhodopsin II; OLPVRII: Organic Lake *Phycodnavirus* rhodopsin; PR: proteorhodopsin; RhGC: rhodopsin–guanylyl cyclase; RhoPDE/Rh-PDE: guanylyl cyclase rhodopsin; *Sr*SRI: *Salinibacter ruber* sensory rhodopsin I; VirR1/VirR2: viral rhodopsins; VirR$_{DTS}$: viral rhodopsin

gene (bop) was fused with the signal sequence of IM proteins. BR was functionally expressed in the presence of retinal. This work is a second early example (the other one was described by E. Racker and W. Stoeckenius [36]) for the new field of Synthetic Biology. Here, a genetically transferred light energy converter to the IM of a non-photoactive eukaryotic organism partially diminished glucose uptake under anaerobic conditions.

Cell-free synthesis The cell free synthesis of peptides was first developed by M. Nirenberg and J. Matthaei [94] with the aim to decipher the genetic code [95]. The addition of synthetic nucleotides like polyuridylic acid to a cell-free extract of *E. coli* resulted in poly-phenylalanine. K. Rothschild and colleagues used a wheat germ extract to synthesize bacterioopsin, which was refolded to functional BR [96]. Applying *E. coli* extracts, R. Kalmbach et al. showed that BR can in situ be inserted into unilamellar liposomes [97]. BR is unidirectionally incorporated into this artificial lipid environment, which increases the sensitivity towards biophysical techniques. KR2, a microbial light-driven sodium or proton pump, has been co-translationally inserted into nanodiscs [98]. This latter method enabled the authors to examine the protein with noncovalent mass-spectrometric, electrophysiological, and flash photolysis measurements. State-of-the-art FTIR techniques allowed the in situ observation of a folding BR into nanodiscs [99]. The refolding of membrane proteins has been studied by various laboratories. Quite a few data are available for BR, which served as model system for the study of membrane protein folding (see e.g. [100, 101]).

Incorporation of unnatural amino acids There are several approaches to modify proteins resulting in products carrying altered or unnatural amino acids. The natural repertoire of post-translational modification has not been extensively utilized. The chemical modification of chemically addressable amino acids (like Cys or Lys) played and still plays an important role. Cys, being a relatively rare amino acid, can be modified with e.g. a fluorescent label or with an EPR-sensitive group or other biophysical probes [102]. However, site-directed mutagenesis is preferable because it leads to site-specific modifications. For rhodopsin, the method has been pioneered by G. Korana, which enabled biophysical studies using EPR- [16] and fluorescence spectroscopy (reviewed in [103]). Another example for the use of fluorescence spectroscopy relates to the measurement of binding constants of membrane protein complexes within lipid bilayers. Using FRET measurements the affinity of *Np*SRII to its cognate transducer *Np*HtrII could be determined [104].

Bacteriorhodopsin and other archaeal rhodopsins have been examined extensively. Results from this research included inter alia a topological structure of the *Np*SRII/*Np*HtrII complex [105] and the light activated outward movement of helix F in *Np*SRII, which seems to be a general observation true also for rhodopsins and type 1 rhodopsins (archaeal rhodopsins are reviewed in [106]).

Another, not very often chosen approach, is the replacement of a naturally occurring amino acid with a close analog. T. Ebrey and coworkers replaced largely tyrosine with *m*-fluorotyrosine (F-Tyr) in the growth medium for *H. salinarum* [107]. However, increasing amounts of F-Tyr incorporation decreased the stability of BR. It should be noted, that this method normally does not lead to single-site modifications.

P. Schultz developed a new methodology, which allows to incorporate unnatural amino acids almost at will (for an early review, *see* [108]). The method is based on chemically acylated amber suppressor tRNA (tRNAcua) (for a recent review, *see* [109]). T. Sakmar has introduced this method to study rhodopsins and other GPCRs [110]. Similarly, he and his coworkers employed genetic code expansion technology to analyze pore hydration of channelrhodopsin [111]. The amber-code suppression technology has to rely on the three stop-codons. Recently, the translation system has been expanded by a four-base codon/anticodon pair to generate BR containing a fluorescent amino acid [112].

4 Molecular Biology of Optogenetics

Optogenetics has been introduced in the early 2000 [113–116]. It makes use of the optical and spectroscopic properties of the protein, bioengineering, and genetics with the aim to render cells (e.g. neurons) sensitive to light. The previous section has described methods and techniques to express microbial rhodopsins in order to get access to their structural and functional properties. These data are of fundamental importance for optogenetic tools because they permit the development of tailor-made targets, which fulfill physical and functional requirements posed by the optogenetic aims. This noninvasive methodology allows controlling intracellular signaling cascades, gene expression, neurite outgrowth, cell survival, membrane composition and potential, protein and organelle localization, protein secretion and degradation, or behavior of whole organism (for a review *see* [117]).

Whereas the expression of proteins in a single cell organism is relatively straightforward, the targeting of proteins to specific cells in tissues necessitates special techniques. Furthermore, targeting a protein to subcellular compartment is even more challenging. As mentioned above, already in 1994 Hoffmann et al. described experiments to direct BR to the inner mitochondrial membrane of yeast [93]. In recent years the targeting of mitochondria has been refined (see e.g. [118]).

In the first experiments with *Caenorhabditis elegans*, M. Gottschalk and coworkers could, in a light-dependent manner, control the behavior of the nematode [114]. To direct the neural inhibitor *Np*HR or neural activator ChR2 to the membrane of specific cells appropriate promoters or intersectional promoters have to be used. The method is described in detail in [119]. A lentiviral gene delivery system was used to target ChR2 to mammalian neurons [113].

In general, it affords two major processes to direct proteins to subcellular compartments of specific cells. First, the gene has to be transferred to the specific cell. This can be accomplished by appropriate viral delivery systems. In the next step, expressed proteins have to find their designed destination. This can be accomplished by adding appropriate N- and/or C-terminal signal domains. For example a C-terminal KRas CAAX domain directs the protein to the plasma membrane [120]. The endoplasmic reticulum can be addressed by adding at the N-terminus the signal peptide from calreticulin and at the C-terminus KDEL retention sequence [121]. There are now several targeting protocols, which enable to address e.g. axons, lumen of secretory vesicles, cytosol of presynaptic terminal, or endoplasmic reticulum. A compilation of targeting tools has been published recently [117].

The molecular biology of optogenetic proteins has made tremendous progress and has led to an exponential growth in literature. It encompasses protein chemistry, biophysics, cellular biology, systems biology, synthetic biology, and genetics. For further more detailed information, the reader is referred to a couple of informative reviews [117, 122–129].

5 Outlook

From its early beginning, investigations of type 1 rhodopsins have experienced an astonishing exponential growth. This field of research stimulated the development of biophysical techniques, methods in molecular biology, or biomedical applications. Recent years have shown that this family of proteins as well as other light-activated systems (e.g. LOV domains) have expanded into areas such diverse as restoring hearing [128] and sight [129], Synthetic Biology [127, 130], Systems Biology [130, 131], Bioengineering [130], Cellular Biology [132], or Optogenetic Medicine [133]. Furthermore, BR has been a model for time-resolved serial femtosecond crystallography using X-ray free electron lasers of [134]. Discovering and engineering new optogenetic tools will refine current methodologies and applications. There will be progress in therapeutic applications of optogenetic tools in humans. It is also foreseeable that optogenetics will benefit from CRISPR technology [135]. These encouraging and promising developments will depend not only on technological advances but also on thorough basic research characterizing intimately the structural, functional, and applicable properties of the target of choice.

References

1. Grote M (2019) Membranes to molecular machines: active matter and the remaking of life. The University of Chicago Press, Chicago, IL
2. Boll F (1977) On the anatomy and physiology of the retina (Translated by Ruth Hubbard with the help of Helene Hoffmann). Vis Res 17(11–12):1249–1265. https://doi.org/10.1016/0042-6989(77)90112-2
3. Lanska DJ (2013) Chapter 4 - Optograms and criminology: science, news reporting, and fanciful novels. In: Stiles A, Finger S, Boller F (eds) Progress in brain research, vol 205. Elsevier, Amsterdam, pp 55–84. https://doi.org/10.1016/B978-0-444-63273-9.00004-6
4. Kühne W (1977) Chemical processes in the retina (Translation by R. Hubbard and G. Wald.). Vis Res 17(11–12):1269–1316. https://doi.org/10.1016/0042-6989(77)90114-6
5. Tansley K (1931) The regeneration of visual purple: its relation to dark adaptation and night blindness. J Physiol 71(4):442–458. https://doi.org/10.1113/jphysiol.1931.sp002749
6. Wald G (1938) On rhodopsin in solution. J Gen Physiol 21(6):795–832. https://doi.org/10.1085/jgp.21.6.795
7. Wald G (1968) The molecular basis of visual excitation. Nature 219(5156):800–807. https://doi.org/10.1038/219800a0
8. Oesterhelt D, Stoeckenius W (1971) Rhodopsin-like protein from the purple membrane of *Halobacterium halobium*. Nat New Biol 233(39):149–152
9. Blaurock AE, Stoeckenius W (1971) Structure of the purple membrane. Nat New Biol 233(39):152–155
10. Siebert F, Mäntele W, Kreutz W (1980) Flash-induced kinetic infrared spectroscopy applied to biochemical systems. Biophys Struct Mech 6(2):139–146
11. Eisenstein L, Lin SL, Dollinger G, Odashima K, Ding WD, Nakanishi K (1987) FTIR difference studies on apoproteins; protonation states of aspartic- and glutamic acid residues during the photocycle of bacteriorhodopsin. J Am Chem Soc 109:6860–6862
12. Mathies R, Freedman TB, Stryer L (1977) Resonance Raman studies of the conformation of retinal in rhodopsin and isorhodopsin. J Mol Biol 109(2):367–372
13. Aton B, Doukas AG, Callender RH, Becher B, Ebrey TG (1977) Resonance Raman studies of the purple membrane. Biochemistry 16(13):2995–2999
14. Harbison GS, Smith SO, Pardoen JA, Mulder PP, Lugtenburg J, Herzfeld J, Mathies R, Griffin RG (1984) Solid-state 13C NMR studies of retinal in bacteriorhodopsin. Biochemistry 23(12):2662–2667
15. Engelhard M, Hess B, Emeis D, Metz G, Kreutz W, Siebert F (1989) Magic angle sample spinning ^{13}C nuclear magnetic resonance of isotopically labeled bacteriorhodopsin. Biochemistry 28:3967–3975
16. Altenbach C, Flitsch SL, Khorana HG, Hubbell WL (1989) Structural studies on transmembrane proteins. 2. Spin labeling of bacteriorhodopsin mutants at unique cysteines. Biochemistry 28:7806–7812
17. Mandal T, Hustedt EJ, Song L, Oh KJ (2019) CW EPR and DEER methods to determine BCL-2 family protein structure and interactions: application of site-directed spin labeling to BAK apoptotic pores. In: Gavathiotis E (ed) BCL-2 family proteins. Methods in molecular biology, vol 1877. Humana Press, New York, NY, pp 257–303. https://doi.org/10.1007/978-1-4939-8861-7_18
18. Engelhard M, Gerwert K, Hess B, Kreutz W, Siebert F (1985) Light-driven protonation changes of internal aspartic acids of bacteriorhodopsin: an investigation by static and time-resolved infrared difference spectroscopy using [4-13C] aspartic acid labeling purple membrane. Biochemistry 24:400–407
19. Bamberg E, Apell H-J, Dencher NA, Sperling W, Stieve H, Läuger P (1979) Photocurrents generated by bacteriorhodopsin on planar bilayer membranes. Biophys Struct Mech 5:277–292
20. Nagel G, Möckel B, Büldt G, Bamberg E (1995) Functional expression of bacteriorhodopsin in oocytes allows direct measurement of voltage dependence of light induced H^+ pumping. FEBS Lett 377:263–266
21. Unwin PN, Henderson R (1975) Molecular structure determination by electron microscopy of unstained crystalline specimens. J Mol Biol 94(3):425–440. https://doi.org/10.1016/0022-2836(75)90212-0
22. Henderson R, Unwin PN (1975) Three-dimensional model of purple membrane obtained by electron microscopy. Nature 257(5521):28–32. https://doi.org/10.1038/257028a0
23. Hargrave PA, McDowell JH, Curtis DR, Wang JK, Juszczak E, Fong SL, Rao JK,

Argos P (1983) The structure of bovine rhodopsin. Biophys Struct Mech 9(4):235–244. https://doi.org/10.1007/BF00535659

24. Pebay-Peyroula E, Rummel G, Rosenbusch JP, Landau EM (1997) X-ray structure of bacteriorhodopsin at 2.5 Ångstroms from microcrystals grown in lipidic cubic phases. Science 277(5332):1676–1681
25. Landau EM, Rosenbusch JP (1996) Lipidic cubic phases - a novel concept for the crystallization of membrane proteins. Proc Natl Acad Sci U S A 93(25):14532–14535
26. Deniaud A, Moiseeva E, Gordeliy V, Pebay-Peyroula E (2010) Crystallography of membrane proteins: from crystallization to structure. Methods Mol Biol 654:79–103. https://doi.org/10.1007/978-1-60761-762-4_5
27. Palczewski K, Kumasaka T, Hori T, Behnke CA, Motoshima H, Fox BA, Le Trong I, Teller DC, Okada T, Stenkamp RE, Yamamoto M, Miyano M (2000) Crystal structure of rhodopsin: a G protein-coupled receptor. Science 289(5480):739–745
28. Ovchinnikov Iu A, Abdulaev NG, Feigina M, Artamonov ID, Bogachuk AS (1983) Visual rhodopsin. III. Complete amino acid sequence and topography in a membrane. Bioorg Khim 9(10):1331–1340
29. Ovchinnikov YA, Abdulaev NG, Feigina MY, Kiselev AV, Lobanov NA (1979) The structural basis of the functioning of bacteriorhodopsin: an overview. FEBS Lett 100(2):219–224. https://doi.org/10.1016/0014-5793(79)80338-5
30. Gerber GE, Anderegg RJ, Herlihy WC, Gray CP, Biemann K, Khorana HG (1977) Partial primary structure of bacteriorhodopsin: sequencing methods for membrane proteins. Proc Natl Acad Sci U S A 76:227–231
31. Nathans J, Hogness DS (1983) Isolation, sequence analysis, and intron-exon arrangement of the gene encoding bovine rhodopsin. Cell 34(3):807–814. https://doi.org/10.1016/0092-8674(83)90537-8
32. Dunn R, McCoy J, Simsek M, Majumdar A, Chang SH, Rajbhandary UL, Khorana HG (1981) The bacteriorhodopsin gene. Proc Natl Acad Sci U S A 78(11):6744–6748. https://doi.org/10.1073/pnas.78.11.6744
33. Dassarma S, Rajbhandary UL, Khorana HG (1984) Bacterio-opsin mRNA in wild-type and bacterio-opsin-deficient Halobacterium halobium strains. Proc Natl Acad Sci U S A 81(1):125–129. https://doi.org/10.1073/pnas.81.1.125
34. Oesterhelt D, Stoeckenius W (1973) Functions of a new photoreceptor membrane. Proc Natl Acad Sci U S A 70(10):2853–2857
35. Lozier RH, Bogomolni RA, Stoeckenius W (1975) Bacteriorhodopsin: a light-driven proton pump in *Halobacterium halobium*. Biophys J 15(9):955–962
36. Racker E, Stoeckenius W (1974) Reconstitution of purple membrane vesicles catalyzing light- driven proton uptake and adenosine triphosphate formation. J Biol Chem 249(2):662–663
37. Mitchell P (1979) Keilin's respiratory chain concept and its chemiosmotic consequences. Science 206(4423):1148–1159
38. Hildebrand E, Dencher N (1975) Two photosystems controlling behavioural responses of *Halobacterium halobium*. Nature 257:46–48
39. Bogomolni RA, Spudich JL (1982) Identification of a third rhodopsin-like pigment in phototactic *Halobacterium halobium*. Proc Natl Acad Sci U S A 79:6250–6254
40. Takahashi T, Tomioka H, Kamo N, Kobatake Y (1985) A photosystem other than PS370 also mediates the negative phototaxis of *Halobacterium halobium*. FEMS Microb Lett 28: 161–164
41. Yao VJ, Spudich JL (1992) Primary structure of an archaebacterial transducer, a methyl-accepting protein associated with sensory rhodopsin I. Proc Natl Acad Sci U S A 89: 11915–11919
42. Seidel R, Scharf B, Gautel M, Kleine K, Oesterhelt D, Engelhard M (1995) The primary structure of sensory rhodopsin II: a member of an additional retinal protein subgroup is coexpressed with its transducer, the halobacterial transducer of rhodopsin II. Proc Natl Acad Sci U S A 92:3036–3040
43. Rudolph J, Nordmann B, Storch KF, Gruenberg H, Rodewald K, Oesterhelt D (1996) A family of halobacterial transducer proteins. FEMS Microbiol Lett 139(2–3):161–168
44. Matsuno-Yagi A, Mukohata Y (1977) Two possible roles of bacteriorhodopsin; a comparative study of strains of Halobacterium halobium differing in pigmentation. Biochem Biophys Res Commun 78(1):237–243
45. Mukohata Y, Kaji Y (1981) Light-induced membrane-potential increase, ATP synthesis, and proton uptake in Halobacterium halobium, R1mR catalyzed by halorhodopsin: effects of N,N′-dicyclohexylcarbodiimide, triphenyltin chloride, and 3,5-di-tert-butyl-4-

hydroxybenzylidenemalononitrile (SF6847). Can J Chem 206(1):72–76

46. Schobert B, Lanyi JK (1982) Halorhodopsin is a light-driven chloride pump. J Biol Chem 257(17):10306–10313
47. Engelhard C, Chizhov I, Siebert F, Engelhard M (2018) Microbial halorhodopsins: light-driven chloride pumps. Chem Rev 118: 10629. https://doi.org/10.1021/acs.chemrev.7b00715
48. Soliman GSH, Trüper HG (1982) Halobacterium pharaonis sp. nov., a new, extremely Haloalkaliphilic Archaebacterium with low magnesium requirement. Zentralblatt für Bakteriologie Mikrobiologie und Hygiene: I. Abt Originale C: Allgemeine, angewandte und ökologische Mikrobiologie 3(2):318–329
49. Bivin DB, Stoeckenius W (1986) Photoactive retinal pigments in Haloalkaliphilic bacteria. J Gen Microbiol 132:2167–2177
50. Shimono K, Iwamoto M, Sumi M, Kamo N (1997) Functional expression of pharaonis phoborhodopsin in *Escherichia coli*. FEBS Lett 420(1):54–56
51. Hohenfeld IP, Wegener AA, Engelhard M (1999) Purification of histidine tagged bacteriorhodopsin, *pharaonis* halorhodopsin and *pharaonis* sensory rhodopsin II functionally expressed in *Escherichia coli*. FEBS Lett 442(2–3):198–202
52. Béjà O, Aravind L, Koonin EV, Suzuki MT, Hadd A, Nguyen LP, Jovanovich S, Gates CM, Feldman RA, Spudich JL, Spudich EN, DeLong EF (2000) Bacterial rhodopsin: evidence for a new type of phototrophy in the sea. Science 289(5486):1902–1906
53. Béjà O, Suzuki MT, Heidelberg JF, Nelson WC, Preston CM, Hamada T, Eisen JA, Fraser CM, DeLong EF (2002) Unsuspected diversity among marine aerobic anoxygenic phototrophs. Nature 415:630–633
54. Venter JC, Remington K, Heidelberg JF, Halpern AL, Rusch D, Eisen JA, Wu D, Paulsen I, Nelson KE, Nelson W, Fouts DE, Levy S, Knap AH, Lomas MW, Nealson K, White O, Peterson J, Hoffman J, Parsons R, Baden-Tillson H, Pfannkoch C, Rogers YH, Smith HO (2004) Environmental genome shotgun sequencing of the Sargasso Sea. Science 304(5667):66–74
55. Foster KW, Smyth RD (1980) Light Antennas in phototactic algae. Microbiol Rev 44(4):572–630
56. Foster KW, Saranak J, Patel N, Zarilli G, Okabe M, Kline T, Nakanishi K (1984) A rhodopsin is the functional photoreceptor for phototaxis in the unicellular eukaryote Chlamydomonas. Nature 311(5988):756–759
57. Litvin FF, Sineshchekov OA, Sineshchekov VA (1978) Photoreceptor electric potential in the phototaxis of the alga Haematococcus pluvialis. Nature 271(5644):476–478. https://doi.org/10.1038/271476a0
58. Harz H, Hegemann P (1991) Rhodopsin-regulated calcium currents in *Chlamydomonas*. Nature 351:489–491
59. Braun FJ, Hegemann P (1999) Two light-activated conductances in the eye of the green alga Volvox carteri. Biophys J 76(3):1668–1678. https://doi.org/10.1016/S0006-3495(99)77326-1
60. Nagel G, Ollig D, Fuhrmann M, Kateriya S, Musti AM, Bamberg E, Hegemann P (2002) Channelrhodopsin-1: a light-gated proton channel in green algae. Science 296(5577):2395–2398
61. Sineshchekov OA, Jung KH, Spudich JL (2002) Two rhodopsins mediate phototaxis to low- and high-intensity light in Chlamydomonasreinhardtii. Proc Natl Acad Sci U S A 99(13):8689–8694
62. Nagel G, Szellas T, Huhn W, Kateriya S, Adeishvili N, Berthold P, Ollig D, Hegemann P, Bamberg E (2003) Channelrhodopsin-2, a directly light-gated cation-selective membrane channel. Proc Natl Acad Sci U S A 100(24):13940–13945
63. Kojima K, Shibukawa A, Sudo Y (2020) The unlimited potential of microbial rhodopsins as optical tools. Biochemistry 59(3):218–229. https://doi.org/10.1021/acs.biochem.9b00768
64. Oesterhelt D, Bräuchle C, Hampp N (1991) Bacteriorhodopsin: a biological material for information processing. Q Rev Biophys 24: 425–478
65. Adamantidis A, Arber S, Bains JS, Bamberg E, Bonci A, Buzsaki G, Cardin JA, Costa RM, Dan Y, Goda Y, Graybiel AM, Hausser M, Hegemann P, Huguenard JR, Insel TR, Janak PH, Johnston D, Josselyn SA, Koch C, Kreitzer AC, Luscher C, Malenka RC, Miesenbock G, Nagel G, Roska B, Schnitzer MJ, Shenoy KV, Soltesz I, Sternson SM, Tsien RW, Tsien RY, Turrigiano GG, Tye KM, Wilson RI (2015) Optogenetics: 10 years after ChR2 in neurons--views from the community. Nat Neurosci 18(9):1202–1212. https://doi.org/10.1038/nn.4106
66. Oprian DD, Molday RS, Kaufman RJ, Khorana HG (1987) Expression of a synthetic bovine rhodopsin gene in monkey kidney

cells. Proc Natl Acad Sci U S A 84(24):8874–8878. https://doi.org/10.1073/pnas.84.24.8874

67. Nathans J, Weitz CJ, Agarwal N, Nir I, Papermaster DS (1989) Production of bovine rhodopsin by mammalian cell lines expressing cloned cDNA: spectrophotometry and subcellular localization. Vis Res 29(8):907–914. https://doi.org/10.1016/0042-6989(89)90105-3

68. Kesidis A, Depping P, Lode A, Vaitsopoulou A, Bill RM, Goddard AD, Rothnie AJ (2020) Expression of eukaryotic membrane proteins in eukaryotic and prokaryotic hosts. Methods 180:3. https://doi.org/10.1016/j.ymeth.2020.06.006

69. Bratanov D, Balandin T, Round E, Shevchenko V, Gushchin I, Polovinkin V, Borshchevskiy V, Gordeliy V (2015) An approach to heterologous expression of membrane proteins. The case of bacteriorhodopsin. PLoS One 10(6):e0128390. https://doi.org/10.1371/journal.pone.0128390

70. Tu CH, Yi HP, Hsieh SY, Lin HS, Yang CS (2018) Overexpression of different types of microbial rhodopsins with a highly expressible bacteriorhodopsin from Haloarcula marismortui as a single protein in E. coli. Sci Rep 8(1):14026. https://doi.org/10.1038/s41598-018-32399-x

71. Heymann J, Jäger R, Subramaniam S (1997) Expression of bacteriorhodopsin in sf9 and cos-1 cells. J Bioenerg Biomembr 29(1):55–59

72. Sineshchekov OA, Trivedi VD, Sasaki J, Spudich JL (2005) Photochromicity of anabaena sensory rhodopsin, an atypical microbial receptor with a cis-retinal light-adapted form. J Biol Chem 280(15):14663–14668

73. Shevchenko V, Mager T, Kovalev K, Polovinkin V, Alekseev A, Juettner J, Chizhov I, Bamann C, Vavourakis C, Ghai R, Gushchin I, Borshchevskiy V, Rogachev A, Melnikov I, Popov A, Balandin T, Rodriguez-Valera F, Manstein DJ, Bueldt G, Bamberg E, Gordeliy V (2017) Inward H+ pump xenorhodopsin: mechanism and alternative optogenetic approach. Sci Adv 3(9):e1603187. https://doi.org/10.1126/sciadv.1603187

74. Seki A, Miyauchi S, Hayashi S, Kikukawa T, Kubo M, Demura M, Ganapathy V, Kamo N (2007) Heterologous expression of Pharaonis halorhodopsin in Xenopus laevis oocytes and electrophysiological characterization of its light-driven Cl- pump activity. Biophys J 92(7):2559–2569. https://doi.org/10.1529/biophysj.106.093153

75. Kim K, Kwon SK, Jun SH, Cha JS, Kim H, Lee W, Kim JF, Cho HS (2016) Crystal structure and functional characterization of a light-driven chloride pump having an NTQ motif. Nat Commun 7:12677. https://doi.org/10.1038/ncomms12677

76. Yoshizawa S, Kumagai Y, Kim H, Ogura Y, Hayashi T, Iwasaki W, DeLong EF, Kogure K (2014) Functional characterization of flavobacteria rhodopsins reveals a unique class of light-driven chloride pump in bacteria. Proc Natl Acad Sci U S A 111(18):6732–6737. https://doi.org/10.1073/pnas.1403051111

77. Inoue K, Ono H, Abe-Yoshizumi R, Yoshizawa S, Ito H, Kogure K, Kandori H (2013) A light-driven sodium ion pump in marine bacteria. Nat Commun 4(1678):1–10. https://doi.org/10.1038/ncomms2689

78. Volkov O, Kovalev K, Polovinkin V, Borshchevskiy V, Bamann C, Astashkin R, Marin E, Popov A, Balandin T, Willbold D, Büldt G, Bamberg E, Gordeliy V (2017) Structural insights into ion conduction by channelrhodopsin 2. Science 358(6366):eaan8862. https://doi.org/10.1126/science.aan8862

79. Ritter E, Stehfest K, Berndt A, Hegemann P, Bartl FJ (2008) Monitoring light-induced structural changes of Channelrhodopsin-2 by UV-visible and Fourier transform infrared spectroscopy. J Biol Chem 283(50):35033–35041. https://doi.org/10.1074/jbc.M806353200. M806353200 [pii]

80. Marshel JH, Kim YS, Machado TA, Quirin S, Benson B, Kadmon J, Raja C, Chibukhchyan A, Ramakrishnan C, Inoue M, Shane JC, McKnight DJ, Yoshizawa S, Kato HE, Ganguli S, Deisseroth K (2019) Cortical layer-specific critical dynamics triggering perception. Science 365(6453):eaaw5202. https://doi.org/10.1126/science.aaw5202

81. Wietek J, Broser M, Krause BS, Hegemann P (2016) Identification of a natural green light absorbing chloride conducting channelrhodopsin from Proteomonas sulcata. J Biol Chem 291(8):4121–4127. https://doi.org/10.1074/jbc.M115.699637

82. Sugiyama Y, Inoue T, Ikematsu M, Iseki M, Sekiguchi T (1997) Controlling the orientation of purple membrane fragments on an air/water interface by a new method of direct electric field application during purple membrane spreading. Jpn J Appl Phys 36(9A):5674–5679

83. Oppermann J, Fischer P, Silapetere A, Liepe B, Rodriguez-Rozada S, Flores-Uribe J, Peter E, Keidel A, Vierock J, Kaufmann J, Broser M, Luck M, Bartl F, Hildebrandt P, Wiegert JS, Beja O, Hegemann P, Wietek J (2019) MerMAIDs: a family of metagenomically discovered marine anion-conducting and intensely desensitizing channelrhodopsins. Nat Commun 10(1):3315. https://doi.org/10.1038/s41467-019-11322-6

84. Schmies G, Chizhov I, Engelhard M (2000) Functional expression of His-tagged sensory rhodopsin I in *Escherichia coli*. FEBS Lett 466(1):67–69

85. Kitajima T, Hirayama J, Ihara K, Sugiyama Y, Kamo N, Mukohata Y (1996) Novel bacterial rhodopsins from *Haloarcula vallismortis*. Biochem Biophys Res Commun 220:341–345

86. Bratanov D, Kovalev K, Machtens JP, Astashkin R, Chizhov I, Soloviov D, Volkov D, Polovinkin V, Zabelskii D, Mager T, Gushchin I, Rokitskaya T, Antonenko Y, Alekseev A, Shevchenko V, Yutin N, Rosselli R, Baeken C, Borshchevskiy V, Bourenkov G, Popov A, Balandin T, Buldt G, Manstein DJ, Rodriguez-Valera F, Fahlke C, Bamberg E, Koonin E, Gordeliy V (2019) Unique structure and function of viral rhodopsins. Nat Commun 10(1):4939. https://doi.org/10.1038/s41467-019-12718-0

87. Needham DM, Yoshizawa S, Hosaka T, Poirier C, Choi CJ, Hehenberger E, Irwin NAT, Wilken S, Yung CM, Bachy C, Kurihara R, Nakajima Y, Kojima K, Kimura-Someya T, Leonard G, Malmstrom RR, Mende DR, Olson DK, Sudo Y, Sudek S, Richards TA, DeLong EF, Keeling PJ, Santoro AE, Shirouzu M, Iwasaki W, Worden AZ (2019) A distinct lineage of giant viruses brings a rhodopsin photosystem to unicellular marine predators. Proc Natl Acad Sci U S A 116(41):20574–20583. https://doi.org/10.1073/pnas.1907517116

88. Kovalev K, Volkov D, Astashkin R, Alekseev A, Gushchin I, Haro-Moreno JM, Chizhov I, Siletsky S, Mamedov M, Rogachev A, Balandin T, Borshchevskiy V, Popov A, Bourenkov G, Bamberg E, Rodriguez-Valera F, Buldt G, Gordeliy V (2020) High-resolution structural insights into the heliorhodopsin family. Proc Natl Acad Sci U S A 117(8):4131–4141. https://doi.org/10.1073/pnas.1915888117

89. Lamarche LB, Kumar RP, Trieu MM, Devine EL, Cohen-Abeles LE, Theobald DL, Oprian DD (2017) Purification and characterization of RhoPDE, a Retinylidene/phosphodiesterase fusion protein and potential optogenetic tool from the Choanoflagellate Salpingoeca rosetta. Biochemistry 56(43):5812–5822. https://doi.org/10.1021/acs.biochem.7b00519

90. Yoshida K, Tsunoda SP, Brown LS, Kandori H (2017) A unique choanoflagellate enzyme rhodopsin exhibits light-dependent cyclic nucleotide phosphodiesterase activity. J Biol Chem 292(18):7531–7541. https://doi.org/10.1074/jbc.M117.775569

91. Scheib U, Stehfest K, Gee CE, Korschen HG, Fudim R, Oertner TG, Hegemann P (2015) The rhodopsin-guanylyl cyclase of the aquatic fungus Blastocladiella emersonii enables fast optical control of cGMP signaling. Sci Signal 8(389):rs8. https://doi.org/10.1126/scisignal.aab0611

92. Luck M, Mathes T, Bruun S, Fudim R, Hagedorn R, Tran Nguyen TM, Kateriya S, Kennis JTM, Hildebrandt P, Hegemann P (2012) A photochromic histidine kinase rhodopsin (HKR1) that is bimodally switched by ultraviolet and blue light. J Biol Chem 287(47):40083–40090

93. Hoffmann A, Hildebrandt V, Heberle J, Büldt G (1994) Photoactive mitochondria: *in vivo* transfer of a light- driven proton pump into the inner mitochondrial membrane of *Schizosaccharomyces pombe*. Proc Natl Acad Sci U S A 91(20):9367–9371

94. Nirenberg MW, Matthaei JH (1961) The dependence of cell-free protein synthesis in E. coli upon naturally occurring or synthetic polyribonucleotides. Proc Natl Acad Sci U S A 47:1588–1602. https://doi.org/10.1073/pnas.47.10.1588

95. Nirenberg M (2004) Historical review: deciphering the genetic code - a personal account. Trends Biochem Sci 29(1):46–54

96. Sonar S, Patel N, Fischer W, Rothschild KJ (1993) Cell-free synthesis, functional refolding, and spectroscopic characterization of bacteriorhodopsin, an integral membrane protein. Biochemistry 32:13777–13781

97. Kalmbach R, Chizhov I, Schumacher MC, Friedrich T, Bamberg E, Engelhard M (2007) Functional cell-free synthesis of a seven helix membrane protein: in situ insertion of bacteriorhodopsin into liposomes. J Mol Biol 371(3):639–648

98. Henrich E, Sormann J, Eberhardt P, Peetz O, Mezhyrova J, Morgner N, Fendler K, Dotsch V, Wachtveitl J, Bernhard F, Bamann C (2017) From gene to function: cell-free electrophysiological and optical analysis of ion pumps in nanodiscs. Biophys J

113(6):1331–1341. https://doi.org/10.1016/j.bpj.2017.03.026
99. Baumann A, Kerruth S, Fitter J, Buldt G, Heberle J, Schlesinger R, Ataka K (2016) In-situ observation of membrane protein folding during cell-free expression. PLoS One 11(3):e0151051. https://doi.org/10.1371/journal.pone.0151051
100. Engelman DM, Chen Y, Chin CN, Curran AR, Dixon AM, Dupuy AD, Lee AS, Lehnert U, Matthews EE, Reshetnyak YK, Senes A, Popot JL (2003) Membrane protein folding: beyond the two stage model. FEBS Lett 555(1):122–125. https://doi.org/10.1016/s0014-5793(03)01106-2
101. Tastan O, Dutta A, Booth P, Klein-Seetharaman J (2014) Retinal proteins as model systems for membrane protein folding. Biochim Biophys Acta 1837(5):656–663. https://doi.org/10.1016/j.bbabio.2013.11.021
102. Cornish VW, Benson DR, Altenbach CA, Hideg K, Hubbell WL, Schultz PG (1994) Site-specific incorporation of biophysical probes into proteins. Proc Natl Acad Sci U S A 91:2910–2914
103. Alexiev U, Farrens DL (2014) Fluorescence spectroscopy of rhodopsins: insights and approaches. Biochim Biophys Acta 1837(5):694–709. https://doi.org/10.1016/j.bbabio.2013.10.008
104. Kriegsmann J, Brehs M, Klare JP, Engelhard M, Fitter J (2009) Sensory rhodopsin II/transducer complex formation in detergent and in lipid bilayers studied with FRET. Biochim Biophys Acta 1788(2):522–531. https://doi.org/10.1016/j.bbamem.2008.11.011. S0005-2736(08)00376-3 [pii]
105. Wegener AA, Klare JP, Engelhard M, Steinhoff HJ (2001) Structural insights into the early steps of receptor-transducer signal transfer in archaeal phototaxis. EMBO J 20(19):5312–5319
106. Klare JP, Bordignon E, Engelhard M, Steinhoff HJ (2004) Sensory rhodopsin II and bacteriorhodopsin: light activated helix F movement. Photochem Photobiol Sci 3(6):543–547
107. Hazard ES III, Govindjee R, Ebrey TG, Crouch RK (1992) Biosynthetic incorporation of M-fluorotyrosine into bacteriorhodopsin. Photochem Photobiol 56:929–934
108. Anthony-Cahill SJ, Griffith MC, Noren CJ, Suich DJ, Schultz PG (1989) Site-specific mutagenesis with unnatural amino acids. Trends Biochem Sci 14:400–403
109. Chemla Y, Ozer E, Algov I, Alfonta L (2018) Context effects of genetic code expansion by stop codon suppression. Curr Opin Chem Biol 46:146–155. https://doi.org/10.1016/j.cbpa.2018.07.012
110. Huber T, Naganathan S, Tian H, Ye S, Sakmar TP (2013) Unnatural amino acid mutagenesis of GPCRs using amber codon suppression and bioorthogonal labeling. Methods Enzymol 520:281–305. https://doi.org/10.1016/B978-0-12-391861-1.00013-7
111. Krause BS, Kaufmann JCD, Kuhne J, Vierock J, Huber T, Sakmar TP, Gerwert K, Bartl FJ, Hegemann P (2019) Tracking pore hydration in channelrhodopsin by site-directed infrared-active azido probes. Biochemistry 58(9):1275–1286. https://doi.org/10.1021/acs.biochem.8b01211
112. Ohtsuka T, Neki S, Kanai T, Akiyoshi K, Nomura SM, Ohtsuki T (2011) Synthesis and in situ insertion of a site-specific fluorescently labeled membrane protein into cell-sized liposomes. Anal Biochem 418(1):97–101. https://doi.org/10.1016/j.ab.2011.06.026
113. Boyden ES, Zhang F, Bamberg E, Nagel G, Deisseroth K (2005) Millisecond-timescale, genetically targeted optical control of neural activity. Nat Neurosci 8(9):1263–1268. https://doi.org/10.1038/nn1525
114. Nagel G, Brauner M, Liewald JF, Adeishvili N, Bamberg E, Gottschalk A (2005) Light activation of channelrhodopsin-2 in excitable cells of Caenorhabditis elegans triggers rapid behavioral responses. Curr Biol 15(24):2279–2284. https://doi.org/10.1016/j.cub.2005.11.032
115. Zemelman BV, Lee GA, Ng M, Miesenbock G (2002) Selective photostimulation of genetically chARGed neurons. Neuron 33(1):15–22. https://doi.org/10.1016/s0896-6273(01)00574-8
116. Zemelman BV, Nesnas N, Lee GA, Miesenbock G (2003) Photochemical gating of heterologous ion channels: remote control over genetically designated populations of neurons. Proc Natl Acad Sci U S A 100(3):1352–1357. https://doi.org/10.1073/pnas.242738899
117. Rost BR, Schneider-Warme F, Schmitz D, Hegemann P (2017) Optogenetic tools for subcellular applications in neuroscience. Neuron 96(3):572–603. https://doi.org/10.1016/j.neuron.2017.09.047
118. Venkatachalam V, Cohen Adam E (2014) Imaging GFP-based reporters in neurons

with multiwavelength optogenetic control. Biophys J 107(7):1554–1563. https://doi.org/10.1016/j.bpj.2014.08.020

119. Glock C, Nagpal J, Gottschalk A (2015) Microbial rhodopsin optogenetic tools: application for analyses of synaptic transmission and of neuronal network activity in behavior. Methods Mol Biol 1327:87–103. https://doi.org/10.1007/978-1-4939-2842-2_8

120. Zhang K, Duan L, Ong Q, Lin Z, Varman PM, Sung K, Cui B (2014) Light-mediated kinetic control reveals the temporal effect of the Raf/MEK/ERK pathway in PC12 cell neurite outgrowth. PLoS One 9(3):e92917. https://doi.org/10.1371/journal.pone.0092917

121. Henderson MJ, Baldwin HA, Werley CA, Boccardo S, Whitaker LR, Yan X, Holt GT, Schreiter ER, Looger LL, Cohen AE, Kim DS, Harvey BK (2015) A low affinity GCaMP3 variant (GCaMPer) for imaging the endoplasmic reticulum calcium store. PLoS One 10(10):e0139273. https://doi.org/10.1371/journal.pone.0139273

122. Wiegert JS, Mahn M, Prigge M, Printz Y, Yizhar O (2017) Silencing neurons: tools, applications, and experimental constraints. Neuron 95(3):504–529. https://doi.org/10.1016/j.neuron.2017.06.050

123. Mühlhäuser WW, Fischer A, Weber W, Radziwill G (2017) Optogenetics - bringing light into the darkness of mammalian signal transduction. Biochim Biophys Acta 1864(2):280–292. https://doi.org/10.1016/j.bbamcr.2016.11.009

124. Govorunova EG, Sineshchekov OA, Li H, Spudich JL (2017) Microbial rhodopsins: diversity, mechanisms, and optogenetic applications. Annu Rev Biochem 86:845–872. https://doi.org/10.1146/annurev-biochem-101910-144233

125. Deisseroth K (2015) Optogenetics: 10 years of microbial opsins in neuroscience. Nat Neurosci 18:1213. https://doi.org/10.1038/nn.4091. https://www.nature.com/articles/nn.4091#supplementary-information

126. Dagliyan O, Hahn KM (2019) Controlling protein conformation with light. Curr Opin Struct Biol 57:17–22. https://doi.org/10.1016/j.sbi.2019.01.012

127. Mansouri M, Strittmatter T, Fussenegger M (2019) Light-controlled mammalian cells and their therapeutic applications in synthetic biology. Adv Sci 6(1):1800952. https://doi.org/10.1002/advs.201800952

128. Dombrowski T, Rankovic V, Moser T (2019) Toward the optical cochlear implant. Cold Spring Harb Perspect Med 9(8):a033225. https://doi.org/10.1101/cshperspect.a033225

129. Simunovic MP, Shen W, Lin JY, Protti DA, Lisowski L, Gillies MC (2019) Optogenetic approaches to vision restoration. Exp Eye Res 178:15–26. https://doi.org/10.1016/j.exer.2018.09.003

130. Kichuk TC, Carrasco-Lopez C, Avalos JL (2020) Lights up on organelles: optogenetic tools to control subcellular structure and organization. Wiley Interdiscip Rev Syst Biol Med 13:e1500. https://doi.org/10.1002/wsbm.1500

131. Hartmann J, Krueger D, De Renzis S (2020) Using optogenetics to tackle systems-level questions of multicellular morphogenesis. Curr Opin Cell Biol 66:19–27. https://doi.org/10.1016/j.ceb.2020.04.004

132. Zhang K, Cui B (2015) Optogenetic control of intracellular signaling pathways. Trends Biotechnol 33(2):92–100. https://doi.org/10.1016/j.tibtech.2014.11.007

133. Ye H, Fussenegger M (2019) Optogenetic medicine: synthetic therapeutic solutions precision-guided by light. Cold Spring Harb Perspect Med 9(9):a034371. https://doi.org/10.1101/cshperspect.a034371

134. Wickstrand C, Nogly P, Nango E, Iwata S, Standfuss J, Neutze R (2019) Bacteriorhodopsin: structural insights revealed using X-ray lasers and synchrotron radiation. Annu Rev Biochem 88:59–83. https://doi.org/10.1146/annurev-biochem-013118-111327

135. Mathony J, Hoffmann MD, Niopek D (2020) Optogenetics and CRISPR: a new relationship built to last. Methods Mol Biol 2173:261–281. https://doi.org/10.1007/978-1-0716-0755-8_18

Chapter 3

Rhodopsin-Based Optogenetics: Basics and Applications

Alexey Alekseev, Valentin Gordeliy, and Ernst Bamberg

Abstract

Optogenetics has revolutionized not only neuroscience but also had an impact on muscle physiology and cell biology. Rhodopsin-based optogenetics started with the discovery of the light-gated cation channels, called channelrhodopsins. Together with the light-driven ion pumps, these channels allow light-mediated control of electrically excitable cells in culture tissue and living animals. They can be activated (depolarized) or silenced (hyperpolarized) by light with incomparably high spatiotemporal resolution. Optogenetics allows the light manipulation of cells under electrode-free conditions in a minimally invasive manner. Through modern genetic techniques, virus-induced transduction can be performed with extremely high cell specificity in tissue and living animals, allowing completely new approaches for analyzing neural networks, behavior studies, and investigations of neurodegenerative diseases. First clinical trials for the optogenetic recovery of vision are underway.

This chapter provides a comprehensive description of the structure and function of the different light-gated channels and some new light-activated ion pumps. Some of them already play an essential role in optogenetics while others are supposed to become important tools for more specialized applications in the future.

At the moment, a large number of publications are available concerning intrinsic mechanisms of microbial rhodopsins. Mostly they describe *Cr*ChR2 and its variants, as *Cr*ChR2 is still the most prominent optogenetic tool. Therefore, many biophysically and biochemically oriented groups contributed to the overwhelming mass of information on this unique ion channel's molecular mechanism. In this context, the function of new optogenetic tools is discussed, which is essential for rational optimization of the optogenetic approach for an eventual biomedical application. The comparison of the effectivity of ion pumps versus ion channels is discussed as well.

Applications of rhodopsins-based optogenetic tools are also discussed in the chapter. Because of the enormous number of these applications in neuroscience, only exemplary studies on cell culture neural tissue, muscle physiology, and remote control of animal behavior are presented.

Key words Optogenetics, Microbial rhodopsins, Channelrhodopsins, Neuroscience, Behavior

1 Optogenetics

Optogenetics allows the control of electrically excitable cells in culture, tissue, and living animals with high before unknown spatiotemporal resolution. This method has revolutionized, in particular, neuroscience enabling experiments on animal behavior and

Valentin Gordeliy (ed.), *Rhodopsin: Methods and Protocols*, Methods in Molecular Biology, vol. 2501,
https://doi.org/10.1007/978-1-0716-2329-9_3,

reading-out the current condition of an animal brain on spatial and millisecond time scale.

Although modern optogenetics is generally based on microbial rhodopsins, the first attempts to control neurons were made with other types of proteins. In 2002, a group of scientists presented the first genetically encoded three-component system, which allowed the control of group neurons by type, but not by spatial arrangement [1]. They coexpressed ten proteins that are responsible for Drosophila photo-sensing in Xenopus oocytes. Using patch-clamp, they detected photocurrents in oocytes with a magnitude around 1 μA. Excluding one by one the proteins that had no significant effect on photocurrents, they came up with chARGe—the combination of three proteins (arrestin-2, rhodopsin, and the alpha subunit of cognate heterotrimeric G-protein). They were sufficient to support the light-induced effect of the same magnitude. They also showed that rat hippocampal neurons with chARGe were photoactive. It was the first genetically encoded tool for optical stimulation of neurons.

In 1984 Foster et al. demonstrated that the phototactic reactions of the unicellular alga *Chlamydomonas reinhardtii* are based on a rhodopsin-like photoreceptor [2]. First electrophysiological studies on the unicellular green algae were conducted during the last decade of the twentieth century [3–6]. In these studies, light-activated depolarization in *Haematococcus pluvialis* and *Chlamydomonas reinhardtii* was shown. The interpretation of the electrical photocurrent was the activity of an unknown ion channel assisted by a rhodopsin. In 1993 and 1994 *Hs*BR was functionally expressed in the plasma membrane and the inner mitochondrial membrane of the eukaryotic cell *Schizosaccharomyces pombe* [7, 8]. The yeast cells were transformed into a light-sensitive organism that uses light energy instead of glucose. In line with this experiment, *Hs*BR was functionally expressed in the plasma membrane of oocytes from *Xenopus leavis.* That allowed the electrophysiological characterization of the voltage dependence of the pump describing the underlying molecular mechanism [9, 10]. These studies can be considered as the precursors for optogenetics as it is applied nowadays.

In 2002 and 2003, a team of scientists published that two rhodopsins from *Chlamydomonas reinhardtii* act as light-gated ion channels, which then were named **Channelrhodopsin-1 and 2** (*Cr*ChR1,2) [4, 5]. The article on *Cr*ChR2 indicated the beginning of the revolution in optogenetics and neuroscience. Unexpectedly, this microbial rhodopsin functioned as a light-gated cation channel showing large inwardly directed photocurrents and depolarizing photovoltages (Fig. 1a, b). It was shown that a single gene encoding the opsin part of *Cr*ChR2 with its seven transmembrane helix motif is the functional unit with superior properties with respect to kinetics and genetic manipulations. It took 2 more

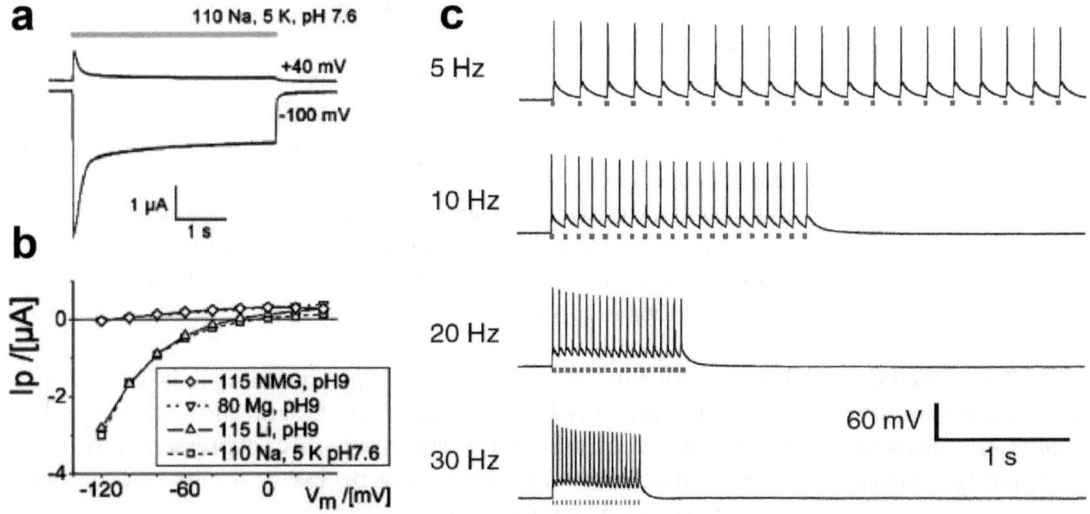

Fig. 1 CrChR2 is a light-gated cation channel capable of firing neurons. (**a**) Voltage-clamp recording from oocytes expressing CrChR2. (**b**, **c**) Current-voltage relationship of stationary photocurrents for one representative oocyte. (The figure is a compilation of the panels of the following publications [11, 12])

years to implement this unique single-component optogenetic tool in neurons and in muscle cells of a living animal [12–14]. Already the first applications resulted in precise activation of neuronal cells with high spatiotemporal resolution (Fig. 1c) [12]. The data suggested that *Cr*ChR2 did not significantly affect the cell health and in absence of light the electrical properties of the plasma membrane. *Cr*ChR2 is still the most frequently used optogenetic tool, i.e. the working horse for the light stimulation of electrically excitable cells. Together with the hyperpolarizing, inhibiting light-driven ion pumps or anion channels they comprise an optimal optogenetic toolbox for the study of cell culture tissues and living animals, as they allow the multimodal control of electrical excitability [15]. Genetically encoded readout fluorescence of specific microbial rhodopsins for the registration of the action potentials [16, 17] or to monitor changes in Ca^{2+} concentrations [18] complete the optogenetic approach for an electrode-free and a pure light-based minimally invasive method.

2 Microbial Rhodopsins and Corresponding Optogenetic Technologies

Various optogenetic applications require different tools. Moreover, the widespread use of optogenetics resulted in the rapid development of synergetic technologies. Such technologies are necessary to perform specific experiments, especially in living animals. Namely, viral technologies, genetic technologies (e.g. genetic recombination) and reporting technologies (GECI and GEVI) are described in Subheading 2.3.

However, microbial rhodopsins play a central role in optogenetic research. Therefore, the absence of the optimal properties of the respective microbial rhodopsin often limits the optogenetic application. Since 2005 scientists increased the diversity of optogenetic tools extensively. However, both basic science and biomedical applications require further development of appropriate optogenetic tools, because out of the great number of available tools only a small number is applied.

2.1 Basic Properties of Channelrhodopsins and Modifications for Different Applications

For rhodopsins to be applied in optogenetics, it is beneficial to know its functional properties. These data allow deriving how many copies of respective rhodopsin are expressed and, more importantly, how many copies are necessary for its optimal functioning as an optogenetic tool.

The transport properties of the light-driven ion pumps are tightly coupled to the kinetics of a photocycle, usually pumping one elementary charge per successful photocycle [19]. For the light-gated ion channels, the ion transport depends on the single-channel conductance, the photocycle duration, and the channel's closing time, which is usually the rate-limiting step of the photocycle, and on the concentration of the transported ion.

The single-channel conductance values are typically extremely small, around 50 fS under physiological conditions. Those values could only be determined by noise analysis [20]. Since the single-channel conductance and the quantum efficiency are almost invariable for different channelrhodopsins, the light sensitivity can be varied only by channel open time adjustment or/and by optimizing heterologous expression.

Channels with fast closing kinetics transport a lower amount of ions per one absorbed photon, whereas slow channels are correspondingly permeable for a larger number of ions per photocycle. In other words, slow channels need fewer photons than fast channels to produce photocurrents of the same magnitude.

These considerations should be taken into account for the design of new appropriate optogenetic tools. For instance, optogenetic activation of neurons with extremely high spiking frequencies requires the use of channelrhodopsins with very fast closing kinetics (open lifetime of about 2 ms). An increased expression level must compensate for a single channel's lower light sensitivity in such a case.

For the ion pumps, the situation is different. Efficient pumps with large turnover numbers up to 1000 s^{-1} require light of correspondingly higher intensity for optimal photocurrents or photovoltages.

2.2 Microbial Rhodopsins and Alternative Optogenetic Approaches

The first reported microbial rhodopsin is the well-known bacteriorhodopsin (*Hs*BR) light-driven proton pump [21]. In 1971 the protein was isolated from the purple membranes [22]. Since then, bacteriorhodopsin was a subject of a considerable number of biophysical experiments [23].

Despite the great interest of the scientific community, only a few microbial rhodopsin genes were identified by 1999, and all of them were archaeal proteins. Thanks to genomic analysis, the breakthrough happened when the rhodopsin gene was found in bacteria. The protein named proteorhodopsin turned out to be a new bacterial light-driven proton pump [24].

In the early 2000s, several groups independently submitted sequences of two microbial rhodopsins from *Chlamydomonas reinhardtii* after large-scale expressed sequence tag (EST) sequencing of the organism [25–27]. These two rhodopsins are expressed in the eye-spot being responsible for phototaxis of the algae. As described above the two rhodopsins act as ion channels: *Cr*ChR1 as a proton-selective channel [28] and more important *Cr*ChR2 as a nonselective cation channel [11].

In a short time, the era of optogenetics began with the first usage of *Cr*ChR2 as a light-sensitive actuator for the neurons and muscle cells [12–14]. These reports significantly increased the interest of the scientific community in microbial rhodopsins. Although CrChR2 is the most used tool for optogenetic manipulations, it has several limitations for special applications. Due to the blue-light absorption spectrum, the light is strongly scattered in living animals' tissue and organs, making transmission low compared to red light. For cells with extremely high spiking frequencies up to 1000 Hz CrChR2 has too slow closing kinetics (15 ms).

The main parameters of the optogenetic tool to be optimized are absorption spectrum, on and off kinetics, the level of the functional expression, ion selectivity, and very important selective expression in the plasma membrane of different cell types of the corresponding organ. The growing interest in optogenetics and the development of genome sequencing technologies led to the discovery of multiple microbial rhodopsin genes. Researchers found them across all domains of life, and they cover broad spectra of desirable properties. Besides that, researchers assembled a lot of mutants and chimera constructs, optimizing rhodopsins for optogenetics.

2.2.1 Optogenetic Actuators

Very soon after the discovery of *Cr*ChR2, researchers made multiple variants of it and obtained actuators with different properties. In 2005 the possibility to evoke muscle contractions in *Caenorhabditis elegans* was successfully investigated [13] (that belongs to the first reports of optogenetics working in vivo; another early example is optogenetic manipulations in embryonic chick spinal cords [14]). The authors performed mutation H134R (the position corresponding to a proton donor in HsBR), which enhanced the magnitude of the light-driven effect of CrChR2, albeit this mutation led to deceleration of channels closing kineticsKinetics as well. Later T159C mutation was shown to increase *Cr*ChR2 photocurrents in a factor of 10, due to the increased lifetime and expression level of the channel [29].

Nevertheless, neuroscience needs faster control of activation. Several chimeric channelrhodopsins comprising domains from *Cr*ChR1 and *Cr*ChR2 were designed [30]. They combined it with site-directed mutagenesis and obtained several promising channelrhodopsin variants. The variants were different in terms of current magnitude, desensitization level, and closing kinetics. The best variant called ChIEF (Ch for the channel, I for I170V mutation, EF for chimera crossing site in EF-loop) had almost six times higher stationary currents and 1.5 times faster closing kinetics than *Cr*ChR2. ChIEF could reliably evoke action potentials in neurons with frequency up to 25 Hz.

Red-shifted spectrum was always a preferable variant for optogenetics since red-light can penetrate deeper in living tissue. The first channelrhodopsin variant with a red-shifted absorption spectrum (maximum at 545 nm) was extracted from a different algae named *Volvox carteri* [31]. Later chimeric rhodopsins comprising domains of channelrhodopsins from *Volvox carteri* were reported. C1V1(T/T) channelrhodopsin was able to fire neurons with frequency up to 40 Hz utilizing 560 nm light [32]. Presented later, ReaChR (ChIEF/VChR1/VChR2 chimera with L171I mutation) had much better plasma membrane localization, and consequently, larger photocurrents [33]. Besides that, its spectral maximum was ~600 nm. Later, ReaChR was modified for enhanced expression in the long-range projections of neurons as well as accelerated channel kinetics, resulting in bReaChES [34].

Meanwhile, a bunch of papers was published describing *Cr*ChR2 optimized by mutagenesis for different purposes. It was shown that DC-pair (C128 and D156 positions in *Cr*ChR2) have a significant impact on channel closing kinetics [32, 35]. CrChR2-C128T/A/S and *Cr*ChR2-C128S-D156A variants called step-function-opsin (SFO) and stabilized-step-function-opsin (SSFO) can be used for prolonged activation of neurons (closing time constant up to ~30 min, Fig. 2g). They can be triggered on and off by short light pulses (470 nm—on, 590 nm—off). Later SSFO-mutations were combined with T159C making SFO opsin variant with higher light-sensitivity, called SOUL [43].

In opposite, another mutation in close proximity of retinal Schiff base E123T significantly accelerates closing kinetics of the channel [36]. The protein called ChETA had a closing time constant ~4.4 ms allowing ultrafast neuronal manipulations up to 200 Hz (Fig. 2a, b).

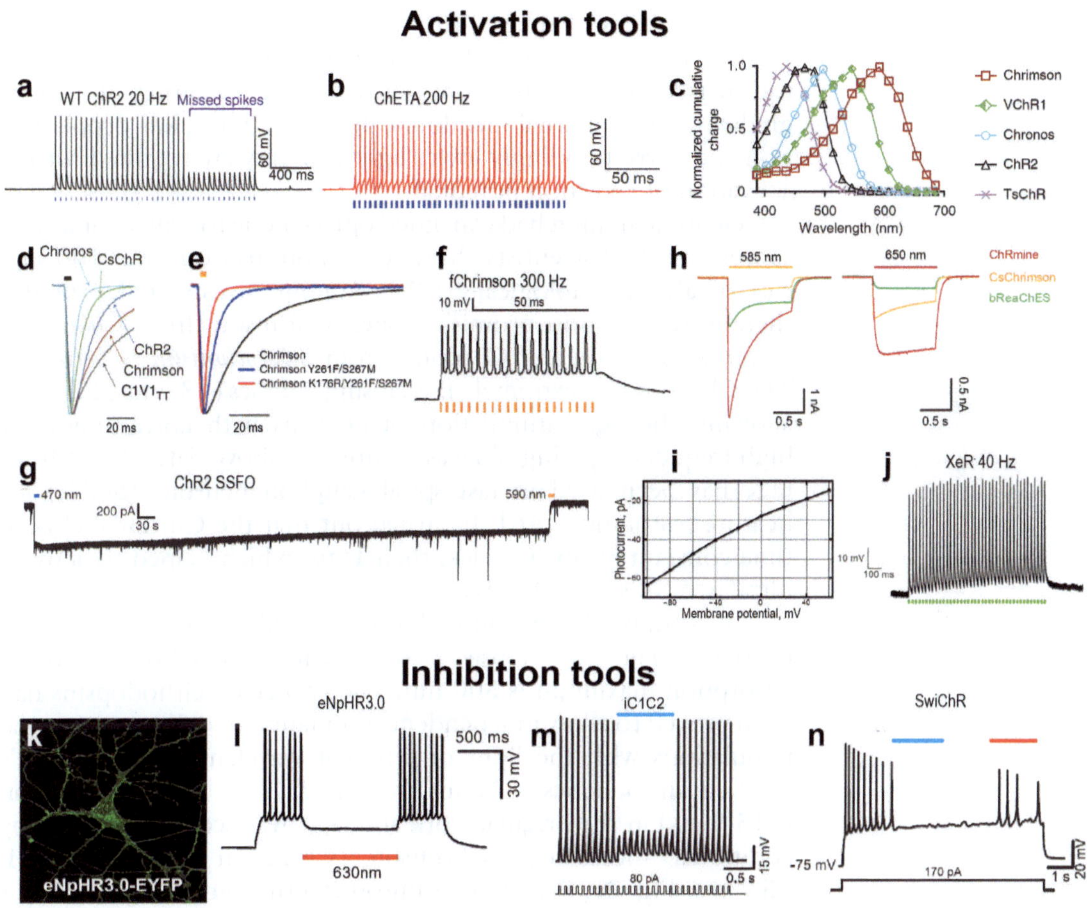

Fig. 2 Different optogenetic tools. (**a**) CrChR2 fires action potential in neurons at 20 Hz and misses spikes. (**b**) Fast mutant called ChETA reliably fires interneurons at 200 Hz. (**c**) Spectra channelrhodopsins obtained by genome mining. Chrimson has a red-shifted spectrum. (**d**) Closing kinetics of channelrhodopsins obtained by genome mining. Chronos has fast off-kinetics. (**e**) Optimization of Chrimson. Mutations accelerated red-shifted Chrimson: fChrimson (Chrimson Y261F/S267M, ~5.7 ms) and vfChrimson (Chrimson K176R/Y261F/S267M, ~2.7 ms). (**f**) fChrimson drives interneurons to their frequency limit of 300 Hz. (**g**) CrChR2 stable step-function mutant has closing kinetics of 30 min. (**h**) ChRmine is a bacteriorhodopsin-like channelrhodopsin capable of inducing photocurrents over 1 nA utilizing far-red 650 nm light. (**i**) Inward proton pump NsXeR current-voltage relation without reversal. (**j**) NsXeR is capable of driving rat hippocampal neurons at 40 Hz. (**k**) eNpHR3.0 is a version of inward chloride pump halorhodopsin in neurons with optimized plasma membrane localization. (**l**) eNpHR3.0 can reliably silence neurons. (**m**) Engineered ACR iC1C2 is capable of neuron silencing. (**n**) Step-function ACR SwiChR turns neuronal firing on and off. (The figure is a compilation of the panels from the following publications [32, 36–42])

Besides closing kinetics, mutagenesis allowed to alter the selectivity properties of the channels. For instance, a single mutation E90K/R could convert *Cr*ChR2 into a light-sensitive chloride-selective channel [44]. Amazingly, a single mutation L132C in *Cr*ChR2 (CatCh) increased relative permeability for calcium by a factor of 6 and indirectly increased light sensitivity by a factor of

70 without loss of speed [45]. The transported Ca^{2+} supposedly binds to the cytoplasmic side of the membrane. That, in turn, shifts the surface potential to more positive values resulting in a lower gradient of the internal membrane potential. Thus, less depolarization and correspondingly lower light intensity are required for the activation of an action potential.

Genome mining had expanded optogenetic toolkit significantly as well. In 2014 scientists characterized opsins from over 100 species of alga physiologically [37]. Among investigated proteins, there were two most promising ones: Chrimson (from *Chlamydomonas noctigama*) and Chronos (from *Stigeoclonium helveticum*). The Chronos had extremely fast closing kinetics (~3.2 ms, Fig. 2d), allowing the light stimulation of neurons with correspondingly high frequency spiking. As was mentioned above, later, this advantage has been used in fast spiral ganglion neurons (SGNs) for hearing restoration [46]. It turned out that the Chronos' closing time constant at 36 °C is less than 1 ms, which helped to activate ultrafast SNGs at 1000 Hz.

In contrast, Chrimson had a red-shifted absorption spectrum (600 nm, Fig. 2c). It was used together with Chronos, whose absorption maximum is 500 nm. The two channelrhodopsins had been proved to allow independent stimulation of different neuronal populations with the light of different wavelengths. Chrimson's slow closing kinetics (~21 ms) was sped up by K176R mutation (~15.8 ms) in the original work and later was accelerated more by additional mutations: fChrimson (Chrimson Y261F/S267M, ~5.7 ms, Fig. 2e, f) and vfChrimson (Chrimson K176R/Y261F/S267M, ~2.7 ms, Fig. 2e) [38].

In 2018, a group of researchers performed topological inversion of some channelrhodopsins in the membrane by fusing with additional N-terminal domain of Neurexin 1B-delta [47]. Such inversion affected the selectivity and function of rhodopsin. Inverted *Cr*ChR2 functioned as a cation pump; thus, it could be used for neuronal silencing. Inverted Chrimson had a significant effect on selectivity: relative permeability for potassium increased seven times. This approach might help in the future to find light-gated channels or pumps with new properties without the metagenomic search procedure.

In 2016 another clade of rhodopsins was shown to function as light-sensitive cation channels [48, 49]. Unlike known channelrhodopsins, they share high functional residues similarity to those of archaeal proton pumps. Later, a member of this clade ChRmine was shown to induce high depolarizing photocurrent (~4 nA, Fig. 2h) in neurons and fire them at frequencies up to 40 Hz [39].

Besides channelrhodopsins, an alternative approach to activate neurons was reported [40]. Proteins of newly found rhodopsin clade [50] were shown to function as light-sensitive inward proton pumps (Fig. 2i) [40, 51]. *Ns*XeR was capable of driving spikes in hippocampal neurons with a frequency of at least 40 Hz (Fig. 2j). Recently, metagenomic searches resulted in the discovery of Schizorhodopsins [52]. These rhodopsins function as inward proton pumps as well [53]. Thus, they also can expand a variety of optogenetic actuators.

2.2.2 Optogenetic Inhibitors

The other mandatory need for neuroscience was to hyperpolarize the cell membrane. The idea of using rhodopsins to hyperpolarize neurons evolved after the big success of channelrhodopsin. In 2005 there were already several known microbial rhodopsins that perform a desirable function: proton pumps and chloride pumps. Since then, multiple attempts have been made to build reliable optogenetic inhibitors.

Natronomonas pharaonis halorhodopsin (*Np*HR) is an inward chloride pump [54]. It was one of the first proteins to be successfully used as light-sensitive neuronal silencer [15, 55]. *Np*HR was complementary to the CrChR2 silencing tool since it had a red-shifted spectral maximum. However, photocurrents in neurons were not sufficient (~50 pA) to use it reliably in a variety of possible applications. Therefore, the *Np*HR gene has undergone several significant optimizations [41, 56]. In brief, the target expression of *Np*HR in the plasma membrane of neurons was optimized by fusion of its gene with several signal peptides: N-terminal signal peptide ensured membrane insertion, ER export, and Golgi export signal peptides, and membrane trafficking signal to ensure surface expression. The final best variant called e*Np*HR3.0 induced photocurrents of ~700 pA on average and was capable of hyperpolarizing neurons for −100 mV when illuminated with 593 nm light (Fig. 2k, l). Besides, e*Np*HR3.0 could sufficiently depolarize neurons under illumination with light of far red-shifted wavelength up to 680 nm.

Meanwhile, proton pumps were also shown to have optogenetic potential. For example, the light-driven pump bacteriorhodopsin was optimized similarly to *Np*HR [41]. Although optimization made *Hs*BR capable of hyperpolarizing neurons up to −10 mV, it still was a much less effective tool than e*Np*HR3.0. That is explained by the lower turnover rate of 30 s^{-1} compared to 800 s^{-1} for *Np*HR [57]. Several proton pumps from different organisms were compared to *Np*HR [58]. Three of them had turnover rates similar to that of *Np*HR. Still, thanks to better expression in the plasma membrane they produced larger photocurrents: Arch (Halorubrum sodomense), Mac (Leptosphaeria maculans), and cruxrhodopsin-1 (Haloarcula). Moreover, Mac could be used simultaneously with *Np*HR utilizing light of different wavelengths (for instance, Mac and *Np*HR can absorb light of 470 nm and 630 nm, respectively). Later, genome mining helped to find an even better archaerhodopsin variant called ArchT. It was three times more sensitive to light [59], which allowed silencing of a twice larger brain tissue volume.

In 2013 the first light-driven non-proton cation pump KR2 was described [60]. KR2 pumped sodium ions, which was unexpected because the newly discovered microbial rhodopsins always functioned as proton or chloride pumps until that time. The sodium pump might be an attractive optogenetic tool as it had new selectivity. However, molecular mechanisms of sodium pumping have to be investigated for a deeper understanding of its applicability and for future engineering of new rhodopsins. The KR2 structure was reported by two groups independently, and both groups showed the possibility to engineer a light-driven potassium pump [61, 62]. This possibility is of uttermost importance for optogenetics. Voltage-gated potassium channels are usually in charge of repolarization of neuronal membranes, therefore silencing by a potassium pump might minimize possible side-effects. The KR2 was also shown to silence neurons with sodium ions, however, with limited efficacy, likely due to an impaired expression level. Optimizations of KR2 lead to engineering eKR2 with much better plasma membrane localization [63]. Besides that, it was reported that one mutation in R109Q makes KR2 passively permeable for potassium, although with residual sodium pumping activity [64]. Passive-K^+ to active-Na^+ transports ratio further increased by additional S70A mutation [64].

However, it is of great interest if cation-conducting channelrhodopsins could be converted into anion-selective channels. The primary physiological anion is chloride, which has its Nernst equilibrium potential at around −80 mV. Anion-conducting channelrhodopsins with a high level of plasma membrane expression might have been used for stable neuronal silencing of neurons with higher efficiency than hyperpolarizing ion pumps. As mentioned above, structure-guided rational mutagenesis revealed that single residue replacement E90 with positively charged residues converts *Cr*ChR2 into chloride-conducting channel [44]. Meanwhile, another group of scientists conducted extensive mutagenesis of chimera C1C2 and came up with other variants of engineered ACRs called iC1C2 and SwiChR [42]. The iC1C2 was a ninefold mutant of C1C2 (Fig. 2m) chimera, while SwiChR had additional "step-function" mutation (Fig. 2n). Both proteins had current-voltage relation reversed at around −65 mV, which was enough to keep neurons from firing. Besides that, SwiChR had slower closing kinetics (~7.3 s compared to 24 ms for iC1C2) and could be activated and deactivated with short light pulses similar to SFO channels. Later, additional mutations named iC++ and SwiChR++, resulted in 15-fold increased photocurrents compared to iC1C2 and SwiChR [65]. However, engineered ACRs had small residual cation conductance, which increased reversal potential and thus lowered its silencing efficacy.

After full genome sequencing of the cryptophyta *Guillardia theta*, two genes coding anion-conducting channelrhodopsins (ACR) were discovered [66]. The photocurrent of the natural anion channelrhodopsin *Gt*ACR2 in cultured pyramidal neurons reversed exactly at Nernst potential for chloride. *Gt*ACR2 photocurrents exceeded the maximal photocurrent of Arch at thousandth lower light intensities. In comparison with engineered chloride channels, *Gt*ACR2 had much higher light sensitivity while maintaining fast kinetics. In order to clarify whether the improved light sensitivity arises from the intrinsic channel property or from the higher expression level, it is necessary to determine the single channel conductance. The introduction of the single E68R mutation in *Gt*ACR1 yielded an interesting variant, which is open in the dark and is closed by the application of light [67]. Later, structural-based rational mutagenesis helped to engineer *Gt*ACR1 double mutant R83Q/N239Q called FLASH, which was capable of effective silencing of neurons with fast closing kinetics [68]. Further research for natural ACRs revealed MerMAIDs: intensely desensitized ACRs [69].

2.2.3 Other Types of Optogenetic Cell Manipulation

Besides modulation of electrically active cells, microbial rhodopsins are prospective candidates for other types of genetic interventions.

Light-driven proton pump Arch was targetly expressed and used for acidification of lysosomes and synaptic vesicles [70]. The generation of electrochemical potential on the synaptic vesicle membrane is crucial for the accumulation of neurotransmitters inside them. Therefore, this approach could indirectly influence signal transmission by chemical synapses. On the other hand, acidification of lysosomes could have effects on the fundamental processes of any cell in the organisms (not only electrically active). Moreover, such tools can speed up research concerning physiological aspects of pH maintenance in different cell compartments.

Another notable example is the target-mitochondrial expression of rhodopsin for control of mitochondrial membrane potential [71]. By controlling the mitochondrial membrane potential, it became possible to control its calcium uptake indirectly. Moreover, it was shown that one could optogenetically control beatings of rat cardiomyocytes via this technique [71].

Interestingly, recently *Cr*ChR2 was also used as a depolarizing tool to investigate H^+-ATPase in *Arabidopsis* plants [72]. Thus, optogenetics appears suitable for noninvasive control of electrical potential in the plants' plasma membrane.

However, microbial rhodopsins allow not only electrogenic control of biological processes. The first microbial rhodopsin capable of triggering cellular cascades with light was reported in 2012. Such rhodopsins are called enzymerhodopsins. They consist of two main parts: N-terminal light-sensitive rhodopsin domain and

C-terminal effector domain. At the moment, several enzymerhodopsins are known. Based on their function, they are divided into three classes: histidine kinase rhodopsins (HKR) [73], rhodopsin guanylyl cyclases (RhGC) [74], and rhodopsin phosphodiesterases (RhoPDE) [75]. Despite the great and obvious potential of enzymerhodopsins, they are still to be integrated into optogenetic research.

Search for new rhodopsin genes is productive, as never before. Each year thousands of new genes are discovered. Investigations of their functions take a considerable amount of time. Currently, there is no reliable approach to predict their functions and properties. However, the history of optogenetics shows that some of them might find their application in it.

2.3 Development of Optogenetic Techniques Synergetic to Rhodopsins

The progress of optogenetics relies on the development of specific technical aspects as well as search for rhodopsins with new properties.

First of all, neural research needs precise delivery of optogenetic tools. Modern working spatial precision is not just cellular but subcellular. In order to obtain reliable expression of certain rhodopsin in neuronal dendrites and axons or cellular organelles, researchers have to develop new methods of targeting. As already mentioned above, the fusion of rhodopsin with signal peptides plays an essential role in target expression [41, 63].

Besides that, modern viral technologies allow researchers to increase the spatial precision of stimulation [76]. First, scientists achieve opsin expression in the desired tissue of wild-type animals by placing genes under the cell-type specific promoters (Fig. 3a). Moreover, one can exclusively stimulate neurons, which project their axons in a specific brain region, placing stimulation fiber above the axonal projection (Fig. 3c). By combining, one can stimulate only neurons specified by a particular promoter and projecting their axons to a particular region.

Second, Cre-Lox recombination is widely used in order to specify cell types and regions in the brain. In this technology Cre-recombinase-dependent opsin gene is delivered to a desirable region (Fig. 3b). The opsin expression occurs either in the Cre-positive or Cre-negative neurons. Genes in Double-floxed Inverted Orientation (DIO or Cre-on) or Double-floxed Orientation (DO or Cre-off) are used for that, respectively. Double-floxed means opsin gene is placed between two different pairs of Lox-sequences, which allows expressing the gene in an **inverted** orientation in the presence of Cre (Cre-on), and in **forward** orientation in the absence of Cre (Cre-off). A cell-specific promoter regulates the expression of the Cre-recombinase gene. The gene can either already exist in transgenic animals or be delivered by the second virus (Fig. 3d left).

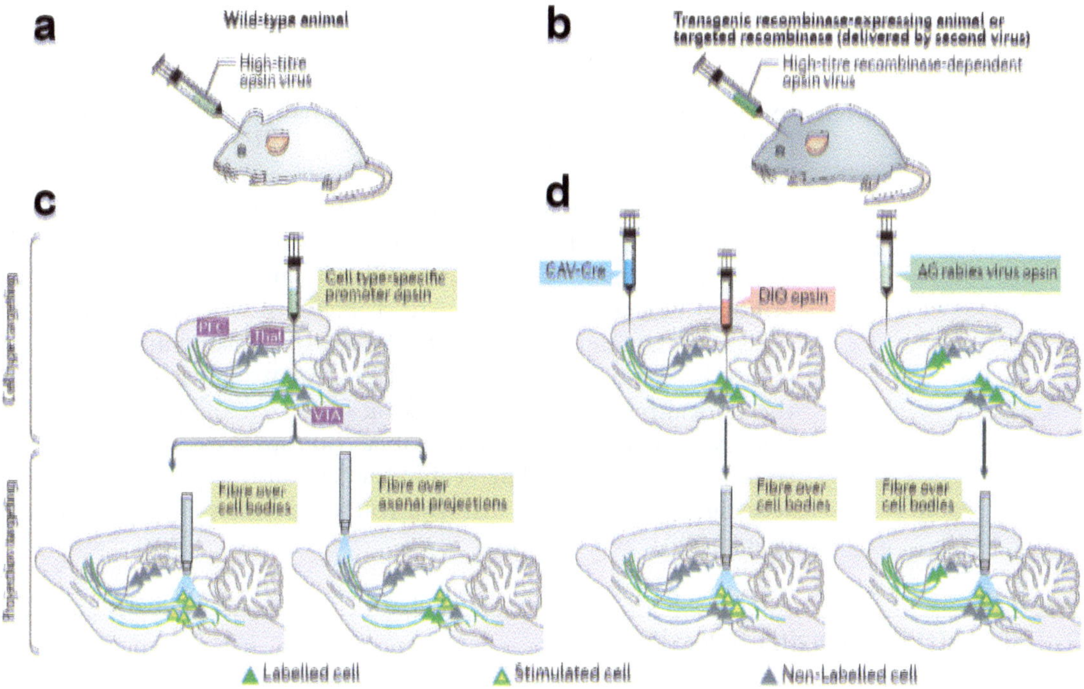

Fig. 3 Cell-specific viral technologies for opsin gene delivery. (**a**, **b**) The high-titer virus is delivered to the brain region of interest. Cell specificity is achieved by using either cell-specific promoter in wild-type animal or recombinase-dependent virus in transgenic animal or by the second injection of virus, carrying recombinase gene. (**c**) AAVs carrying opsin gene under control of the cell-specific promoter are injected into the region of interest. The stimulating light can either be delivered over cell bodies (down-left) or be delivered over neuronal axons projecting to the desired brain region (down-right). (**d**) Projecting region-dependent opsin expression is achieved by injection of DIO opsin gene in combination with retrograde CAV virus, carrying Cre-recombinase (left). Another approach is to inject the retrograde rabies virus, carrying the opsin gene. (The figure is adapted from [76])

Third, retrograde viruses can be used in order to label neurons with projections to a specific region (Fig. 3d right). Insertion of the Cre-recombinase gene into the genome of the retrograde virus allows researchers to label neurons of defined type and brain region.

To sum up, advances in genome editing, virus gene delivery, and light-targeting have a significant impact on optogenetic research.

Brain discoveries also benefit a lot from the development of complementary to optogenetics technologies. For instance, a combination of in vivo electrophysiological recording with optogenetics allowed closed-loop optogenetic interventions when optical stimulation is driven by ongoing activity. Indeed, researchers detected initiation of epileptic seizures in the thalamus and, at the moment, silenced thalamocortical neurons with halorhodopsin in order to terminate the epileptic activity [77].

Genetically encoded calcium indicators (GECI) are the primary optogenetic readout technique. Synergetic use of GECI with microbial rhodopsins allows inducing and reading specific actions at the same time. GECI are developing gradually in parallel with the invention of new optogenetic approaches. The first GECI to be used for calcium imaging in living animals was invented in 1997 [18]. However, the first successful use of GECI in mammals was in 2004 [78]. Genetically modified mice expressing GECI in smooth muscles were created to monitor postsynaptic signals in the muscles. However, GECIs have too slow kinetics to resolve action potentials in vivo. It may be possible to use genetically encoded **voltage** indicators (GEVI) instead. Nevertheless, optimization of spectral parameters (to separate them from opsins) and kinetics is still to be done for GEVI as well [79]. An attractive alternative for the optical recording of the spiking is the use of voltage-sensitive fluorescence of engineered microbial rhodopsin with high temporal resolution [16, 17]. The technique was also expanded for drug screening, for instance, with the voltage-gated sodium channels as targets [80].

3 Microbial Rhodopsins Diversity

Massive genome sequencing reveals thousands of rhodopsins each year. Nowadays, the superfamily of rhodopsins is considered as the most diverse and widespread family of light-harvesting proteins among organisms on Earth [81].

Discoveries of proteorhodopsins [24] and heliorhodopsins [82] significantly expanded the phylogenetic tree of rhodopsins (Fig. 4). The current conventional opinion is that rhodopsins have been evolving for a long time. Now they are capable of performing an enormously diverse set of biological functions. Furthermore, rhodopsins are abundant among all domains of life and also in viruses [83].

In some cases, like the case of *Hs*BR [21] or channelrhodopsins, function and biological role are already explored. Outward proton pumps provide cells with energy when the channelrhodopsins provide unicellular organisms with a kind of "eye" for phototaxis [26].

In other cases, like for xenorhodopsins, researchers have already only revealed molecular function [40, 51]. Although their function as inward proton pumping is already used for optogenetics, the biological role is still to be explored. Another rhodopsins clade Schizorhodopsins were proven to be inward proton pumps as well [53], which shows an example of divergent evolution of this property. Thus, inward proton pumps play an important biological role, which is still to be revealed.

Viral rhodopsins comprise another group of rhodopsins without explored biological roles [83]. Giant viruses, carrying genes of rhodopsins, are widely distributed in the oceans [84]. They infect

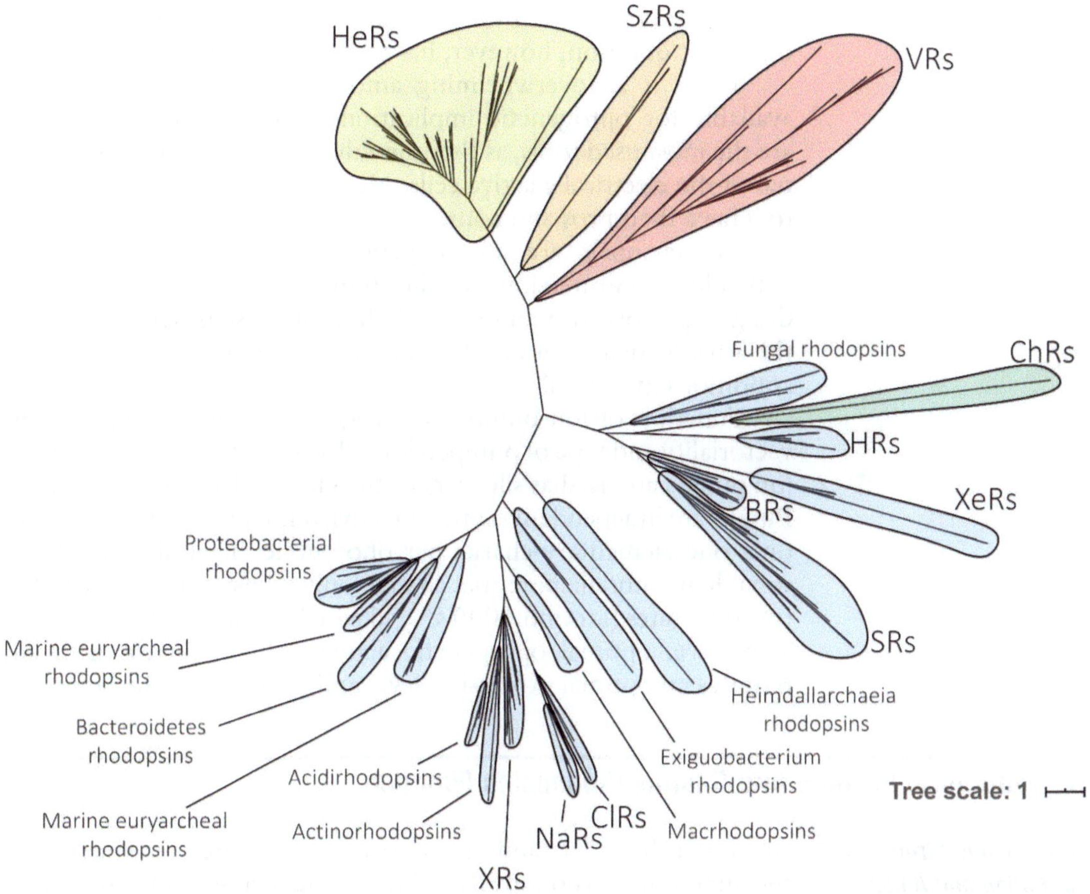

Fig. 4 Phylogenetic tree of microbial rhodopsins. An unrooted maximum-likelihood tree of microbial rhodopsins of type-1 (including channelrhodopsins and viral rhodopsins), heliorhodopsins, and schizorhodopsins. (Protein sequences were adapted from Bulzu et al. [52])

different organisms independent on their initial light-sensitivity. At the moment, X-ray structures of VirR$_{DTS}$ and OLPVR1 (viral rhodopsins group 1), and OLPVR2 are available (viral rhodopsins group 2) [84–86]. However, the biological role of these rhodopsins is not yet clear. VirR$_{DTS}$ and OLPVR2 act as proton pumps, whereas recent electrophysiological studies show cation channel activity of VirChR1 (viral rhodopsin group 1) [86].

Heliorhodopsins constitute a new class of microbial rhodopsins discovered in 2018 [82]. However, neither their function nor their biological role is disclosed. Nevertheless, the wide distribution and diversity of heliorhodopsins lead to the belief that some of them might have a potential for applications, including optogenetics.

Taking together, nowadays, a large variety of microbial rhodopsins are available as optogenetic tools. Nevertheless, the genomic discoveries of new rhodopsins might have a meaningful impact not only in the field of optogenetics. The biodiversity of rhodopsins raises an even more critical question of their effect on global ecology. That question, however, has to be answered in the future.

There is an overwhelming amount of microbial rhodopsins available for optogenetic implications. Ion channels and pumps are the most used tools, as they directly influence membrane potential of the electrically active cells. However, both types of modulators have their pros and cons.

Ion channels are advantageous because of the life time-dependent transmission of the transported ions. That allows designing tools with an extremely high light sensitivity. However, the ion selectivity is not perfect and they are dependent on the ionic conditions in the cell.

Function of ion pumps as optogenetic tools depends on the vectoriality and type of pumped ion. However, the great majority of ion pumps are used as silencers of the electrical activity of the cells. Pumps are independent on the ion environment, but they transport only one elementary charge per photocycle. That allows get reasonable membrane potential change only by using pumps with fast turnover rates (around 1000 s^{-1}, like in Arch and *Np*HR) [57–59] Therefore, optimal optogenetic activity of ion pumps generally requires an optimal expression level and saturating light intensity.

4 Structural Insights into Channelrhodopsins Function

4.1 X-Ray Structures of Cation and Anion Conducting Channelrhodopsins

An in-depth understanding of structural features responsible for the function of rhodopsins is interesting per se and can help to engineer new optogenetic tools with optimized properties. As already mentioned, many optogenetic tools were improved by using X-ray structure-based mutagenesis. A deep investigation of only one molecule of C1C2 resulted in the engineering of new tools for both activation and inhibition of neural cells, with increased sensitivity and optimized kinetics.

Until 2012 many efforts for the optimization of *Cr*ChR2 were based on knowledge of the structure of *Hs*BR and some analogs [87]. Before the first high-resolution structure was published, a low-resolution cryo-EM structure confirmed the dimeric form of *Cr*ChR2 [88]. With the same technique, the *Cr*ChR2 C128T (SFO variant with a long-living open state) was investigated under light and dark conditions. The illumination of the 2D crystals resulted in movement of three helices, which probably indicated the open state of the channel [89]. There were, however, considerable efforts to obtain a high-resolution X-ray structure of channelrhodopsin. The first successful attempt was obtained in 2012 with the chimera of *Cr*ChR1 and *Cr*ChR2 [87]. In this study, structural insights were confirmed by extensive mutagenesis and electrophysiological experiments. At the moment, three X-Ray structures of cation conducting channelrhodopsins and two X-Ray structures of

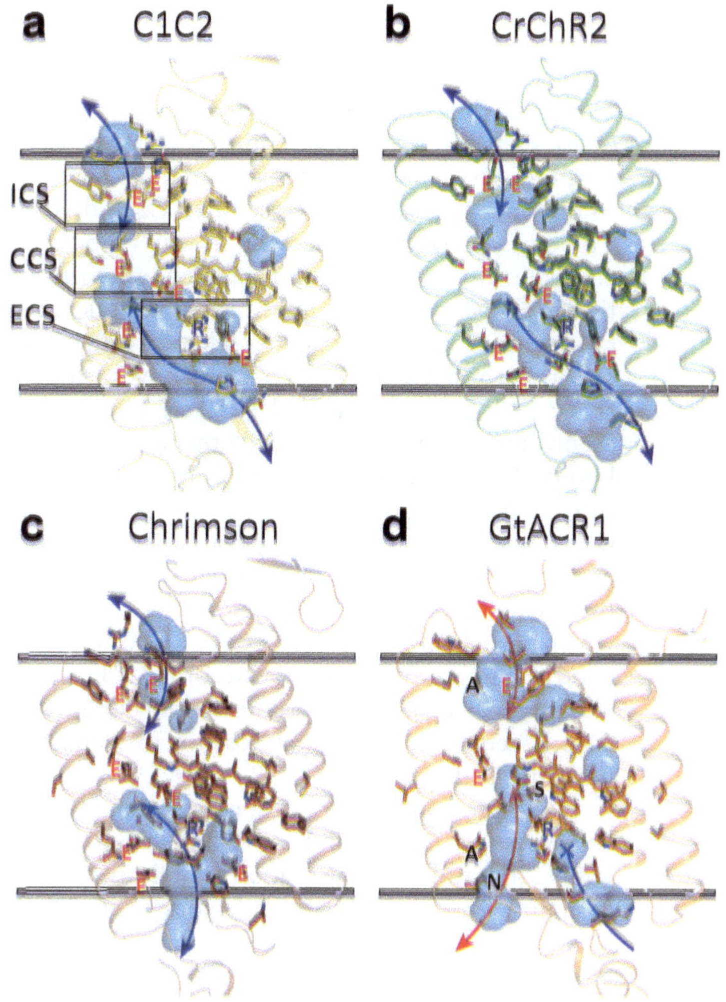

Fig. 5 Overall X-ray crystal structures of channelrhodopsins. Blue arrows indicate a putative cation-conducting pathway. Conservative glutamates, lining cation pathway, are indicated in each structure as "E" in red (namely, E82, E83, E90, E97, E101, E123, E235, according to CrChR2 numeration). R120 is shown as "R" in blue. Red arrows indicate a putative anion conducting pathway in the GtACR1. Gray lines represent two surfaces of the lipid membranes. (**a**) The structure of the chimera C1C2. (**b**) The structure of the CrChR2. (**c**) The structure of the red-shifted channelrhodopsin Chrimson. (**d**) The structure of the anion channelrhodopsin GtACR1; alterations of glutamates are indicated in black

anion conducting channelrhodopsins are available (Fig. 5) [68, 87, 90–92]. In this paragraph, we concentrate only on channelrhodopsins' structures, as they are used in optogenetics. More high-resolution structures of other microbial rhodopsins are also discussed in the "Rhodopsins" chapter of this issue.

In brief, channelrhodopsins, as rhodopsins in general, comprising of 7-transmembrane helices and cofactor retinal covalently bonded to specific lysine in G-helix. Presumably, channelrhodopsins have a pore protruding through the complete protein in the open state, passing near the retinal Schiff base. However, at the moment, there is no high-resolution structure of the open state of any channelrhodopsin.

Meanwhile, in the ground state, one can either see several cavities separated by constriction sites (case of *Cr*ChR2 and Chrimson, Fig. 5b, c) or complete pore going from extracellular space up to the central gate region (case of C1C2 and *Gt*ACR1, Fig. 5a, d). The surfaces of the pores are shown to be electronegative and electropositive for cation and anion conducting rhodopsins, respectively. Thus, the pores might serve as selectivity filters for the channelrhodopsins. For such a selectivity, the putative ion translocation path in CCRs is lined with six very conservative Glutamic acids: E82, E82, E90, E97, E101, and E235 (hereafter numbering according to *Cr*ChR2, unless otherwise stated). Considering all known X-ray structures of channelrhodopsins, the channels have three constriction sites along the ion pathway (Fig. 5a): extracellular, central, and intracellular sites (ECS, CCS, ICS). However, the gates are formed a bit differently in each case. The unique feature of the *Gt*ACR1 is that ECS completely blocks ion translocation way, which is usually used in CCRs. *Gt*ACR1 has its own pore, approaching the CCS from the other side (Fig. 5). This pores' surface is electropositive, presumably making it selective for anions. This is achieved due to the absence of some of the conservative glutamic acids present in CCRs.

An interesting observation concerning DC-pair (C128/D156) comes from high-resolution X-ray structures. As was discussed above, mutations in this region significantly alter channel kinetics [32, 35]. But the DC-pair is not located in the putative ion translocation path. Thus, the understanding of its role remains elusive.

4.2 Photocycle of CrChR2

As all microbial rhodopsins undergo a photocycle, it is interesting to correlate the transport activity with the kinetics of the photointermediates. For example, in *Hs*BR, the formation of the M intermediate and the concomitant deprotonation of the Schiff base, and the M decay followed by the reprotonation are the main electrogenic events in the photocycle. In the case of *Cr*ChR2, the situation is quite different, as can be seen in Fig. 6. The M state lasts only for few microseconds. The channel activity occurs in the protonated state of the Schiff base with intermediate lifetimes of 2 and 15 ms [11, 93, 95].

Channelrhodopsins have complex photocycles. It was a challenge to explain all effects that occur during *Cr*ChR2 function: light/dark adaptation, desensitization, and different kinetics in the presence of different ions. Flash experiments, together with the determination of the time-resolved photocurrents, resulted in the identification of the photo products, which represent the open state of the channel. It was shown that a light-adapted and a dark-adapted form with an absorption maximum occur at 470 nm during the photocycle. By this the splitting of the cycle could be

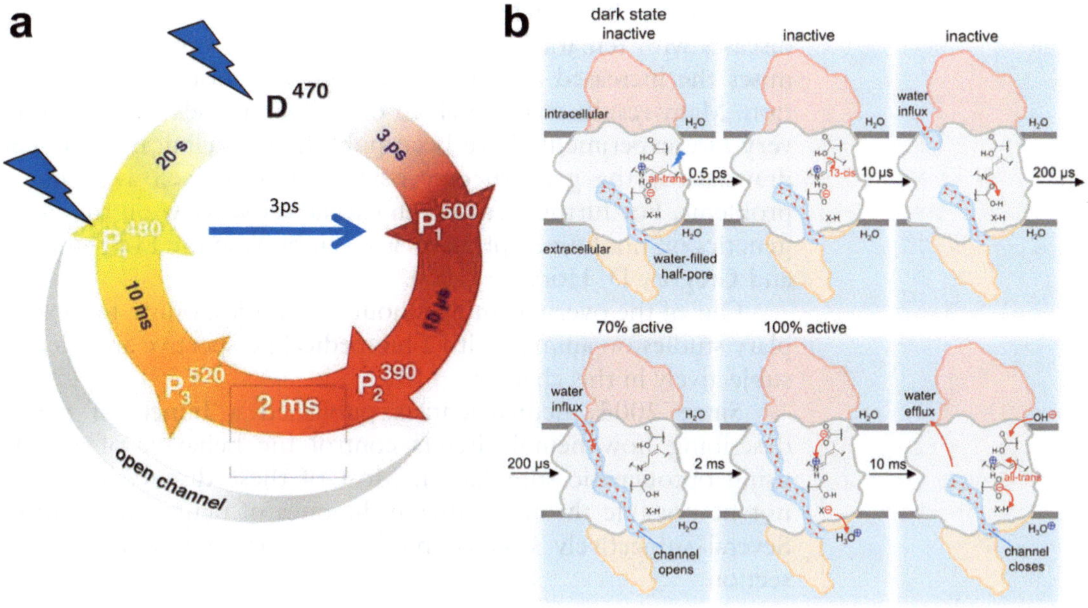

Fig. 6 The photocycle of the channelrhodopsin *Cr*ChR2. (**a**) The second photon absorption accelerates the photocycle under continuous light conditions, whereas flashlight experiments show the decay time of around 20 s. (**b**) Schematic representation of gating mechanisms in *Cr*ChR2 under continuous light conditions. (Figure is adapted from publications [93, 94])

explained. A single-photon activation results in a rate-limiting step of several seconds, namely the decay from the light-adapted form to the dark-adapted form, the ground state. In contrast, in electrophysiological experiments, the lifetime of the channel was determined to 10 ms. The discrepancy can be explained that a second photon activates the light-adapted form, which short circuits the photocycle (Fig. 6a). However, the desensitization of the photocurrents could not be explained.

There is a large number of time-resolved infrared spectroscopy (FTIR) studies. Many of them deal with protonation and deprotonation reactions of functionally important amino acid residues [96, 97]. However, others allow determining *Cr*ChR2 conformational changes during its photocycle (Fig. 6b) [94, 98].

Besides, 2D Cryo-EM and DEER analysis helped to analyze the helices movement involved in the channel's opening and closing; namely, second, third, and sixth helices movements participate in the process [89]. More detailed description of the *Cr*ChR2 photocycle is given in [99].

5 Exemplary Optogenetic Applications with Biomedical Background

There is a wide variety of options for stimulation or inhibition of neural circuits by the use of extracellular neural electrode stimulation. Those techniques are used to cure or suppress neural diseases with remarkable success. Nevertheless, optogenetics promises the increased specificity and higher spatiotemporal resolution. Many results were obtained on animal models but until now very rare experiments have been published on clinical trials. The drawback is the gene therapy rising ethical as well as technical problems. For further reading an excellent overview on the optogenetics on nonhuman primates is given by Yasmine El-Shamayleh and Gregory D. Horwitz [100].

Out of the overwhelming amount of studies, only a few exemplary studies on animals with a biomedical perspective are selected subjectively in this chapter.

Since 2005, neuroscientists published a bunch of papers describing how neural circuits control the behavior of animals using optogenetic stimulation. Most of these discoveries would not be possible through other techniques of neural modulation. Several subjectively selected publications are mentioned in this section.

5.1 Optogenetic Memory Retrieval

Alzheimer's disease is the most common variant of dementia of aged people. It has a wide range of symptoms: at the beginning there will be difficulty in remembering recent events. As the disease advances, it affects language, mood, motivation and behavior, and spatial orientation. In the later stages, the disease leads to self-care disability. This disease has a dramatic social effect and significant economic effects in developed countries.

Currently, there is no competent medical cure for Alzheimer's disease. However, a prolonged survey of the scientific group from MIT in 2016 led to the first optogenetic approach for extracting memories in AD mice.

Scientists from the Tonegawa laboratory in 2012 managed to optogenetically extract fear memories in healthy mice [101]. They labeled engram cells in mice (which were active during training in fear context, where mice are electrically shocked) with channelrhodopsin-2 (Fig. 7a, b). Mice with labeled engrams froze in neutral conditions when they were optically stimulated. This experiment showed the possibility to extract memories optogenetically. More importantly, this experiment showed that the activation of engram cells is sufficient to retrieve the memory, which had a tremendous impact on neuroscience.

In 2013 the same group of scientists used a similar methodology for making false fear memories in living animals [103]. They labeled mice engrams activated during neutral context A with channelrhodopsin-2. Later they stimulated these engrams in distinct fear context B. This made mice brain encode fear into engrams corresponding to neutral context A as well as into engrams of context B. After this procedure mice froze in context A without any optical stimulation (in control neutral context C mice were still

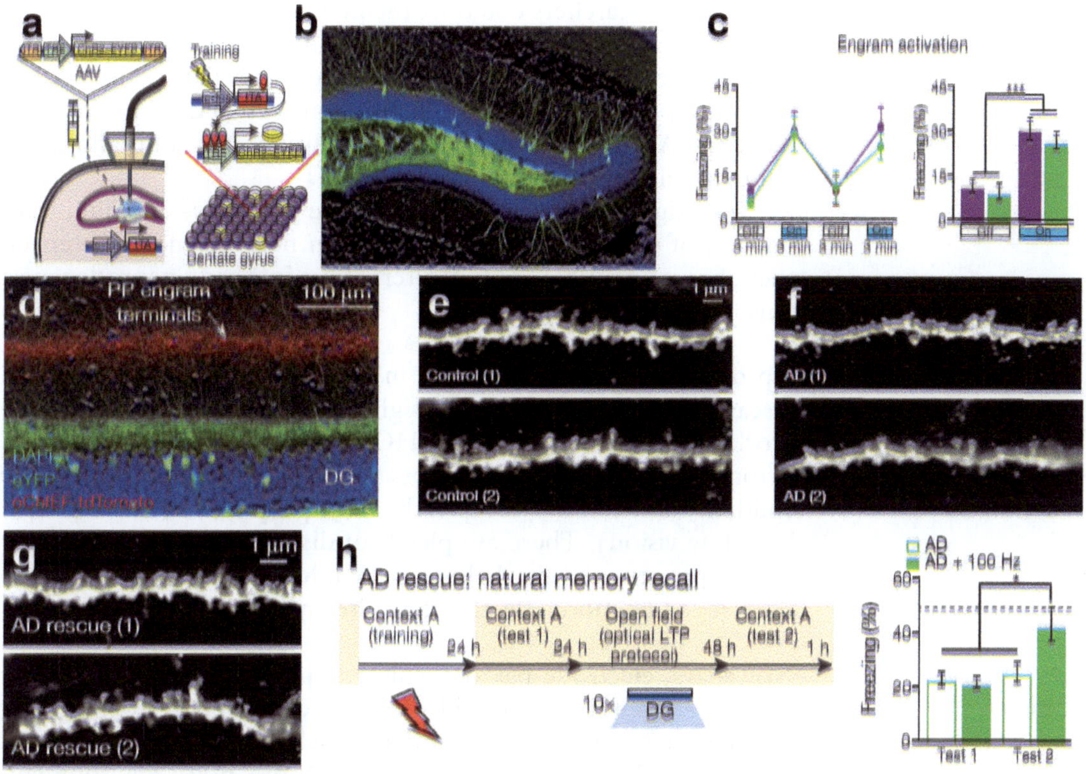

Fig. 7 Optogenetic memory retrieval in AD mice. (**a**) Two-virus technology for activity-dependent neuron labeling. (**b**) DG engram labeled with CrChR2 (green). (**c**) Memory retrieval by optogenetic activation of DG engrams. (**d**) EC engram labeled with ChIEF (red) and DG engram labeled with eYFP (yellow). (**e**) Dendrite spines in healthy mice DG. (**f**) Dendrite spines in AD mice DG. (**g**) Recovered dendrite spines in AD mice DG after optogenetic activation of long-term potentiation. (**h**) Memory rescue in AD mice. Behavioral schedule (left) and freezing restoration. Dashed line represents control mice freezing. (The figure is a compilation of the panels of the following publications [101, 102])

active). Therefore, this methodology allowed the authors to create a false memory.

In 2014 another group of scientists showed that activation of engram neurons is also necessary for memory retrieval [104]. They conducted a similar experiment except for expressing silencing archaerhodopsin-3 in engrams. The silencing of these neurons significantly reduced the recall of the fear memory. Therefore, the theorem: "Neurons of dentate gyrus activated during contextual fear conditioning are both necessary and sufficient for fear memory retrieval" was proven. Q.E.D.

Using accumulated knowledge, Tonegawa's group reached the goal [102]. They proved the possibility to retrieve memory in healthy and AD mice optogenetically by activation of fear-engrams (Fig. 7c). Importantly, their study showed no difference in optogenetic memory recall between healthy and AD mice, which means AD impairs recall of memory rather than the encoding of memory.

Moreover, researchers could fix the dendritic spines deficit in early AD mice by long-term potentiation induced by light-activation of ChIEF-labeled engrams in the entorhinal cortex (EC) (Fig. 7d, e, f, g). The spines' restoration, in its turn, rescued term memory retrieval: AD mice froze in fear conditions without additional light stimulation (Fig. 7h).

Investigations described above are examples of significant impacts of optogenetic interventions on neuroscience. Moreover, they showed possible medical alternatives in curing neurodegenerative diseases.

5.2 Optogenetic Sight and Hearing Restoration

Optogenetics can be applicable in curing hereditary and acquired diseases depriving a person of sight or hearing. According to the World Health Organization (WHO), over 5% of the world's population are people with hearing loss (~466 mln, ~34 mln of them are children) and 39 mln completely blind people (~246 mln people have low vision). There are plenty of diseases causing such disorders. Optogenetics may help to treat at least some of them. Moreover, it can be more effective than already existing techniques.

5.2.1 Hearing Restoration

Nowadays, electrode-based cochlear implants are widely used for hearing restoration. Although they can effectively restore hearing for speech comprehension under noiseless conditions, it is not enough for listening to music and talk in a noisy environment. Electrical implants stimulate the spiral ganglion neurons (SGNs) in the inner ear, depending on their placement on the cochlea, responsible for different sound wave frequencies.

A broad spread of electrical activation allows using only 10–20 stimulation channels along the cochlear [105] (Fig. 8a). Moreover, the loudness growth function typical for electrical stimulation has several times more narrow dynamic range than for the natural stimulation [106]. These facts lead to low frequency and intensity resolution. Optogenetic stimulation of SGNs can solve these problems (Fig. 8b). Optical stimulation can be effectively focused; thus, the sound frequency can be precisely transformed into the activation of corresponding SGNs. Spatial precision also has impact on intensity resolution. During natural listening, loudness increase leads to activation of neighboring SNGs in addition to frequency-defined ones. Besides that, intensity increase leads to activation of SNGs with high spontaneous firing rates, which can also be mimicked by different channelrhodopsin expression levels in different SGNs.

Successful experiments on mice were reported showing that *Cr*ChR2 can be used for cochlear implants [107].

However, *Cr*ChR2 is not an optimal optogenetic tool for hearing restoration. SGNs are among the fastest neurons in mammals, which can fire with up to 1000 Hz rate. Therefore, rhodopsins with fast closing kinetics are needed. Chronos is a fast-closing

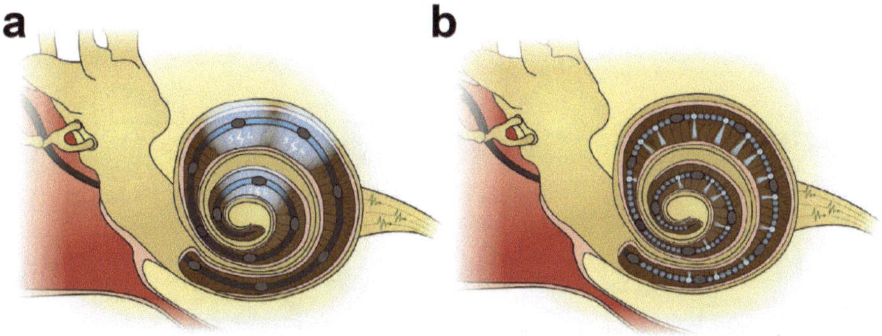

Fig. 8 Schematic overview of cochlear operating principles. (**a**) A broad spread of activation by electrical cochlear implants. (**b**) Optogenetic activation of SGNs promises a significant increase in frequency resolution. (The figure is adapted from [105])

blue-light absorbing naturally existing channelrhodopsin [37]. It was reported to elicit auditory brainstem responses even at 1000 Hz of stimulation frequency [46]. Besides, a red-shifted ultrafast channelrhodopsin variant, derived from Chrimson, was obtained by rational optimization and shown to be capable of restoring hearing in deaf mice [38].

5.2.2 Sight Restoration

Microbial opsins can also be used for vision restoration. It is especially convenient due to the presence of human rhodopsin in the eye. There are several possible optogenetic strategies of sight restoration, depending on the severity of the disease.

In brief, photoreception in the eye is arranged as follows. There are several different types of cells present in the retina: cones, rods, on and off bipolar neurons, ganglion neurons, amacrine, and horizontal cells (Fig. 9a). All cells communicate with each other through neurotransmitters, which are released in response to changes in membrane potential. Thus, the illumination of the retina in a healthy state results in depolarization and hyperpolarization of specific types of cells. In particular, rods and cones, as well as off-bipolar cells and off-ganglion neurons, are hyperpolarized in response to light. On the contrary, on-bipolar neurons and on-ganglion cells are hyperpolarized under the illumination of the retina. However, only cones and rods are light-sensitive: they absorb light and transmit the signal to bipolar cells, which in turn transmit the signal to ganglion neurons. Axons of ganglion cells assemble into the optic nerve that transmits visual information to the brain. Horizontal cells and amacrine cells modulate signal transmission from photosensitive cells to bipolar cells and from bipolar cells to ganglion neurons, respectively.

Already in 2006 one of the first works concerning optogenetic eye treatment in mice was published [109].

Optogenetic tools can be expressed in the retina in different ways corresponding to the state of the disease [110]. For instance,

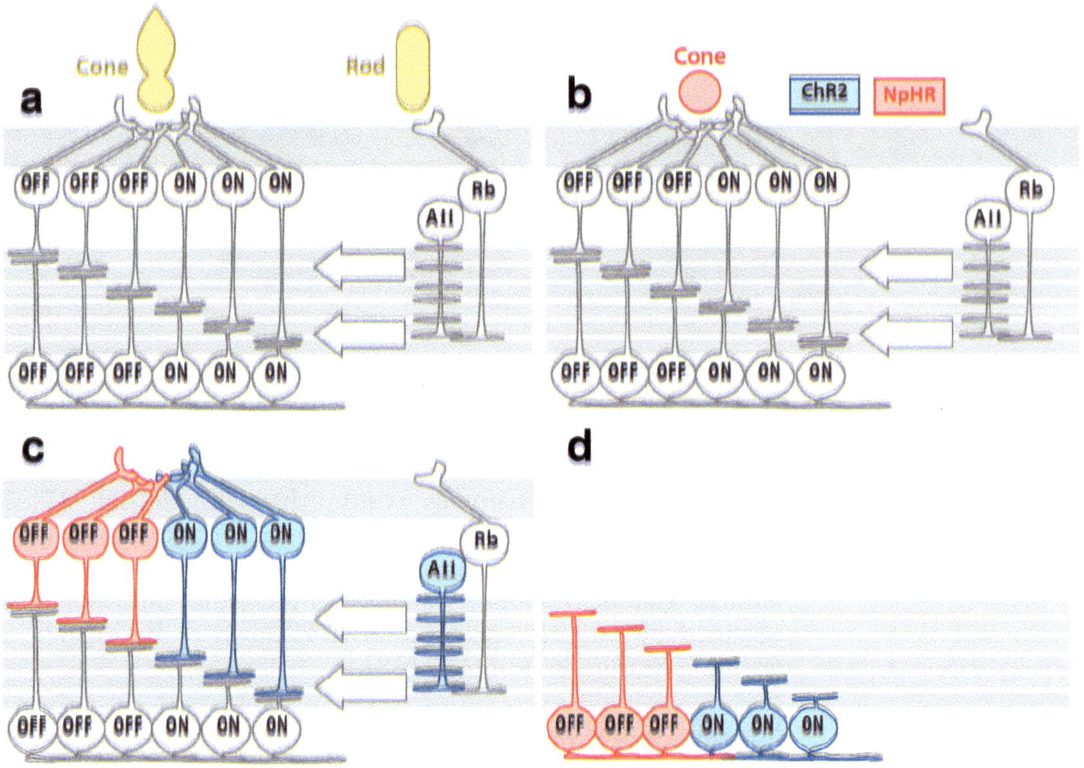

Fig. 9 Stages of retinal degradation and optogenetic therapy for blindness. Blue/red colors show expression of CrChR2/NpHR in certain types of cells, respectively, as a therapy. (**a**) A healthy retina consists of cones and rods, on/off bipolar cells, and ganglion neurons. (**b**) Rods degenerated, cones outer segments lost: restoration of cones. (**c**) Cones degenerated: optogenetic intervention into bipolar cells of amacrine cells. (**d**) Bipolar cells degenerated: optogenetic intervention into ganglion cells. (The figure is adapted from [108])

if only cones' outer photosensitive segments are degenerated, then optogenetic inhibitors can be expressed in cones for sight restoration [111, 112] (Fig. 9b). If cones are fully degenerated, then optogenetic activators can be expressed in on-bipolar cells, and the inhibitors in off-bipolar cells or optogenetic activators can be expressed in amacrine cells (Fig. 9c). If the layer of the bipolar cells also degenerates, then optogenetic activation/suppression of on/off ganglion cells can help [113] (Fig. 9d).

Indeed, in 2021 the first optogenetic therapy in a human was reported [114]. The red-light absorbing optogenetic actuator Chrimson K176R [37] was targeted to the foveal retinal ganglion cells by injecting an adeno-associated virus. The researchers used specially engineered goggles to optimize the light intensity range for the rhodopsin. This approach partially recovered visual function in a treated eye: the blind patient could locate, count, and touch different objects while wearing the goggles.

However, an efficient application of these approaches requires future improvements [108]. Optogenetic tools respond to a much narrower intensity range of light than healthy cones and rods. Although it can be solved by using special goggles that adjust light intensity, it is preferable to have more sensitive optogenetic tools. Besides that, all approaches described above cannot restore color vision: one has to express several rhodopsins with different spectra at a time. Finally, retina cells membrane potentials are finely modulated in a healthy state, which can hardly be controlled using only one optogenetic inhibitor/actuator at a time. At least some of the described problems can be solved by the rational design of optogenetic tools.

5.3 Deep Brain Stimulation Therapy Using Microbial Rhodopsins

Deep brain stimulation (DBS) is a neurosurgical procedure. An electrical neurostimulator is implanted into the patient brain. It sends electrical impulses to defined brain regions and thereby intervene in neural work. Nowadays, DBS is one of the primary methods used to treat neuronal disorders such as Parkinson's disease (PD), dystonia, obsessive-compulsive disorder, and epilepsy.

Despite the great importance and wide distribution, the exact mechanism underlying DBS is still not known. For instance, DBS can treat many symptoms of PD, including tremor, rigidity, and bradykinesia—but not others, such as speech impairment, depression, and dementia. It is evident that conventional DBS methodology lacks cell-type specificity, spatial resolution, and is invasive. In 2009 researchers made progress in the understanding of DBS treatment of PD using optogenetics [115]. They stimulated and inhibited distinct neural circuits in parkinsonian rodents. The authors observed that neither inhibition with e*Np*HR nor excitation with *Cr*ChR2 of subthalamic nucleus (STN, the brain region linked to movement impairments in PD patients) neurons did not affect rodents' behavior. Therefore, then they tested if effects of DBS arise from activation of axonal projections that enter the STN. Such activation led to the complete restoration of the normal healthy behavior of the parkinsonian rodents. Thus, optogenetics has contributed to the understanding of DBS and PD symptoms.

In 2018 a group of scientists invented a noninvasive near-infrared optogenetic solution for DBS [116]. They expressed channelrhodopsin-2 in mice neurons of interest and stimulated them using 980 nm laser. This highly red-shifted light penetrated through the brain well enough without neurosurgical implanting of fiber-optics. However, the wavelength was inappropriate for the activation of channelrhodopsin-2. To deal with the problem, scientists used upconversion nanoparticles (UCNPs). These particles can absorb near-infrared light (980 nm) and emit blue light instead (450 and 475 nm). This approach allowed scientists to introduce DBS without neurosurgical interventions.

Nevertheless, this methodology still needs to deliver non-natural compounds (UCNPs) inside the brain. Engineering of rhodopsin channels with a highly red-shifted spectrum and sufficient quantum efficiency would allow making DBS without any additional particles. Rational mutagenesis of rhodopsins and bioinformatic approaches have already optimized optogenetic working spectral range. However, optogenetic tools absorbing in far-infrared are desirable.

6 Conclusions and Outlook

There is an overwhelming number of publications on optogenetics, showing that the method opens completely new approaches in neuroscience, muscle physiology, and cell biology. In many of these studies, the authors claim the biomedical impact. Already after the first applications in 2005, it became apparent that this new method might be promising for the cure of neurodegenerative diseases indeed. Exciting studies on the analysis of neural circuits, behavior, and animal disease models have been performed mainly in rodents and also in nonhuman primates. These studies resulted in new insights into the nervous system. However, clinical trials are scarce. According to our knowledge, only the most obvious application, gene therapy for the recovery of vision after the destruction of the retina's photoreceptor cells, are in preliminary clinic phases.

In this review, the authors restricted the manuscript to the principles of rhodopsin-based optogenetics, knowing that many interesting aspects were deleted. We have focused our interest on the interplay of physics biochemistry and physiology of microbial rhodopsins, which has been proven to give a solid basis for the application of optogenetics.

Up to now most of optogenetic manipulations are done through the control of cell plasma membrane potential. We believe that one of the future exciting developments of optogenetics might be connected with the control of cell organelles like mitochondria, lysosomes, and endoplasmic reticulum. First experiments are in progress.

Acknowledgment

This work was supported by Russian Science Foundation (RSF) Project 21-64-00018.

References

1. Zemelman BV, Lee GA, Ng M et al (2002) Selective photostimulation of genetically ChARGed neurons. Neuron 33:15–22
2. Foster KW, Saranak J, Patel N et al (1984) A rhodopsin is the functional photoreceptor for phototaxis in the unicellular eukaryote Chlamydomonas. Nature 311:756–759
3. Sineshchekov OA, Litvin FF, Keszthelyi L (1990) Two components of photoreceptor potential in phototaxis of the flagellated green alga Haematococcus pluvialis. Biophys J 57:33–39
4. Harz H, Hegemann P (1991) Rhodopsin-regulated calcium currents in Chlamydomonas. Nature 351:489–491
5. Holland EM, Braun FJ, Nonnengässer C et al (1996) The nature of rhodopsin-triggered photocurrents in Chlamydomonas. I. Kinetics and influence of divalent ions. Biophys J 70:924–931
6. Nonnengässer C, Holland EM, Harz H et al (1996) The nature of rhodopsin-triggered photocurrents in Chlamydomonas. II. Influence of monovalent ions. Biophys J 70:932–938
7. Hildebrandt V, Fendler K, Heberle J et al (1993) Bacteriorhodopsin expressed in Schizosaccharomyces pombe pumps protons through the plasma membrane. Proc Natl Acad Sci 90:3578–3582
8. Hoffmann A, Hildebrandt V, Heberle J et al (1994) Photoactive mitochondria: in vivo transfer of a light-driven proton pump into the inner mitochondrial membrane of Schizosaccharomyces pombe. Proc Natl Acad Sci 91: 9367–9371
9. Nagel G, Möckel B, Büldt G et al (1995) Functional expression of bacteriorhodopsin in oocytes allows direct measurement of voltage dependence of light induced H+ pumping. FEBS Lett 377:263–266
10. Geibel S, Friedrich T, Ormos P et al (2001) The voltage-dependent proton pumping in bacteriorhodopsin is characterized by optoelectric behavior. Biophys J 81:2059–2068
11. Nagel G, Szellas T, Huhn W et al (2003) Channelrhodopsin-2, a directly light-gated cation-selective membrane channel. Proc Natl Acad Sci 100:13940–13945
12. Boyden ES, Zhang F, Bamberg E et al (2005) Millisecond-timescale, genetically targeted optical control of neural activity. Nat Neurosci 8:1263–1268
13. Nagel G, Brauner M, Liewald JF et al (2005) Light activation of channelrhodopsin-2 in excitable cells of Caenorhabditis elegans triggers rapid behavioral responses. Curr Biol 15: 2279–2284
14. Li X, Gutierrez DV, Hanson MG et al (2005) Fast noninvasive activation and inhibition of neural and network activity by vertebrate rhodopsin and green algae channelrhodopsin. Proc Natl Acad Sci 102:17816–17821
15. Zhang F, Wang L-P, Brauner M et al (2007) Multimodal fast optical interrogation of neural circuitry. Nature 446:633–639
16. Kralj JM, Douglass AD, Hochbaum DR et al (2012) Optical recording of action potentials in mammalian neurons using a microbial rhodopsin. Nat Methods 9:90–95
17. Hochbaum DR, Zhao Y, Farhi SL et al (2014) All-optical electrophysiology in mammalian neurons using engineered microbial rhodopsins. Nat Methods 11:825–833
18. Miyawaki A, Llopis J, Heim R et al (1997) Fluorescent indicators for Ca2+ based on green fluorescent proteins and calmodulin. Nature 388:882–887
19. Ishizuka T, Kakuda M, Araki R et al (2006) Kinetic evaluation of photosensitivity in genetically engineered neurons expressing green algae light-gated channels. Neurosci Res 54:85–94
20. Feldbauer K, Zimmermann D, Pintschovius V et al (2009) Channelrhodopsin-2 is a leaky proton pump. Proc Natl Acad Sci 106: 12317–12322
21. Oesterhelt D, Stoeckenius W (1973) Functions of a new photoreceptor membrane. Proc Natl Acad Sci 70:2853–2857
22. Oesterhelt D, Stoeckenius W (1971) Rhodopsin-like protein from the purple membrane of Halobacterium halobium. Nat New Biol 233:149–152
23. Lanyi JK (2004) Bacteriorhodopsin. Annu Rev Physiol 66:665–688
24. Béjà O, Aravind L, Koonin EV et al (2000) Bacterial rhodopsin: evidence for a new type of phototrophy in the sea. Science 289: 1902–1906
25. Hegemann P, Fuhrmann M, Kateriya S (2001) Algal sensory photoreceptors. J Phycol 37:668–676
26. Sineshchekov OA, Jung K-H, Spudich JL (2002) Two rhodopsins mediate phototaxis to low- and high-intensity light in Chlamydomonas reinhardtii. Proc Natl Acad Sci 99: 8689–8694
27. Suzuki T, Yamasaki K, Fujita S et al (2003) Archaeal-type rhodopsins in Chlamydomonas: model structure and intracellular localization. Biochem Biophys Res Commun 301: 711–717
28. Nagel G, Ollig D, Fuhrmann M et al (2002) Channelrhodopsin-1: a light-gated proton

channel in green algae. Science 296: 2395–2398
29. Berndt A, Schoenenberger P, Mattis J et al (2011) High-efficiency channelrhodopsins for fast neuronal stimulation at low light levels. Proc Natl Acad Sci 108:7595–7600
30. Lin JY, Lin MZ, Steinbach P et al (2009) Characterization of engineered channelrhodopsin variants with improved properties and kinetics. Biophys J 96:1803–1814
31. Zhang F, Prigge M, Beyrière F et al (2008) Red-shifted optogenetic excitation: a tool for fast neural control derived from Volvox carteri. Nat Neurosci 11:631–633
32. Yizhar O, Fenno LE, Prigge M et al (2011) Neocortical excitation/inhibition balance in information processing and social dysfunction. Nature 477:171–178
33. Lin JY, Knutsen PM, Muller A et al (2013) ReaChR: a red-shifted variant of channelrhodopsin enables deep transcranial optogenetic excitation. Nat Neurosci 16:1499–1508
34. Rajasethupathy P, Sankaran S, Marshel JH et al (2015) Projections from neocortex mediate top-down control of memory retrieval. Nature 526:653–659
35. Berndt A, Yizhar O, Gunaydin LA et al (2009) Bi-stable neural state switches. Nat Neurosci 12:229–234
36. Gunaydin LA, Yizhar O, Berndt A et al (2010) Ultrafast optogenetic control. Nat Neurosci 13:387–392
37. Klapoetke NC, Murata Y, Kim SS et al (2014) Independent optical excitation of distinct neural populations. Nat Methods 11: 338–346
38. Mager T, Lopez de la Morena D, Senn V et al (2018) High frequency neural spiking and auditory signaling by ultrafast red-shifted optogenetics. Nat Commun 9:1750
39. Marshel JH, Kim YS, Machado TA et al (2019) Cortical layer–specific critical dynamics triggering perception. Science 365: eaaw5202
40. Shevchenko V, Mager T, Kovalev K et al (2017) Inward H+ pump xenorhodopsin: mechanism and alternative optogenetic approach. Sci Adv 3:e1603187
41. Gradinaru V, Zhang F, Ramakrishnan C et al (2010) Molecular and cellular approaches for diversifying and extending optogenetics. Cell 141:154–165
42. Berndt A, Lee SY, Ramakrishnan C et al (2014) Structure-guided transformation of channelrhodopsin into a light-activated chloride channel. Science 344:420–424
43. Gong X, Mendoza-Halliday D, Ting JT et al (2020) An ultra-sensitive step-function opsin for minimally invasive optogenetic stimulation in mice and macaques. Neuron 107:38
44. Wietek J, Wiegert JS, Adeishvili N et al (2014) Conversion of channelrhodopsin into a light-gated chloride channel. Science 344:409
45. Kleinlogel S, Feldbauer K, Dempski RE et al (2011) Ultra light-sensitive and fast neuronal activation with the Ca2+-permeable channelrhodopsin CatCh. Nat Neurosci 14:513–518
46. Keppeler D, Merino RM, Lopez de la Morena D et al (2018) Ultrafast optogenetic stimulation of the auditory pathway by targeting-optimized Chronos. EMBO J 37:e99649
47. Brown J, Behnam R, Coddington L et al (2018) Expanding the optogenetics toolkit by topological inversion of rhodopsins. Cell 175:1131–1140.e11
48. Govorunova EG, Sineshchekov OA, Spudich JL (2016) Structurally distinct cation channelrhodopsins from cryptophyte algae. Biophys J 110:2302–2304
49. Sineshchekov OA, Govorunova EG, Li H et al (2017) Bacteriorhodopsin-like channelrhodopsins: alternative mechanism for control of cation conductance. Proc Natl Acad Sci 114: E9512–E9519
50. Ugalde JA, Podell S, Narasingarao P et al (2011) Xenorhodopsins, an enigmatic new class of microbial rhodopsins horizontally transferred between archaea and bacteria. Biol Direct 6:52
51. Inoue K, Ito S, Kato Y et al (2016) A natural light-driven inward proton pump. Nat Commun 7:1–10
52. Bulzu P-A, Andrei A-Ş, Salcher MM et al (2019) Casting light on Asgardarchaeota metabolism in a sunlit microoxic niche. Nat Microbiol 4:1129–1137
53. Inoue K, Tsunoda SP, Singh M et al (2020) Schizorhodopsins: a family of rhodopsins from Asgard archaea that function as light-driven inward H+ pumps. Sci Adv 6:eaaz2441
54. Schobert B, Lanyi JK (1982) Halorhodopsin is a light-driven chloride pump. J Biol Chem 257:10306–10313
55. Han X, Boyden ES (2007) Multiple-color optical activation, silencing, and desynchronization of neural activity, with single-spike temporal resolution. PLoS One 2:e299
56. Gradinaru V, Thompson KR, Deisseroth K (2008) eNpHR: a Natronomonas halorhodopsin enhanced for optogenetic applications. Brain Cell Biol 36:129–139
57. Kleinlogel S, Terpitz U, Legrum B et al (2011) A gene-fusion strategy for stoichiometric and co-localized expression of light-gated membrane proteins. Nat Methods 8: 1083–1088

58. Chow BY, Han X, Dobry AS et al (2010) High-performance genetically targetable optical neural silencing by light-driven proton pumps. Nature 463:98–102
59. Han X, Chow BY, Zhou H et al (2011) A high-light sensitivity optical neural silencer: development and application to optogenetic control of non-human primate cortex. Front Syst Neurosci 5:18
60. Inoue K, Ono H, Abe-Yoshizumi R et al (2013) A light-driven sodium ion pump in marine bacteria. Nat Commun 4:1678
61. Gushchin I, Shevchenko V, Polovinkin V et al (2015) Crystal structure of a light-driven sodium pump. Nat Struct Mol Biol 22: 390–395
62. Kato HE, Inoue K, Abe-Yoshizumi R et al (2015) Structural basis for Na+ transport mechanism by a light-driven Na+ pump. Nature 521:48–53
63. Grimm C, Silapetere A, Vogt A et al (2018) Electrical properties, substrate specificity and optogenetic potential of the engineered light-driven sodium pump eKR2. Sci Rep 8:9316
64. Vogt A, Silapetere A, Grimm C et al (2019) Engineered passive potassium conductance in the KR2 sodium pump. Biophys J 116: 1941–1951
65. Berndt A, Lee SY, Wietek J et al (2016) Structural foundations of optogenetics: determinants of channelrhodopsin ion selectivity. Proc Natl Acad Sci 113:822–829
66. Govorunova EG, Sineshchekov OA, Janz R et al (2015) Natural light-gated anion channels: a family of microbial rhodopsins for advanced optogenetics. Science 349:647–650
67. Sineshchekov OA, Govorunova EG, Li H et al (2015) Gating mechanisms of a natural anion channelrhodopsin. Proc Natl Acad Sci 112: 14236–14241
68. Kato HE, Kim YS, Paggi JM et al (2018) Structural mechanisms of selectivity and gating in anion channelrhodopsins. Nature 561: 349–354
69. Oppermann J, Fischer P, Silapetere A et al (2019) MerMAIDs: a family of metagenomically discovered marine anion-conducting and intensely desensitizing channelrhodopsins. Nat Commun 10:1–13
70. Rost BR, Schneider F, Grauel MK et al (2015) Optogenetic acidification of synaptic vesicles and lysosomes. Nat Neurosci 18:1845–1852
71. Tkatch T, Greotti E, Baranauskas G et al (2017) Optogenetic control of mitochondrial metabolism and Ca2+ signaling by mitochondria-targeted opsins. Proc Natl Acad Sci 114:E5167–E5176
72. Reyer A, Häßler M, Scherzer S et al (2020) Channelrhodopsin-mediated optogenetics highlights a central role of depolarization-dependent plant proton pumps. Proc Natl Acad Sci 117:20920–20925
73. Luck M, Mathes T, Bruun S et al (2012) A photochromic histidine kinase rhodopsin (HKR1) that is bimodally switched by ultraviolet and blue light. J Biol Chem 287: 40083–40090
74. Avelar GM, Schumacher RI, Zaini PA et al (2014) A rhodopsin-guanylyl cyclase gene fusion functions in visual perception in a fungus. Curr Biol 24:1234–1240
75. Yoshida K, Tsunoda SP, Brown LS et al (2017) A unique choanoflagellate enzyme rhodopsin exhibits light-dependent cyclic nucleotide phosphodiesterase activity. J Biol Chem 292:7531–7541
76. Kim CK, Adhikari A, Deisseroth K (2017) Integration of optogenetics with complementary methodologies in systems neuroscience. Nat Rev Neurosci 18:222–235
77. Paz JT, Davidson TJ, Frechette ES et al (2013) Closed-loop optogenetic control of thalamus as a tool for interrupting seizures after cortical injury. Nat Neurosci 16:64–70
78. Ji G, Feldman ME, Deng K-Y et al (2004) Ca2+-sensing transgenic mice postsynaptic signaling in smooth muscle. J Biol Chem 279:21461–21468
79. Vogt N (2015) Voltage sensors: challenging, but with potential. Nat Methods 12:921–924
80. Zhang H, Reichert E, Cohen AE (2016) Optical electrophysiology for probing function and pharmacology of voltage-gated ion channels. elife 5:e15202
81. Gómez-Consarnau L, Raven JA, Levine NM et al (2019) Microbial rhodopsins are major contributors to the solar energy captured in the sea. Sci Adv 5:eaaw8855
82. Pushkarev A, Inoue K, Larom S et al (2018) A distinct abundant group of microbial rhodopsins discovered using functional metagenomics. Nature 558:595–599
83. Yutin N, Koonin EV (2012) Proteorhodopsin genes in giant viruses. Biol Direct 7:34
84. Needham DM, Yoshizawa S, Hosaka T et al (2019) A distinct lineage of giant viruses brings a rhodopsin photosystem to unicellular marine predators. Proc Natl Acad Sci 116: 20574–20583
85. Bratanov D, Kovalev K, Machtens J-P et al (2019) Unique structure and function of viral rhodopsins. Nat Commun 10:4939
86. Zabelskii D, Alekseev A, Kovalev K et al (2020) Viral rhodopsins 1 are an unique

family of light-gated cation channels. Nat Commun 11:5707
87. Kato HE, Zhang F, Yizhar O et al (2012) Crystal structure of the channelrhodopsin light-gated cation channel. Nature 482: 369–374
88. Müller M, Bamann C, Bamberg E et al (2011) Projection structure of channelrhodopsin-2 at 6 Å resolution by electron crystallography. J Mol Biol 414:86–95
89. Müller M, Bamann C, Bamberg E et al (2015) Light-induced helix movements in channelrhodopsin-2. J Mol Biol 427:341–349
90. Volkov O, Kovalev K, Polovinkin V et al (2017) Structural insights into ion conduction by channelrhodopsin 2. Science 358: eaan8862
91. Oda K, Vierock J, Oishi S et al (2018) Crystal structure of the red light-activated channelrhodopsin Chrimson. Nat Commun 9:1–11
92. Kim YS, Kato HE, Yamashita K et al (2018) Crystal structure of the natural anion-conducting channelrhodopsin Gt ACR1. Nature 561:343–348
93. Lórenz-Fonfría VA, Resler T, Krause N et al (2013) Transient protonation changes in channelrhodopsin-2 and their relevance to channel gating. Proc Natl Acad Sci 110: E1273–E1281
94. Lórenz-Fonfría VA, Bamann C, Resler T et al (2015) Temporal evolution of helix hydration in a light-gated ion channel correlates with ion conductance. Proc Natl Acad Sci U S A 112:E5796–E5804
95. Bamann C, Kirsch T, Nagel G et al (2008) Spectral characteristics of the photocycle of channelrhodopsin-2 and its implication for channel function. J Mol Biol 375:686–694
96. Eisenhauer K, Kuhne J, Ritter E et al (2012) In channelrhodopsin-2 Glu-90 is crucial for ion selectivity and is deprotonated during the photocycle. J Biol Chem 287:6904–6911
97. Ito S, Kato HE, Taniguchi R et al (2014) Water-containing hydrogen-bonding network in the active center of channelrhodopsin. J Am Chem Soc 136:3475–3482
98. Radu I, Bamann C, Nack M et al (2009) Conformational changes of channelrhodopsin-2. J Am Chem Soc 131:7313–7319
99. Kuhne J, Vierock J, Tennigkeit SA et al (2019) Unifying photocycle model for light adaptation and temporal evolution of cation conductance in channelrhodopsin-2. Proc Natl Acad Sci 116:9380–9389
100. El-Shamayleh Y, Horwitz GD (2019) Primate optogenetics: progress and prognosis. Proc Natl Acad Sci 116:26195–26203
101. Liu X, Ramirez S, Pang PT et al (2012) Optogenetic stimulation of a hippocampal engram activates fear memory recall. Nature 484: 381–385
102. Roy DS, Arons A, Mitchell TI et al (2016) Memory retrieval by activating engram cells in mouse models of early Alzheimer's disease. Nature 531:508–512
103. Ramirez S, Liu X, Lin P-A et al (2013) Creating a false memory in the hippocampus. Science 341:387–391
104. Denny CA, Kheirbek MA, Alba EL et al (2014) Hippocampal memory traces are differentially modulated by experience, time, and adult neurogenesis. Neuron 83:189–201
105. Jeschke M, Moser T (2015) Considering optogenetic stimulation for cochlear implants. Hear Res 322:224–234
106. Zeng F-G, Rebscher S, Harrison W et al (2008) Cochlear implants: system design, integration, and evaluation. IEEE Rev Biomed Eng 1:115–142
107. Hernandez VH, Gehrt A, Reuter K et al (2014) Optogenetic stimulation of the auditory pathway. J Clin Invest 124:1114–1129
108. Sahel J-A, Roska B (2013) Gene therapy for blindness. Annu Rev Neurosci 36:467–488
109. Bi A, Cui J, Ma Y-P et al (2006) Ectopic expression of a microbial-type rhodopsin restores visual responses in mice with photoreceptor degeneration. Neuron 50:23–33
110. Busskamp V, Picaud S, Sahel JA et al (2012) Optogenetic therapy for retinitis pigmentosa. Gene Ther 19:169–175
111. Busskamp V, Duebel J, Balya D et al (2010) Genetic reactivation of cone photoreceptors restores visual responses in retinitis pigmentosa. Science 329:413–417
112. Chuong AS, Miri ML, Busskamp V et al (2014) Noninvasive optical inhibition with a red-shifted microbial rhodopsin. Nat Neurosci 17:1123–1129
113. McGregor JE, Godat T, Dhakal KR et al (2020) Optogenetic restoration of retinal ganglion cell activity in the living primate. Nat Commun 11:1–9
114. Sahel J-A, Boulanger-Scemama E, Pagot C et al (2021) Partial recovery of visual function in a blind patient after optogenetic therapy. Nat Med 1–7
115. Gradinaru V, Mogri M, Thompson KR et al (2009) Optical deconstruction of Parkinsonian neural circuitry. Science 324:354–359
116. Chen S, Weitemier AZ, Zeng X et al (2018) Near-infrared deep brain stimulation via upconversion nanoparticle–mediated optogenetics. Science 359:679

Chapter 4

Searching Metagenomes for New Rhodopsins

Francisco Rodriguez-Valera, Alina Pushkarev, Riccardo Rosselli, and Oded Béjà

Abstract

Most microbial groups have not been cultivated yet, and the only way to approach the enormous diversity of rhodopsins that they contain in a sensible timeframe is through the analysis of their genomes. High-throughput sequencing technologies have allowed the release of community genomics (metagenomics) of many habitats in the photic zones of the ocean and lakes. Already the harvest is impressive and included from the first bacterial rhodopsin (proteorhodopsin) to the recent discovery of heliorhodopsin by functional metagenomics. However, the search continues using bioinformatic or biochemical routes.

Key words High-throughput sequencing, Photic zone, Aquatic habitats, Heliorhodopsin, Proteorhodopsin

1 Introduction to Metagenomics

One major discovery of the twentieth century has been the gigantic scale of the diversity of prokaryotic microbes. Although discovered in the seventeenth century, their small size and morphological monotony prevented scientific advance till the mid-nineteenth century when the discovery of pure culture (monoclonal populations derived from a single cell) allowed major advances in the study of their role in human health and ecology [1]. In spite of the tremendous advance that some models like *Escherichia coli* or its phages T4 and λ facilitated [2], pure culture remained the workhorse of microbiology, and it provided what at the time was thought a rather broad representation of the prokaryotic world [1]. However, the first studies of molecular chronometers (mostly ribosomal components) during the 1970s [3] started to shake the foundations of this self-satisfaction. The polyphyletic nature of prokaryotic cells was exposed [4]. They were composed of two different kinds at the level of basic molecular biology mechanisms, later called bacteria and archaea. By the mid-1980s, the same molecular chronometers

Valentin Gordeliy (ed.), *Rhodopsin: Methods and Protocols*, Methods in Molecular Biology, vol. 2501,
https://doi.org/10.1007/978-1-0716-2329-9_4,

could be cloned (first) [5] and PCR amplified (later) [6] and provided a vertiginous increase in the scale of the diversity of these two kinds of cells. Many major microbial groups had remained hidden by their reluctance to grow in laboratory conditions and be retrieved as pure cultures, thus they had to be studied only by their sequence. This was the era of the "molecular approach" when several new major taxa were described and proven to be much more relevant for the ecology of the planet than the "classical" pure culture-derived prokaryotic clades [7]. Painstakingly some of the major newly described microbes could be eventually cultured and thus representatives of *Prochlorococcus* [8, 9] and *Pelagibacter* [10] probably the most abundant prokaryotic cells on Earth could be obtained. It turns out that we were lucky with pathogenic bacteria because they are used to grow in the rich environment of animal bodies, but in nature microbes tend to be much more conservative of resources, so their growth is slow and their requirements are very specific and often unpredictable [11]. Therefore, the culturing avenue (although useful and important) has a very low yield. Meanwhile, the development of high-throughput sequencing led to the birth of the new science of genomics. By the mid-1990s the first bacterial and archaeal genomes were released, and they revealed yet another unexpected boost in the hidden diversity of prokaryotes. If the sequence of molecular chronometers showed the existence of multiple novel clades, comparative genomics revealed the pangenome [12, 13]. In higher organisms, animals, and plants, the taxonomy of one individual (say a *Homo sapiens*) allows a quite precise prediction of the genes this individual will contain (although of course allelic variation would provide diversity of phenotypes). In bacterial or archaeal strains (say an *E. coli* strain) the species only allows prediction between 30 and 50% of the genes present, the other half will be strain-dependent. This is the pangenome, the pool of genes contained within a prokaryotic species, that already with 2000 *E. coli* strains reaches the mind boggling figure of 100,000 different gene families (more than three times as many genes as a human contains), and still every new *E. coli* strain sequenced adds another 300 genes to the pool with no end in sight [14]. This complexity within complexity illustrates two facts: (1) the sequence diversity in the prokaryotic world is of astronomical proportions and (2) even if we cultivate a strain of each of the large numbers of yet uncultivated prokaryotic species, we would only get a tiny fraction of the real diversity contained within that species.

The bold approach of metagenomics has a methodological and a conceptual components. Methodologically it is very simple; by using high-throughput approaches it is possible to sequence a whole community of prokaryotes [15]. Conceptually, the sequence should allow the reconstruction of the lifestyle, metabolism, and biotechnological potential of the sampled community

[16]. Furthermore, in theory full pangenomes are accessible. There are two caveats: (1) the completeness (coverage) of the sequenced community might be limited or lost for the rare components and (2) since all sequencing methods available presently provide only restricted sequence length, the contiguity (linkage) of the sequence within a cell remains unknown. The first is in principle easy within reasonable limits by increasing coverage or sequencing enrichments. The second can be solved by binning sequencing reads that are assumed to belong to the same species and then assembly of the so-called metagenome assembled genomes (MAGs) [17]. At this point in time the major limitations of metagenomics derive from assembly and the extraordinary microdiversity that often characterizes prokaryotic populations [18]. Some microbes (often the most abundant) do not assemble well due to the heterogeneity of the reads covering the same sections of different clonal lineages. The flexible genome is particularly resilient to assembly since it is extraordinarily variable from one clone to another within the same population and often contains the most interesting genes from the point of view of the ecology and biotechnology [19]. But long read sequencing approaches might come to the rescue in the immediate future. Regardless, and supplemented by the availability of large amounts of single cell genomes (single amplified genomes, SAGs), the gigantic potential of prokaryotic diversity is at the tip of our fingers. Lots of interesting work for the next generations of microbiologists.

2 Discovery of Novel Rhodopsins via Different Metagenomic Approaches

After the initial step of sampling the environment and extracting DNA, metagenomics could be divided into two approaches; (a) "wet metagenomics": by cloning into (1) single copy artificial bacterial chromosomes (BACs) with ~100–150 kbp inserts, or (2) using in vitro packaging to generate smaller (40 kbp inserts) fosmids with either single copy number or inducible copy number to ~60 copies, or (b) "dry metagenomics": direct sequencing using (1) fragmentation and sequencing using the various approaches available (lately mostly Illumina reads that are 100–300 nucleotides but with high coverage) or (2) long-read sequencing methods like PacBio or Oxford nanopore, and reconstructing MAGs by bioinformatic assembly.

The introduction of BAC and fosmid (F1 origin-based cosmid vector) cloning vectors already in 1992 has dramatically improved genomic cloning efforts of HMW DNA and was instrumental for the development of metagenomics. In 2000 the first two metagenomic BAC libraries were reported from soil [20] and marine [21] environments. The richness of the marine libraries and their huge potential was fully appreciated only when a novel light-driven proton pump (proteorhodopsin, bacteriorhodopsin homolog) neighboring a bacterial rRNA gene was uncovered on a 150 kb BAC fragment originating from the uncultured gammaproteobacterial SAR86 group. The new rhodopsin was detected via Blast searches, and had less than 30% protein identity to the already known haloarchaeal rhodopsins (detected in the seventies within cultured halophilic archaea [22–25]). Proteorhodopsin was new to science and was completely unexpected to be found in a bacterial genome, certainly not in an abundant marine bacterium. Since their initial discovery in 2000, proteorhodopsins have been found in different environments and in the genomes of widely divergent microbial groups (bacteria, archaea and eukarya) and even viruses (*see* below) with current estimates of 70% and above of microbes in the photic zone containing these rhodopsins [26–28].

In 2012, in the course of comprehensive comparative genomic analysis of giant viruses, proteorhodopsin-like genes where detected in giant viruses which infect unicellular aquatic eukaryotes [29]. These viruses include Organic Lake Phycodnaviruses and a virus infecting *Pheocystis globosa* (a photosynthetic marine microalgae). The sequences of these proteins were up to 28% identical to proteorhodopsins. Comparison of the putative viral rhodopsins to metagenomic datasets revealed numerous highly similar sequences. Interestingly, and differently to proteorhodopsins, the new viral rhodopsins lack some key conserved amino acids which precluded assigning function at that time [29, 30]. It was only recently that their function as light-driven proton pumps [31] or channels [32, 33] was suggested.

Metagenomics is usually performed using different bioinformatic tools; however, in 2007 functional metagenomics was suggested for the search of new rhodopsins. In functional metagenomics the library is used not only as a means to stabilize and sequence long fragments of HMW DNA, but also phenotypes are screened for by expressing the library in a heterologous host, such as *E. coli*. Martínez et al. [34] screened marine environmental fosmid clones for the characteristic phenotype; (pinkish/orange) color of *E. coli* colonies which express microbial rhodopsins. Three colonies were scored positive (based on their color), and were found to contain a proteorhodopsin gene as well as a gene cassette coding for a retinal biosynthetic pathway resembling cassettes already known based on "dry" metagenomics [35]. In another screen, proton pumping phenotype was used to screen for rhodopsins, with pH microelectrodes, and more microbial rhodopsins (mostly proteorhodopsins) were reported in the marine environment [36]. Recent functional metagenomic screening for colored *E. coli* cell pellets revealed the presence of a new microbial rhodopsin group, the heliorhodopsin, in freshwater samples [37]. This new and distinct group showed less than 15% identity to known microbial rhodopsins, hence was missing from metagenomics in Blast screens (at that time), and showed inverted membrane topology compared to all other known rhodopsin families. The new heliorhodopsin family is found in bacteria, archaea, eukarya, and viruses. The phenotypic functional screen proved itself as a powerful tool of searching for new rhodopsins in the world of unannotated and miss-annotated databases.

3 Novel Rhodopsins by Bioinformatic Searches

The identification of novel proteins is one of the major objectives of genomic and metagenomic studies. Traditional pairwise algorithms, such as BLAST [38, 39] or the most recent developed DIAMOND [40], allow to compare a protein of interest (so-called query) against a repository of sequences with known function (defined subjects). These comparison approaches evaluate the optimal alignment between amino acids and make possible the selection of subjects that correlate the query over a certain percent of sequence identity or expected value (e-value). However, two major issues may arise from the identification of novel rhodopsins based on pairwise sequence similarity searches. The first problem is merely computational, and concerns computing power and capacity of data storage. It might represent a limiting factor especially for metagenomic studies, where the number of predicted proteins that need classification can reach hundreds of thousands of peptides even for a single experiment. Not to mention the storage space which must be allocated to save the reference database that will be used, and the time needed to process all data. The second problem is related to the rhodopsin amino acid sequence. Pairwise algorithms lose their capacity to identify homologous proteins when the sequence identity is below 30% [41, 42]. Whether BLAST % identities between proton pumping rhodopsins often fall within this "twilight zone," identity values between rhodopsins with different functions are often close or below the resolution provided by such sequence similarity searches. This is the case of proteorhodopsins and sodium- or chloride-pumps (~25%), or proteorhodopsins and viral rhodopsins (~21%).

An alternative approach that yielded positive results in identifying novel rhodopsin from genomic and metagenomic data, relies on the usage of Hidden Markow Models (HMMs) [43]. The implementation of Markow chains for protein search algorithms such as HMMER software (hmmer.org) is based on the assumption that amino acids at each position of related proteins fit a probability distribution which constitute a model or, in other words, a profile. Under the operative point of view, it can be argued that profile-based algorithms follow an inverse logic with respect to pairwise algorithms for sequence similarity searches. While BLAST and DIAMOND utilize the query to identify similar proteins in a given database, HMMs use, the same model to scan proteins for which the function needs to be determined.

The whole method consists of three main steps, (1) definition of the profile HMM; (2) screening of the dataset in which novel rhodopsins need to be found (e.g., predicted proteins from metagenomic contigs); and (3) cleaning the result and certifying the candidate rhodopsins.

1. The first step of a HMM-based search is the definition of the model. The HMM is usually built (trained) based on multiple alignments of related proteins, and it is performed by accessory programs within the HMM-search main algorithm (*hmmbuild*, in the case of HMMER). The accuracy of the process depends on the quality of the model, and the definition of related proteins represents a crucial step that might affect the whole protocol. In the case of rhodopsins, protein clusters can be defined based on conserved amino acid residues which have been demonstrated to correlate with the function. For example, protein with proton pumping motif DTT or DTE (D86-T89-D96 in bacteriorhodopsin numbering) can be grouped together, as well as sodium-pumps with NDQ motifs or chloride-pumps NTQ. These functional clusters can constitute the protein groups which will be used to train the model. A different type of clustering relies on sequence relationships between proteins. It can be performed by following a phylogenetic approach, through the definition of a tree-like topology. Distances between tree nodes can be used to define protein-clades, and each clade can be the template for an HMM profile.
2. The second step is performed by the "*core*" program of the HMM-search tool (*hmmsearch*). The model is queried against the dataset of target proteins, and the output shows matching proteins which fit each profile HMM included in the analysis. The screening is fast, and thousands of proteins can be scanned in a very short time and with little computer requirements.
3. Results quality filtering and identification of novel proteins. Each HMM-matching hit is associated with an e-value and a quality score, which can be used to filter out nonspecific matches. After the first screening based on match quality indexes, candidate rhodopsins can be finally retrieved basied on other sequence features such as the presence of 7TM domains or key-amino acids which are known to be conserved among distantly related proteins. Programs for the prediction of transmembrane domains such as Phobius [44], TMHMM [45] and Philius [46] are often based on different applications of HMMs, and can be run in local computer using standalone versions and implemented to process large datasets.

The successful identification of rhodopsins through HMM-based screening led to the description of novel proteins with interesting biotechnology applications such as the recently characterized channelrhodopsins MerMAIDs [47], or the light gated cation channels belonging to the viral rhodopsin group-1 [33]. These studies indicate that complementary approaches to the pairwise sequence similarity searches can be further implemented and updated with the aim of allowing even more accurate identifications of proteins of biological interest and potential biotechnology application.

Acknowledgment

This work was supported by Russian Science Foundation (RSF) Project 21-64-00018.

References

1. Stanier RY, Doudoroff M, Adelberg EA (1971) General microbiology, 3rd edn. Prentice Hall
2. Krisch HM, Comeau AM (2008) The immense journey of bacteriophage T4—from d'Herelle to Delbruck and then to Darwin and beyond. Res Microbiol 159(5):314–324
3. Magrum L et al (1975) Corrections in the catalogue of oliogonucleotides produced by digestion of *Escherichia coli* 16S rRNA with T1 RNase. Nature 257(5525):423–426
4. Woese CR, Fox GE (1977) Phylogenetic structure of the prokaryotic domain: the primary kingdoms. Proc Natl Acad Sci U S A 74(11): 5088–5090
5. Pace NR et al (1986) The analysis of natural microbial populations by ribosomal RNA sequences. Adv Microb Ecol 9:1–55
6. Giovannoni SJ et al (1990) Genetic diversity in Sargasso Sea bacterioplankton. Nature 345(6270):60–63
7. Hugenholtz P, Goebel BM, Pace NR (1998) Impact of culture-independent studies on the emerging phylogenetic view of bacterial diversity. J Bacteriol 180(18):4765–4774
8. Chisholm SW et al (1992) Prochlorococcus marinus nov. gen. nov. sp.: an oxyphototrophic marine prokaryote containing divinyl chlorophyll a and b. Arch Microbiol 157:297–300
9. Rippka R et al (2000) Prochlorococcus marinus Chisholm et al. 1992 subsp. pastoris subsp. nov. strain PCC 9511, the first axenic chlorophyll a2/b2-containing cyanobacterium (Oxyphotobacteria). Int J Syst Evol Microbiol 5:1833–1847
10. Rappé MS et al (2002) Cultivation of the ubiquitous SAR11 marine bacterioplankton clade. Nature 418(6898):630–633
11. Giovannoni SJ, Cameron Thrash J, Temperton B (2014) Implications of streamlining theory for microbial ecology. ISME J 8(8):1553–1565
12. Welch RA et al (2002) Extensive mosaic structure revealed by the complete genome sequence of uropathogenic *Escherichia coli*. Proc Natl Acad Sci U S A 99(26): 17020–17024
13. Tettelin H et al (2005) Genome analysis of multiple pathogenic isolates of Streptococcus agalactiae: implications for the microbial "pan-genome". Proc Natl Acad Sci U S A 102(39):13950–13955
14. Land M et al (2015) Insights from 20 years of bacterial genome sequencing. Funct Integr Genomics 15(2):141–161
15. Suzuki MT et al (2004) Phylogenetic screening of ribosomal RNA gene-containing clones in Bacterial Artificial Chromosome (BAC) libraries from different depths in Monterey Bay. Microbial Ecol 48:473–488
16. Venter JC et al (2004) Environmental genome shotgun sequencing of the Sargasso Sea. Science 304:66–74
17. Kunin V et al (2008) A bioinformatician's guide to metagenomics. Microbiol Mol Biol Rev 72(4):557–578, Table of Contents

18. Haro-Moreno JM et al (2020) Ecogenomics of the SAR11 clade. Environ Microbiol 22(5): 1748–1763
19. Rodriguez-Valera F, Martin-Cuadrado AB, Lopez-Perez M (2016) Flexible genomic islands as drivers of genome evolution. Curr Opin Microbiol 31:154–160
20. Rondon MR et al (2000) Cloning the soil metagenome: a strategy for accessing the genetic and functional diversity of uncultured microorganisms. Appl Environ Microbiol 66(6):2541–2547
21. Béjà O et al (2000) Construction and analysis of bacterial artificial chromosome libraries from a marine microbial assemblage. Environ Microbiol 2(5):516–529
22. Oesterhelt D, Stoeckenius W (1971) Rhodopsin-like protein from the purple membrane of Halobacterium halobium. Nat New Biol 233(39):149–152
23. Matsuno-Yagi A, Mukohata Y (1977) Two possible roles of bacteriorhodopsin; a comparative study of strains of Halobacterium halobium differing in pigmentation. Biochem Biophys Res Commun 78(1):237–243
24. Spudich JL, Bogomolni RA (1984) Mechanism of colour discrimination by a bacterial sensory rhodopsin. Nature 312(5994): 509–513
25. Jung KH, Trivedi VD, Spudich JL (2003) Demonstration of a sensory rhodopsin in eubacteria. Mol Microbiol 47(6):1513–1522
26. Finkel OM, Béjà O, Belkin S (2012) Global abundance of microbial rhodopsins. ISME J 7:448–451
27. Gómez-Consarnau L et al (2019) Microbial rhodopsins are major contributors to the solar energy captured in the sea. Sci Adv 5:eaaw8855
28. Fuhrman JA, Schwalbach MS, Stingl U (2008) Proteorhodopsins: an array of physiological roles? Nat Rev Microbiol 6:488–494
29. Yutin N, Koonin EV (2012) Proteorhodopsin genes in giant viruses. Biol Direct 7:34
30. Philosof A, Béjà O (2013) Bacterial, archaeal and viral-like rhodopsins from the Red Sea. Environ Microbiol Rep 5:475–482
31. Needham DM et al (2019) A distinct lineage of giant viruses brings a rhodopsin photosystem to unicellular marine predators. Proc Natl Acad Sci U S A 116(41):20574–20583
32. Bratanov D et al (2019) Unique structure and function of viral rhodopsins. Nat Commun 10(1):4939
33. Zabelskii D et al (2020) Viral rhodopsins 1 are an unique family of light-gated cation channels. Nat Commun 11:5707
34. Martínez A et al (2007) Proteorhodopsin photosystem gene expression enables photophosphorylation in heterologous host. Proc Natl Acad Sci U S A 104:5590–5595
35. Sabehi G et al (2005) New insights into metabolic properties of marine bacteria encoding proteorhodopsins. PLoS Biol 3:e173
36. Pushkarev A, Béjà O (2016) Functional metagenomic screen reveals new and diverse microbial rhodopsins. ISME J 10:2331–2335
37. Pushkarev A et al (2018) A distinct abundant group of microbial rhodopsins discovered using functional metagenomics. Nature 558: 595–599
38. Altschul SF et al (1990) Basic local alignment search tool. J Mol Biol 215:403–410
39. Gish W, States DJ (1993) Identification of protein coding regions by database similarity search. Nat Genet 3:266–272
40. Buchfink B, Xie C, Huson DH (2015) Fast and sensitive protein alignment using DIAMOND. Nat Methods 12:59–60
41. Brenner SE, Chothia C, Hubbard TJP (1998) Assessing sequence comparison methods with reliable structurally identified distant evolutionary relationships. Proc Natl Acad Sci U S A 95:6073–6078
42. Park J et al (1998) Sequence comparisons using multiple sequences detect three times as many remote homologues as pairwise methods. J Mol Biol 284:1201–1210
43. Krogh A et al (1994) Hidden Markov Models in computational biology applications to protein modeling. J Mol Biol 235:1501–1531
44. Käll L, Krogh A, Sonnhammer EL (2004) A combined transmembrane topology and signal peptide prediction method. J Mol Biol 338(5): 1027–1036
45. Sonnhammer EL, von Heijne G, Krogh A (1998) A hidden Markov model for predicting transmembrane helices in protein sequences. In: ISMB '98: proceedings of the 6th international conference on intelligent systems for molecular biology. ACM, pp 175–182
46. Reynolds SM et al (2008) Transmembrane topology and signal peptide prediction using dynamic bayesian networks. PLoS Comput Biol 4:e1000213
47. Oppermann J et al (2019) MerMAIDs: a novel family of metagenomically discovered, marine, anion-conducting and intensely desensitizing channelrhodopsins. Nat Commun 10:3315

Chapter 5

E. coli Expression and Purification of Microbial and Viral Rhodopsins

Taras Balandin, Dmytro Volkov, Alexey Alekseev, Kirill Kovalev, Dmitry Bratanov, and Valentin Gordeliy

Abstract

Microbial rhodopsins have become an indispensable tool for neurobiology. Of thousands of identified microbial rhodopsins, a minute fraction has been studied so far and they have shown remarkable functional diversity suggesting more great promises that this large family holds. Effective production of recombinant microbial and viral rhodopsins is a prerequisite for the success of functional and structural studies of these proteins. *Escherichia coli* (*E. coli*) are suitable for high yield expression of many of microbial and viral rhodopsins and they facilitate rapid exploration of this large protein family.

Key words Rhodopsin, Membrane protein, Recombinant, Production, *Escherichia coli*

1 Introduction

Rhodopsin field is a hot and important filed of current biology and biophysics [1]. Rhodopsins are in the core of optogenetics [2]. Among over 10,000 identified opsin genes only a small number of them have been heterologically expressed and studied. Recombinant production of microbial rhodopsin is a prerequisite for the success of functional and structural studies of these proteins. Expression of these proteins in *Escherichia coli* (*E. coli*) facilitates and accelerates research on rhodopsins. However, it is not always easy to express these membrane proteins, in particular, in sufficient amounts required by some of the studies, for instance, structural studies. An example of the expression problem we should mention is bacteriorhodopsin (bR). For many years, bacteriorhodopsin was functionally expressed in the only host, *Halobium salinarum*. This approach takes a long time, and only in 2015 the problem of functional expression of bR in *E. coli* was solved [3].

Recombinant protein production in *Escherichia coli* has already been covered in this book series in general [4] and for membrane

Valentin Gordeliy (ed.), *Rhodopsin: Methods and Protocols*, Methods in Molecular Biology, vol. 2501,
https://doi.org/10.1007/978-1-0716-2329-9_5,

proteins in particular [5]. What is specific about rhodopsins is that to acquire proper structure and function they need a cofactor, retinal, and usually (the only exception is heliorhodopsin) rhodopsins are oriented in cell membrane so that *N*-terminus faces outside and *C*-terminus faces inside of the cell. The features should be taken into account in the design of the recombinant constructions for expression and in the purification procedure.

In this chapter we provide information and efficient protocols for microbial rhodopsin expression, which were used in the recent studies [6–11].

2 Materials

2.1 Bacteria Cultivation Materials

Stock solution to prepare growth media are listed in Table 1.

Compositions of preliminary culture media (PCM), base media for induction turning (BIM), and basic auto-inducing media (AIM) are given in Tables 2, 3, and 4.

To prepare agar plates mix 1.5% agar, 1% yeast extract, 1% tryptone, 0.9% NaCl, 0.5% glycerol in water.

Boil once shortly by microwaving to dissolve agar, autoclave 15 min at 120 °C, mix while cooling down.

When the mixture reaches 50 °C, add 40% sterile glucose to final concentration of 0.4%, ampicillin to 150 mg/L or kanamycin to 50 mg/L, mix well. Pour 20 mL per 9 cm plate, let them set and dry, keep at 4 °C, and use within 2 weeks.

Commonly used DNA plasmids for *E. coli* expression are generally suitable to produce rhodopsins. A few are listed in Table 5.

E. coli expression strains BL21(DE3)T1, C41(DE3), C43(DE3), and NovaBlue (DE3) designed for T7 expression system suit for rhodopsins.

2.2 Bacteria Cultivation Equipment

100 mL, 250 mL, and 2000 mL baffled Erlenmeyer flasks.

Orbital shaker-incubator for 18–37 °C suitable for 100 mL, 250 mL, and 2000 mL Erlenmeyer flasks.

Ultrasonicator with 3 mm tip.

2.3 Protein Purification Solutions

Protease inhibitor cocktail cOmplete™ EDTA-free: 50× stock solution in water, store at −20 °C.

2 M sodium 6-aminohexanoate (6-AHA), pH 7.5: 1000× stock solution in water, store at 4 °C.

25% Triton-X100 water solution.

Cell lysis buffer LB: 20 mM Na_2HPO_4, 150 mM NaCl, 2 mM 6-AHA, 1× cOmplete™, pH 7.0.

Table 1
Basic growth media components

Solution	Stock conc.	Working conc.	Sterilization	Storage
Water			Autoclaving	RT
Glycerol	50%	0.5–1.5%	Autoclaving	RT
Glucose	40%	0.05–0.5%	Filtration	RT
α-Lactose	10%	0.02–0.2%	Filtration	RT
Di-sodium succinate	1 M	10 mM	Filtration	RT
$MgSO_4$	1 M	2 mM	Filtration	RT
$MgCl_2$	1 M	2 mM	Filtration	RT
Ampicillin	100 g/L	100–150 mg/L	Filtration	−20 °C
Kanamycin	100 g/L	50–100 mg/L	Filtration	−20 °C
IPTG	1 M	0.2–1 mM	Filtration	−20 °C
Tween-20	25%	0.005%	Filtration	RT
All-trans-retinal	50 mM in ethanol	10 μM	–	−20 °C
"NPS" salt mixture [12]	1 M Na_2HPO_4 1 M KH_2PO_4 0.5 M $(NH_4)_2SO_4$ pH 6.8	50 mM 50 mM 25 mM	Autoclaving	RT
"5052" sugar mixture [12]	25% glycerol 2.5% glucose 10% α-lactose	0.5% 0.05% 0.2%	Filtration	RT
TY media base	1% yeast extract 1% tryptone	1% 1%	Autoclaving	RT

Table 2
Preliminary culture media PCM

Component	Stock dilution factor	Final conc.
TY media base	–	1% yeast extract 1% tryptone
NPS mixture	20	50 mM Na_2HPO_4 50 mM KH_2PO_4 25 mM $(NH_4)_2SO_4$
1 M $MgSO_4$	500	2 mM
50% glycerol	50	1%
40% glucose	80	0.5%
1 M di-sodium succinate	100	10 mM

Just before use, add 150 mg/L ampicillin or 100 mg/L kanamycin depending on the culture resistivity

Table 3
Base media for induction turning BIM

Component	Stock dilution factor	Final conc.
TY media base	–	1% yeast extract 1% tryptone
NPS mixture	20	50 mM Na_2HPO_4 50 mM KH_2PO_4 25 mM $(NH_4)_2SO_4$
1 M $MgSO_4$	500	2 mM
50% glycerol	100	0.5%
40% glucose	800	0.05%

Add 150 mg/L ampicillin or 100 mg/L kanamycin depending on the culture resistivity

Table 4
Basic auto-inducing media AIM

Component	Stock dilution factor	Final conc.
TY media base	–	1% yeast extract 1% tryptone
NPS mixture	20	50 mM Na_2HPO_4 50 mM KH_2PO_4 25 mM $(NH_4)_2SO_4$
1 M $MgSO_4$	500	2 mM
5052 mixture	50	0.5% glycerol 0.05% glucose 0.2% α-lactose

Add 150 mg/L ampicillin or 100 mg/L kanamycin depending on the culture resistivity

Table 5
Expression vectors

Name	Length	Ori	Selection	IaqIQ
pET15	5708	ColEI	Amp	+
pET28	5369	ColEI	Kan	+
pSCodon1.2	5840	ColEI	Amp	+
pBKT7R	3907	pBR322	Kan	+
pBKT7	2275	pBR322	Kan	–

NiNTA buffer base: as LB, but pH adjusted to 7.8.

NiNTA equilibration S0: NiNTA buffer base, 0.05% DDM, 20 mM imidazole, pH 7.8.

NiNTA wash A1: NiNTA buffer base, 0.1% DDM, pH 7.8.

NiNTA wash A2: NiNTA buffer base, 0.1% DDM, 30–50 mM imidazole, pH 7.8.

NiNTA elution A3: NiNTA buffer base, 0.2% DDM, 300 mM imidazole, pH 7.8.

SEC buffer: 10 mM Na_2HPO_4, 100 mM NaCl, 2 mM 6-AHA, 0.5 mM EDTA, 0.5 mM EGTA, pH 7.0.

2.4 Protein Purification Equipment

Microfluidizer M-110P.

Empty columns for gravity flow or low-pressure liquid chromatography (e.g. GE Healthcare C10/10, XK16/20).

Immobilized metal affinity chromatography (IMAC) resin of choice (e.g. NiNTA).

Millipore centrifugal ultrafilter units.

Size-exclusion chromatography column (e.g. GE Healthcare Superdex 200 Increase GL 10/300).

3 Methods

3.1 Preparation of Producer Cells

3.1.1 Design of the Expression Constructs

1. Carefully select target protein sequences (*see* **Note 1**). When full-size sequence does not result in a reasonable protein yield, consider truncation of C-terminus closer to membrane and to the last well-ordered segment, usually seventh helix.
2. When sequence analysis strongly suggests presence of a short *N*-terminal secretory signal sequence [13], for expression in *E. coli* consider replacing it with an *E. coli* native secretory signal [14].
3. Optimize coding DNA sequence for codon usage frequency in the expression host and for mRNA stability [15] (*see* **Note 2**).
4. Proper membrane insertion is important for folding and thereby stability and resistance to proteolytic degradation of nascent opsins. The selected sequences might not intrinsically insert or orient in lipid membranes very well; employing membrane insertion and/or orientation drivers might improve these properties. Thermostabilized *N*-terminal full size apocytochrome b562RIL [16, 17] can be tested at first as probably one of the simplest drivers (*see* **Note 3**).
5. To facilitate purification add oligohistidine tag, usually at the *C*-termini (*see* **Note 4**). Mind the tag accessibility, when the target protein *C*-termini is predicted to be very close to lipid membrane, add few hydrophilic amino acids between the target protein and the His-tag to expose it better (*see* **Note 5**). When

downstream applications require removal of the tag, add specific protease recognition site between rhodopsin and the tag (*see* **Note 6**). Refrain from multiplying number of histidines in the tag significantly (*see* **Note** 7).

6. Add 5′ and 3′ DNA restriction enzyme recognition sites to facilitate insertion of the target sequence into plasmid DNA vector (*see* **Note 8**). Extend the recognition sites with six random nucleotides to let enzymes recognize and bind their sites better.

3.1.2 Synthesis of the Designed Sequences

Once desired coding sequence is established, have it synthesized by a commercial service (*see* **Note 9**).

3.1.3 Insertion of the Synthesized DNA Fragment into Plasmid DNA Vector of Choice

A simple medium copy number DNA plasmid supporting T7 polymerase-driven expression, like pET15 or pET28, is usually sufficient for rhodopsins production in *E. coli*. We typically use pET15, pET28, pSCodon1.2, pBKT7, and pBKT7R (*see* **Note 10**). The latter two are minimal plasmids assembled in our lab featuring ColE1 replication origin, Kan resistance, T7 promoter/MCS/terminator cassette. Plasmid pBKT7R additionally has lacIq gene (*see* **Note 11**).

3.1.4 Transformation of E. coli *Expression Strains with the Prepared Plasmids*

C41(DE3) is a very robust host that can deliver very high rhodopsin expression yield and can be tested at first. BL21(DE3)T1 are better to deal with formation of insoluble inclusions (*see* **Note 12**). NovaBlue(DE3) allow improving solubilization efficiency even further but is not as sturdy as the first two and may be considered when target protein is clearly expressed and incorporated into lipid membranes well but is hardly amenable to solubilization.

1. Use C41(DE3) and BL21(DE3)T1 competent cells obtained or prepared in advance to transform with the expression plasmids bearing target genes (*see* **Note 13**). After transformation spread cells over 9-cm cultural 1.5% agar plate with appropriate antibiotic (Subheading 2.1) and let it grow overnight for 16 h at 37 °C. Do not let the plate overgrow, especially for C41.
2. After colonies appear, it is convenient to shortly propagate few of them in 5 mL shaking culture in PCM media (Table 2) with appropriate antibiotic to OD600 of 2 to 4 AU, mix 0.8 mL portion of the culture with 0.5 mL of sterile 50% glycerol in a cryotube, flash-freeze the mixture in liquid nitrogen and keep the prepared glycerol stock for later use at −80 °C.

3.2 Small-Scale Test Expression

In T7-expression system, biosynthesis of the target protein in the producer cell culture is triggered by unblocking of the *lacUV5* operator in λDE3 lysogen. For that, IPTG or lactose is usually used. IPTG penetrates cell membranes and can be detected inside the cells in seconds after addition to the cell culture. Then, cell biosynthesis machinery is instantly and intensely engaged for the target protein production. While for many rhodopsins it works perfectly fine, some fold better when induction intensity is lower and protein synthesis is slower. Then lactose might be a better option. Lactose does not penetrate cell membranes and require transporters to do their work for intracellular lactose level to increase gradually and for transcription to start. The transporter dependence allows additional control over the time when the induction starts. This had been used to formulate growth media recipes enabling auto-induction at the optimal grow phase [4, 18]. We routinely test both approaches.

3.2.1 IPTG as Inducer

1. Prepare a fresh plate: either do a fresh transformation or use a −80 °C glycerol stock prepared earlier (*see* **Notes 14–16**).
2. Use 5 mL of PCM media and a disposable cell scrapper or spreader to resuspend cells on the plate. Collect the suspension into a separate tube. Measure OD600 of the obtained suspension (*see* **Notes 17–19**).
3. Use the suspension to inoculate 25 mL of PCM media supplemented with the appropriate antibiotic in 250 mL baffled Erlenmeyer flask. Adjust dilution so that the culture starts from OD600 of 0.1 AU.
4. Shake the culture at 37 °C until its OD600 reaches 2.5–4.0 AU (*see* **Notes 20–23**).
5. Use the preliminary culture to inoculate 120 mL of base induction media BIM and make OD600 of 0.1 AU. Similarly, inoculate 20 mL of fresh PCM media as a control for promoter leakage. Distribute the BIM cell suspension by 20 mL aliquots over six 100 mL baffled Erlenmeyer flasks.
6. Shake them intensely at 37 °C. Follow **Notes 22** and **23**.
7. When OD600 is 0.8 AU, in five of six flasks with induction media add 0.05, 0.1, 0.2, 0.5, 1.0 mM IPTG from 1 M stock solution and in all flasks add 10 μM retinal from 10 mM stock solution solubilized in 5% Tween-20 and continue shaking (*see* **Notes 24** and **25**).
8. 1, 3, and 16 h after induction measure OD600 of the cultures and take representative samples for SDS-PAGE and western blotting (WB) (*see* **Note 26**).
9. Analyze the expression level of the target protein in the test cultures depending on the expression time and inducer concentration by western blotting; load 10 μL of the prepared samples per 3 × 1 mm^2 SDS-PAGE well. WB bands' intensities indicate expression levels *per cell*, therefore take into account the corresponding culture density to account for the target protein synthesis level *per liter of culture*.

3.2.2 Lactose as Inducer

1–5. Follow **steps 1–5** in Subheading 3.2.1, but in addition, in **step 5**, in five of six flasks with base induction media BIM add 0.02%, 0.05%, 0.1%, and 0.2% lactose from 10% sterile stock.

6. Shake the flasks at 20 °C following **Notes 22, 23**, and **27**.
7. When OD600 reaches 0.8 AU add 10 μM retinal (*see* **Notes 24** and **25**) and continue shaking overnight.
8. 16–18 h later, measure OD600 of the cultures and take representative samples for SDS-PAGE and WB (*see* **Note 26**).
9. Follow **step 9** in Subheading 3.2.1.

3.2.3 Choice of the Inducer and Its Concentration

Thanks to retinal, cells expressing very high level of correctly folded rhodopsins can be easily recognized already by appearance (*see* **Note 28**). When this is not the case, use WB results to choose minimal IPTG or lactose concentration and cultivation time that reliably results in a maximal yield of the target protein. Pay attention to the proteolytic stability of the target protein over the cultivation period. Small-scale culture tests are mostly useful to rule out induction approach that results in very weak or nor expression at all. When IPTG and lactose induce expression well, it makes sense to upscale both type of cultures as they might behave different in purification.

3.3 Upscale Expression Culture Volume

Upscaling of the expression cultures is rather straightforward.

1. Start with a fresh plate as above and resuspend colonies from the plate in 5 mL media.
2. Inoculate 250 mL of PCM media in 2 L baffled Erlenmeyer flask to make OD600 of 0.1 AU and vigorously shake it to OD600 of 3–4 AU (*see* **Notes 20, 22**, and **23**).
3. Use the preliminary culture to inoculate four to twelve 250–300 mL aliquots of base induction media BIM or AIM in 2 L baffled Erlenmeyer flasks.
4. Follow **steps 6–8** of Subheading 3.2.2 or Subheading 3.2.1 (and use chosen IPTG concentration).
5. Harvest grown cultures at 4 °C, 4000 × *g*, for 20 min. Typically, IPTG-induced cultures are ready for harvesting 3–4 h after induction and lactose-induced cultures are after 16- to 18-h cultivation. Collect cell pellet and proceed to purification or distribute collected cell pellet in LN2-resistant plastic bags, 5–10 g of cells per bag, freeze them in LN2 and keep at −80 °C for up to a few weeks.

3.4 Protein Isolation and Purification

We start with outlaying the typical basic workflow for rhodopsin isolation and purification from bacteria. At first, it can be followed as it is. Although for the best efficiency, some steps in this workflow require optimization in each given case that can be done iteratively.

3.4.1 Basic Steps

These steps provide general guidelines that can be adjusted to the available facilities following explanation in corresponding Notes (*see* Subheading 4).

1. In 100 mL of +4 °C cold cell lysis buffer LB add 1 mL of fresh 10 mg/mL lysozyme and 1 mL of fresh 1 mg/mL DNAse I solutions in the lysis buffer LB, set it for vigorous stirring (*see* **Notes 29** and **30**).
2. Then add 10 g of the frozen cell pellet broken in chunks, let them thaw for 10–15 min, watch them to not block stirring (*see* **Note 31**).
3. Homogenize the mixture with Dounce homogenizer using loose pestle.
4. Mix the homogenized suspension on ice for 20–30 min for lysozyme and DNAse to work.
5. Break cells by passing the suspension through microfluidizer M-110P at 90 MPa 2–3 times.
6. Optionally, add 0.3% TX-100 as 25% stock solution, mix vigorously on ice and pass two more times through the microfluidizer and mix for 20 min longer (*see* **Note 32**).
7. Collect insoluble cell fraction by centrifugation for 1 h at 100k × *g* (50k × *g* is sufficient if **step 6** was followed) (*see* **Notes 33** and **34**).
8. Discard the supernatant (rhodopsins are typically observed in the insoluble cell fraction) and resuspend the pellet in 50 mL of membrane solubilization buffer MSB.
9. Homogenize the suspension with Dounce homogenizer using tight pestle.
10. Set the suspension for vigorous stirring and add 0.5 g DDM as 20% stock solution, do not let foam formation. Continue stirring overnight (1 h is often sufficient; this has to be tested in a separate experiment) (*see* **Notes 35** and **36**).
11. Remove detergent-insoluble material by centrifugation for 30 min at 50k × *g*.
12. Add 20 mM imidazole in the supernatant from 2 M stock solution and adjust pH to 7.8.
13. Mix the sample with 2 mL of NiNTA resin equilibrated with S0 buffer, gently roll it for 30 min at +4 °C (*see* **Note 37**).
14. Gently pellet down the loaded resin by centrifugation at 1000 × *g* for 4 min in 50 mL transparent conical centrifuge tube. Decant supernatant, if its color suggests that there is more target protein in it, mix it with a fresh portion of NiNTA resin and repeat **steps 13** and **14**.

15. Transfer the loaded NiNTA resin into an empty column for chromatography; as gravity flow mode and automated system can be used (automated system with a multi-wavelength detector is preferred).
16. Wash the column with 10 CV of buffer A1 at linear flow rate of 0.20–0.25 cm/min.
17. Then, similarly wash the column with 8 CV of buffer A2, collect fractions of 0.2–0.3 CV each.
18. Elute the target protein with 5–6 CV of buffer A3 at linear flow rate of 0.1 cm/min, collect fractions of about 0.1 CV each.
19. Pick the peak fractions, estimate protein concentration by spectrophotometry.
20. For the downstream size exclusion chromatography (SEC) step, adjust concentration and volume of the sample if needed by ultra-filtration.
21. Clarify the sample by 10 min centrifugation at 12k × *g*. Filter the supernatant through 200 nm membrane to remove large particles.
22. Load the sample onto SEC column equilibrated with 1.5 CV of SEC buffer, inject no more than 2% of the SEC column total volume.
23. Develop the SEC column with 1.5 CV of SEC buffer at 0.5 mL/min (*see* **Note 38**).
24. Repeat **steps 22** and **23** as many times as needed to inject the entire sample collected after NiNTA purification.
25. Combine fractions corresponding to the best $Abs_{\lambda ret}/Abs_{280}$ ratio. Take spectra, concentrate as needed, and flash freeze in liquid nitrogen until later use.

4 Notes

1. Phylogenetically close rhodopsins are expected to exhibit the same function but might have different expression levels. Choose few of them to compare. Consider the length of the full sequence and of its intramembrane part in particular, length and ordering of interhelical loops, predicted structure stability, and membrane insertion scores. Use homology model predictors to obtain topology and identify disordered extra--membrane extensions. Long consecutive repeats of solvent-exposed residues in C-termini can indicate evolutionary rudiments that are unlikely immediately involved in light response; however, it can complicate purification. Consider eliminating them.

2. For efficient CDS optimization use approaches implemented for instance at GenArt service [15].
3. It is convenient to have a set of plasmid DNA vector (s) encoding different extensions to the target sequence in the same translation frame; then, multiple target sequences can be easily tested with various expression, insertion, and orientation drivers at once.
4. Consider expected orientation of the target protein in membrane and, at first, place the His-tag at the termini expected to face cytoplasm. In living *E. coli*, terminal His-tag can bear a significant positive charge that inhibits cytoplasm to periplasm cross-membrane transfer and thereby proper membrane insertion in case of unfortunate placement.
5. For purification, membrane proteins have to be solubilized in membrane mimetics, usually detergents, and His-tag might be buried in a bulk of the detergent micelle.
6. Flank the cleavage site with a few small amino acids for better cleavage site exposition, especially when the target protein C-termini is predicted to be very close to lipid membrane. Stretches like SSG and GSS work well.
7. With proper exposition, 6× His tag provides sufficient IMAC binding, especially for oligomeric proteins. For monomeric integral membrane proteins 8× His tag sometimes improves the binding without a significant trade off of suppressing protein synthesis. While it is sometimes observed that 10× His tag does improve binding in IMAC comparing to 6× His tag, that high number of consecutive histidines still can adversely affect the protein synthesis level. Furthermore, typically used immobilized metal ions are not capable of coordinating that many histidines at once. A reasonable improvement in binding can be achieved simply by better exposition of 6× His or 8× His tags.
8. Surely, any alternative sub-cloning techniques can be used as long as those techniques do not introduce undesired amino acid residues into the planned design of the expression construct.
9. It usually is more affordable to order coding sequences just as "DNA fragments" or "DNA strands" but not as already incorporated into a DNA plasmid. It should not matter which service to use, quality of the delivered product is estimated solely by correctness of the sequence of the synthesized DNA fragment.
10. Although we optimize target sequences for *E. coli* codon usage, sometimes pSCodon1.2, featuring additional genes for "rare" in *E. coli* tRNA, allow significant improvement in expression level.

11. We find that additional transcription control by lacIq gene introduced in pET vector series to allow expression of toxic targets is over-suppressive and unnecessary for some rhodopsins.
12. It is in general important to use T-bacteriophage-resistant derivative of BL21(DE3) to reduce probability of phage contamination of laboratory cultures.
13. Competent cells for the transformation can be commercially obtained from a local vendor or can be self-made. We follow UCC [19] and TSS [20] transformation protocols. Transformation method does not influence expression yields.
14. For some targets, fresh transformation results in higher and better reproducible expression yields and has to be done routinely.
15. A good plate after incubation has 500–1000 separate colonies. Observe their morphology to rule out overgrowth, secondary growth due to insufficient antibiotic selection pressure, phage contamination, and cross-contamination from other microbial cultures used in the lab.
16. To obtain a plate with a reasonable colonies density from a glycerol stock, inoculate a small 2 mL culture and shake it at 37 °C to OD600 a little above 0.1 AU so that it can be measured reliably. Then perform 3 serial, about tenfold, dilutions in 1 mL of fresh media to make culture samples with OD600 corresponding to 1 and 0.1 mAU, spread 50 μL of each suspensions over separate 9 cm plates. After colonies appear, choose the plate with highest density of separate clones.
17. Dense cell suspensions require appropriate dilution for OD measurement. Measured value should fall in range 0.1–0.8 AU600. If dilution necessary to obtain corresponding measured value is above 10×, then few serial weaker dilutions should be done.
18. Portable photometers, comparing to photometers of 'classic' design, sometimes show elevated absorption for the same samples of cell cultures, we saw a 1.4× fold increase. This should be tested and the read value should be appropriately interpreted.
19. We have not justified routine screening for the best expressing separate clone by picking and comparing them individually, and rather prefer chunk sampling.
20. The suspension should be mixed intensely but formation of large amount of foam should be avoided, especially of stable foam, which cover liquid surface and block thereby oxygenation. Revolution velocity is not an appropriate measure of mixing intensity here as the intensity depends very much on the flasks geometry, shape of baffles, their size, amount of

liquid in the flask, shaking orbit. Because anti foam agents are of amphiphilic nature, we avoid them while producing membrane proteins to not affect the composition of bacterial lipid membranes.

21. Do not let the culture overgrow. By the indicated OD600 is reached, glucose should remain at a level above 0.05% in the culture. When it drops, promoter might leak and prematurely expressed target protein might suppress the cells. At the same time, glucose level has to be quite below the initial one of 0.5%, otherwise, after dilution at the next step, glucose level in the induction media turn significantly higher than expected and this would delay the auto induction moment. Glucose level can be verified using for example test stripes; it approaches the optimum when OD600 of the culture reaches 3–4 AU.
22. If the culture requires ampicillin for cultivation, mind its instability. A more stable carbenicillin can replace ampicillin, although it is more expensive. Ampicillin resistant growing culture continuously and rapidly decomposes supplemented ampicillin by b-lactamase secreted in media. As culture density grows, supplemented ampicillin decomposes gradually faster and faster. To maintain the selection pressure use 150 mg/L in the starting culture of 0.1 AU, then add 100 mg/L more when culture density is 0.5 AU, and 100 mg/L more at 1.0 AU.
23. If the culture requires kanamycin for maintenance, note that its effect is compromised by high phosphates concentration in the culture media [12]. So, use 100 mg/L kanamycin supplement with 100 mM phosphates in media (i.e. when diluting 20× NPS 20-folds) and 50 mg/L kanamycin with 20 mM phosphates in media. There is no need to add more kanamycin while the culture is growing.
24. Retinal is hydrophobic and, if added as 1000× or higher concentrated ethanol solution, precipitates instantly in culture and does not distribute evenly in membranes of all cells in the culture. We find it safe for cells and useful for homogeneous distribution in the culture volume to mix 1 part of 50 mM retinal stock solution in ethanol, 1 part of 25% Tween-20, and 3 parts of sterile water and use this mixture as 1000× retinal stock for supplementing bacterial cultures.
25. The supplements should be added directly into liquid of the cultures and immediately distributed in the culture volume by shaking. When drops of a stock solution fall on the dense foam pad floating on the culture surface, it might take long before its concentration in the media reach the desired values.
26. Calculate volume of the culture sample to take as 2[mL*AU]/OD600[AU], collect cells from the sample in 1.5 mL eppy by

1 min centrifugation at 8000 rpm, remove media, resuspend the pellet in 160 μL of water, add 40 μL of 5× SDS-PAGE loading buffer and gently sonicate with 3 mm tip at minimal ultrasonic power for about 20 s until the suspension is homogenous, clear, and not viscous. Keep the tip deep in the sample. Do not overheat the sample by too long treatment. Do not let any foam formation by setting too high sonic power. Freeze the sample until analysis. Concentration of this way prepared sample can be referred to as 10 AU600 which it would have if the cells were not lysed.

27. The lower grow temperature, the slower metabolism and not that vigorous shaking is sufficient.
28. Take into account that retinal is absorbed by cell lipid membranes and renders yellowish appearance to the cells itself when no opsin is expressed by the cells.
29. Preliminary disintegration of the cell walls with lysozyme allow applying lower pressure for cell disruption that is important when the target protein is sensitive to mechanical stress.
30. While mechanical stress applied for cell opening disintegrates large genomic DNA, DNAse reduces viscosity of the mixtures at all the steps below, up to membrane solubilization, that significantly improves efficiency of all the procedures.
31. It is convenient to adjust the total volume of the buffer at the first step and of the cell pellet at the second to the available cell disruptor and ultracentrifuge capacities. With that cell-to-buffer ratio, better not exceed 1:4 for microfluidizer or 1:9 for ultra-sonic disruptor. When the suspension is too dense, cells do not open efficiently and cell membranes precipitate too slow due to viscosity.
32. The amount of the detergent at this washing step is to be determined in a separate experiment.
33. If using swing-out rotor, optionally, underlay the sample with about 3 mL of 20% sucrose per centrifugation tube to ease the next step.
34. Rhodopsins are expected to accumulate in inner bacterial membranes. Sucrose density gradient centrifugation of lysed cells suspension allows separation of soluble cell fraction, inner membranes, outer membranes, and insoluble materials. Yet, in the case of overexpression, the inner membranes are so enriched with the recombinant protein that they sink into denser sucrose layers and mix with the outer membrane fraction making the sucrose density cell fractionation not very useful.
35. Total amount of the detergent for solubilization has to correlate with the total amount of material collected from the

insoluble cell fraction susceptible to solubilization. The latter is not always easy to estimate, then total amount of cells taken for the purification procedure can be used as its indirect measure. About 0.5 g of DDM is usually sufficient to solubilize membrane fraction of insoluble material from 10 g of cell pellet that might contain up to 100 mg of rhodopsin. The optimal amount of detergent can be estimated more accurately in a separate titration experiment.

36. In case the target rhodopsin is not very stable, the solubilization mixture might be supplemented with 5–10 μM retinal.
37. The amount of IMAC resin to be taken is to be optimized in a separate experiment. Too much resin binds too much contaminants in addition to the target protein. Too little resin does not extract the target protein completely. Thanks to the intrinsic color of rhodopsins, this step is usually possible to monitor just by the suspension color intensity.
38. Alternatively, use the flow rate recommended by the manufacturer of the available column and taking into account temperature dependent increase in viscosity of water. Collect fractions of 0.01 CV in the column working range 0.3–0.9 CV. SEC column with separation efficiency above 15,000 N/m is to be used.

Acknowledgments

This work was supported by Deutsche Forschungsgemeinschaft (DFG) project 430170559. V.G. greatly acknowledges the support of Hermann von Helmholtz-Gemeinschaft Deutscher Forschungszentren (HGF) within Helmholtz professorship. This work was supported by Russian Science Foundation (RSF) Project 21-64-00018.

References

1. Kandori H (2020) Biophysics of rhodopsins and optogenetics. Biophys Rev 12:355–361. https://doi.org/10.1007/s12551-020-00645-0
2. Kim CK, Adhikari A, Deisseroth K (2017) Integration of optogenetics with complementary methodologies in systems neuroscience. Nat Rev Neurosci 18:222–235. https://doi.org/10.1038/nrn.2017.15
3. Bratanov D, Balandin T, Round E et al (2015) An approach to heterologous expression of membrane proteins. The case of bacteriorhodopsin. PLoS One 10:e0128390. https://doi.org/10.1371/journal.pone.0128390
4. Studier FW (2014) Stable expression clones and auto-induction for protein production in *E. coli*. Methods Mol Biol 1091:17–32. https://doi.org/10.1007/978-1-62703-691-7_2
5. McIlwain BC, Kermani AA (2020) Membrane protein production in *Escherichia coli*. Methods Mol Biol 2127:13–27. https://doi.org/10.1007/978-1-0716-0373-4_2
6. Shevchenko V, Gushchin I, Polovinkin V et al (2014) Crystal structure of Escherichia coli-expressed Haloarcula marismortui bacteriorhodopsin I in the trimeric form. PLoS One

9:e112873. https://doi.org/10.1371/journal.pone.0112873
7. Gushchin I, Shevchenko V, Polovinkin V et al (2015) Crystal structure of a light-driven sodium pump. Nat Struct Mol Biol 22: 390–395. https://doi.org/10.1038/nsmb.3002
8. Shevchenko V, Mager T, Kovalev K et al (2017) Inward H+pump xenorhodopsin: mechanism and alternative optogenetic approach. Sci Adv 3:e1603187. https://doi.org/10.1126/sciadv.1603187
9. Tu C-H, Yi H-P, Hsieh S-Y et al (2018) Overexpression of different types of microbial rhodopsins with a highly expressible bacteriorhodopsin from haloarcula marismortui as a single protein in E. coli. Sci Rep 8: 14026. https://doi.org/10.1038/s41598-018-32399-x
10. Bratanov D, Kovalev K, Machtens J-P et al (2019) Unique structure and function of viral rhodopsins. Nat Commun 10:4939. https://doi.org/10.1038/s41467-019-12718-0
11. Kovalev K, Volkov D, Astashkin R et al (2020) High-resolution structural insights into the heliorhodopsin family. Proc Natl Acad Sci U S A 117:4131–4141. https://doi.org/10.1073/pnas.1915888117
12. Studier FW (2005) Protein production by auto-induction in high density shaking cultures. Protein Expr Purif 41:207–234. https://doi.org/10.1016/j.pep.2005.01.016
13. Almagro Armenteros JJ, Tsirigos KD, Sønderby CK et al (2019) SignalP 5.0 improves signal peptide predictions using deep neural networks. Nat Biotechnol 37:420–423. https://doi.org/10.1038/s41587-019-0036-z
14. Freudl R (2018) Signal peptides for recombinant protein secretion in bacterial expression systems. Microb Cell Factories 17:52. https://doi.org/10.1186/s12934-018-0901-3
15. Raab D, Graf M, Notka F et al (2010) The GeneOptimizer Algorithm: using a sliding window approach to cope with the vast sequence space in multiparameter DNA sequence optimization. Syst Synth Biol 4:215–225. https://doi.org/10.1007/s11693-010-9062-3
16. Nikkila H, Gennis RB, Sligar SG (1991) Cloning and expression of the gene encoding the soluble cytochrome b562 of Escherichia coli. Eur J Biochem 202:309–313. https://doi.org/10.1111/j.1432-1033.1991.tb16377.x
17. Chu R, Takei J, Knowlton JR et al (2002) Redesign of a four-helix bundle protein by phage display coupled with proteolysis and structural characterization by NMR and X-ray crystallography. J Mol Biol 323:253–262. https://doi.org/10.1016/s0022-2836(02)00884-7
18. Studier FW (2018) T7 expression systems for inducible production of proteins from cloned genes in E. coli. Curr Protoc Mol Biol 124: e63. https://doi.org/10.1002/cpmb.63
19. Inoue H, Nojima H, Okayama H (1990) High efficiency transformation of Escherichia coli with plasmids. Gene 96:23–28. https://doi.org/10.1016/0378-1119(90)90336-p
20. Chung CT, Niemela SL, Miller RH (1989) One-step preparation of competent Escherichia coli: transformation and storage of bacterial cells in the same solution. Proc Natl Acad Sci U S A 86:2172–2175. https://doi.org/10.1073/pnas.86.7.2172

Chapter 6

Crystallization of Microbial Rhodopsins

Kirill Kovalev, Roman Astashkin, Valentin Gordeliy, and Vadim Cherezov

Abstract

Microbial rhodopsins are light-sensitive transmembrane proteins, evolutionary adapted by various organisms like archaea, bacteria, simple eukaryote, and viruses to utilize solar energy for their survival. A complete understanding of functional mechanisms of these proteins is not possible without the knowledge of their high-resolution structures, which can be primarily obtained by X-ray crystallography. This technique, however, requires high-quality crystals, growing of which is a great challenge especially in case of membrane proteins. In this chapter, we summarize methods applied for crystallization of microbial rhodopsins with the emphasis on crystallization in lipidic mesophases, also known as in meso approach. In particular, we describe in detail the methods of crystallization using lipidic cubic phase to grow both large crystals optimized for traditional crystallographic data collection and microcrystals for serial crystallography.

Key words Microbial rhodopsins, Crystallization, X-ray diffraction, Serial crystallography, In meso, Lipidic cubic phase, Membrane fusion, Bicelles

1 Introduction

1.1 Structural Studies of Rhodopsins: Bacteriorhodopsin

Since the discovery of the first microbial rhodopsin, bacteriorhodopsin (BR), from the archaeon *Halobacterium salinarum* in 1971 [1], structural investigations of this unique light-sensitive protein attracted the attention of a large number of research groups around the world. Indeed, H. Michel and colleagues obtained the first 3D crystals of BR already in 1980 [2]. Unfortunately, the crystals were of low quality and did not allow the determination of BR structure. In parallel to crystallization efforts, the first attempts to determine the structure of BR were made using cryo-electron microscopy (cryoEM) and electron diffraction (ED) by R. Henderson and P.N.T. Unwin in 1975 [3]. This resulted in the first low-resolution models of BR, showing that this protein forms a seven-α-helical bundle. Later, the resolution of the model was improved notably by R. Henderson and colleagues to 3.5 Å [4–6]. The structure revealed trimers of BR with key functional amino acid residues resolved. However, in order to understand the

Valentin Gordeliy (ed.), *Rhodopsin: Methods and Protocols*, Methods in Molecular Biology, vol. 2501,
https://doi.org/10.1007/978-1-0716-2329-9_6,

mechanism of proton pumping by BR, high-resolution structures that enable the identification of hydrogen bonds and tightly bound water molecules are required. Despite recent progress in cryoEM, X-ray diffraction (XRD) remains the primary method for obtaining high-resolution structures for such small proteins as BR. XRD requires the growth of high-quality protein crystals, which often presents a great challenge, especially in the case of microbial rhodopsins and many other membrane proteins, since they require a proper lipidic environment during crystallization.

Crystallization of BR had long remained a formidable scientific problem. Although the first 3D crystals were reported as early as in 1980 [2], it took 17 more years to obtain the first high-resolution (2.5 Å) BR structure [7]. This breakthrough was enabled by the development of a new approach to the crystallization of membrane proteins using lipidic cubic phase (LCP), which is also often referred to as in cubo or in meso method [8]. Thus, since 1997, numerous high-resolution structures of BR and other microbial rhodopsins have been reported. At the moment, the highest resolution achieved for BR structures is 1.25 Å [9] and for other microbial rhodopsins is 1.07 Å, reported for archaerhodopsin-3 (AR3) from the archaebacterium *Halorubrum sodomense* [10]; both obtained using the in meso crystallization approach.

It should be noted that recent advances in the development of single particle cryoEM made this method suitable in principle for structural studies of some representatives of the microbial rhodopsin superfamily. In particular, large oligomers, such as pentamers of proteorhodopsins (PRs) or light-driven sodium pumps (NaRs) with a molecular weight of about 150 kDa should be amenable for cryoEM studies. However, by April 2021, there have been neither reported results on such studies nor deposited high-resolution structures of microbial rhodopsin determined using cryoEM. Therefore, at the moment, crystallization is required for the structural characterization of microbial rhodopsins and understanding their functional mechanisms. Below we briefly describe the application of different crystallization methods to structural studies of microbial rhodopsins.

1.2 Vapor Diffusion Crystallization of Protein–Detergent Complexes

Vapor diffusion is the most commonly used method for crystallization of both soluble proteins and membrane proteins solubilized in detergent micelles. Additionally, vapor diffusion technique is at the core of several other approaches to crystallize membrane proteins and microbial rhodopsins in particular, such as the bicelles approach (Subheading 1.4 of this chapter) and the membrane fusion method (Subheading 1.5 of this chapter). LCP crystallization (Subheading 1.3 of this chapter) can also be done using vapor diffusion.

The major shortcoming of the vapor diffusion approach for crystallization of protein–detergent complexes is in the loose packing of solubilized proteins and the low number of intermolecular contacts within a crystal. This is due to the presence of a detergent "cloud" surrounding the hydrophobic regions of membrane proteins and, as a consequence, the absence of specific interactions between the hydrophobic parts of the proteins. This is especially critical in the case of microbial rhodopsins, where the water-soluble regions are limited to the interhelical loops, which are typically short and do not provide many options for mediating strong interactions in crystals. Moreover, geometrical restrains due to the presence of the detergent shell often do not allow forming proper interactions between polar parts of such proteins. These are the main reasons why BR crystals grown by the vapor diffusion approach did not show diffraction better than several angstroms despite many years spent on their optimization. Therefore, vapor diffusion of protein–detergent complexes is the least-used method for crystallization of microbial rhodopsins. Nevertheless, it played an important role in the development of structural studies of these proteins at early stages.

Indeed, H. Michel and colleagues succeeded in using vapor diffusion and obtained the first 3D crystals of BR already in 1980 [2]. This method has then been used extensively to obtain numerous high-resolution structures of other membrane proteins. The first model of BR resolved at better than 3 Å using crystals grown with vapor diffusion method was reported in 1998 [11, 12]. Vapor diffusion was also the first method to obtain the crystals of the D97N mutant of a blue-light-absorbing proteorhodopsin isolated from the Pacific Ocean near Hawaii at a depth of 75 m (HOT75BPR) in 2012 [13]. The authors showed diffraction patterns up to 3 Å and identified five protein molecules in the asymmetric unit using single wavelength anomalous diffraction (SAD) phasing. Later in 2013, the crystallization conditions were optimized, which resulted in the structures of the D97N and D97N/Q105L mutants of HOT75BPR at 2.7 and 2.6 Å, respectively [14].

1.3 In Meso Approach

In meso crystallization revolutionized structural biology of membrane proteins [15]. This method has been successfully applied for crystallization of membrane transporters, enzymes, receptors, including G protein-coupled receptors (GPCRs), etc. However, BR was the first protein crystallized from LCP [7, 8]. The most commonly used comprehensive protocols for LCP crystallization were published in 2009 [16].

LCP provides several principal advantages. First, it consists of a single lipid bilayer, which mimics the native environment for membrane proteins. Indeed, this is a critical point for many unstable proteins, which can change their conformation or unfold during vapor diffusion crystallization trials in detergent micelles. The single lipid bilayer of LCP divides the space into two networks of water channels. Membrane proteins reconstituted in the lipid bilayer of

LCP can diffuse over large distances and interact with other membrane proteins to nucleate a crystal and support its growth. Second, crystals grown in LCP belong to the type I, in which protein molecules are usually packed tighter and the crystal contacts are stronger compared to type II crystals of detergent-solubilized proteins. This typically results in higher-quality diffraction data from the crystals grown using in meso approach. Third, LCP is optically transparent, which allows detecting crystals larger than few micrometers in size and observing their growth using different microscope imaging modalities. Fourth, LCP is tolerant to the addition of inevitable amounts of small amphiphilic molecules and lipids often surrounding the purified proteins [17].

The major disadvantage of LCP is a relatively high-curvature of the lipid bilayer and a limited size of its water channels. This means that membrane proteins with large water-soluble domains often cannot be incorporated in LCP or their mobility within the lipid bilayer is restricted. As a result, such proteins cannot be crystallized by the traditional in meso approach and may require conversion of LCP into a swollen sponge phase [18]. Nevertheless, it remains the preferred method for the crystallization of microbial rhodopsins and relatively small membrane proteins.

The first high-resolution structure of a microbial rhodopsin obtained with the in meso approach was the structure of BR at 2.5 Å; however, crystal twinning was overlooked [7]. In an intense competition by several groups, the resolution was quickly improved to 1.55 Å [19], making it the most-cited BR structure due to its high resolution and a moderate twinning level. In 2002, further efforts to improve resolution were made by the group of J. Lanyi [20, 21]; unfortunately, BR structures in the ground, K, and M_1 states at 1.47, 1.43, and 1.43 Å resolution, respectively, were obtained using perfectly twinned crystals, which affected the quality of the crystallographic data. Finally, a ground-state BR structure at a 1.25 Å resolution was published in 2018 [9]. As mentioned above, the highest-resolution structure of a microbial rhodopsin to date corresponds to that of AR3 reported in 2021 at 1.07 Å [10]. In meso crystallization was also instrumental for obtaining high-resolution structures of many other microbial rhodopsins as summarized in Table 1.

1.4 Crystallization from Bicelles

Bicelles are small planar pieces of lipid bilayers [56]. Their formation is achieved by mixing bilayer-forming phospholipids with small amphiphilic molecules that stabilize the boundaries of the bilayers. The most commonly used bicelles are made of a DMPC/CHAPSO mixture. Bicelles form in an aqueous solution at low temperatures, allowing to pipette them easily. Upon increasing temperature, bicelles transform into a gel-like perforated lamellar phase that supports growth of type I crystals of membrane proteins [57]. The history of the method development for membrane

Table 1
Highest-resolution structures of microbial rhodopsins determined using crystals grown by in meso approach

Protein	Resolution, Å	PDB ID	Year	References
Chloride pump from *Halobacterium salinarum* (*Hs*HR)	1.8	1E12	2000	[22]
Sensory rhodopsin II from *Natronomonas pharaonis* (*Np*SRII)	2.4 2.1	1JGJ 1H68	2001 2001	[23] [24]
*Np*SRII in complex with its cognate transducer (HtrII)	1.93	1H2S	2002	[25]
Anabaena sensory rhodopsin (ASR)	2.0	1XIO	2004	[26]
Eukaryotic proton pump *Acetabularia* rhodopsin II (ARII)	3.2	3AM6	2011	[27]
Cation channelrhodopsin C1C2 from *Chlamydomonas reinhardtii*	2.3	3UG9	2012	[28]
Proteorhodopsin from *Exiguobacterium sibiricum* (ESR)	2.3	4HYJ	2013	[29]
Archaeal proton pump from *Haloarcula marismortui* (*Hm*BRI)	2.5	4PXK	2014	[30]
Acetabularia rhodopsin I (ARI)	1.52	5AX0	2015	[31]
Bacterial chloride pump from *Nonlabens marinus* (ClR, NM-R3)	1.56 1.58	5G28 5B2N	2016 2016	[32] [33]
Archaeal proton pump from *Haloquadratum walsbyi* (*Hw*BR)	1.85	4QI1	2017	[34]
Inward proton pump xenorhodopsin from *Candidatus nanosalina* (*Ns*XeR)	2.5	6EYU	2017	[35]
Cation channelrhodopsin-2 from *Chlamydomonas reinhardtii* (*Cr*ChR2)	2.4	6EID	2017	[36]
Bacteriorhodopsin from *Halobacterium salinarum* (BR)	1.25	5ZIM	2018	[9]
Blue-shifted cation channelrhodopsin Chrimson	2.6	5ZIH	2018	[37]
Anion channelrhodopsin from *Guillardia theta* (*Gt*ACR1)	2.9 2.9	6CSM 6EDQ	2018 2019	[38] [39]
Rhodopsin from *Phaeocystis globosa* virus ($VirR_{DTS}$)	1.6	6JO0	2019	[40]
Rhodopsin II from *Organic Lake phycodnavirus* (OLPVRII)	1.9	6SQG	2019	[41]
Proton pump from *Coccomyxa subellipsoidea* (*Cs*R)	2.0	6GYH	2019	[42]
Archaeal heliorhodopsin from *Thermoplasmatales archaeon* (*Ta*HeR)	2.4	6IS6	2019	[43]
Bacterial 48C12[a] heliorhodopsin from *Actinobacteria bacterium*	1.5	6SU3	2020	[44]
Rhodopsin 1 from *Organic Lake phycodnavirus* (OLPVR1)	1.4	7AKY	2020	[45]
Chloride pump from *Mastigocladopsis repens* (*Mr*HR)	1.9	6K6I	2020	[46]

(continued)

Table 1
(continued)

Protein	Resolution, Å	PDB ID	Year	References
Thermophilic rhodopsin from *Rubrobacter xylanophilus* (*Rx*R)	1.84	6KFQ	2020	[47]
Sodium pump from *Krokinobacter eikastus* (KR2)[b]	2.0	6YC3	2020	[48]
Rhodopsin phosphodiesterase (transmembrane domain)[c] from *Salpingoeca rosetta*	2.6	7CJ3	2021	[49]
Schizorhodopsin 4 from *Asgard group archaeon*	2.1	7E4G	2021	[50]
Archaerhodopsin-3 from *Halorubrum sodomense* (AR3)	1.07	6S6C	2021	[10]

[a] Only the highest-resolution structure of 48C12 heliorhodopsin is presented. Another structure of 48C12 was solved at 2.7 Å resolution (PDB ID: 6UH3 [51])

[b] Only the structure of the pentameric form of KR2 under physiological conditions with the highest resolution is included. 6YC3 model of KR2 supersedes the originally reported in 2019 structure with the resolution of 2.2 Å (PDB ID: 6REW [52]). Other structures of KR2 were obtained for the monomeric form of the protein under acidic and neutral conditions at the resolutions of 1.45 Å (acidic, PDB ID: 4XTL [53]), 2.3 Å (neutral, PDB IDs: 3X3C [54]), and 1.6 Å (neutral, PDB ID: 6TK6 [55])

[c] Only the structure of the transmembrane domain of the rhodopsin phosphodiesterase is included. Another structure of the transmembrane domain and the linker region of the rhodopsin phosphodiesterase was obtained at 3.5 Å resolution (PDB ID: 7D7Q [49])

protein crystallization from bicelles as well as all necessary procedures for the use of this method are excellently summarized in reference 58. In brief, this method relies on the vapor diffusion approach; however, the target protein is reconstituted into bicelles prior to adding precipitant solutions.

Crystallization from bicelles has several advantages. First, the method is relatively simple, since, in principle, it resembles the procedure of traditional vapor diffusion experiments. Second, it does not require the use of highly viscous lipid phases like in the case of the in meso crystallization. The latter facilitates crystal inspection, manipulation, and harvesting. The disadvantage of the method is in a complex phase behavior of bicellar lipid mixtures, low compatibility with a wide range of precipitant solutions, and a high rate of false-positive crystal hits.

BR was the first protein crystallized from bicelles in 2002 [59]. Interestingly, the protein crystallized in the P*21* space group and did not form trimers, as opposed to a native trimeric arrangement of BR in the P*63* space group for crystals grown in LCP. Unfortunately, the P*21* crystals were also twinned. Importantly, the first bicelles-grown crystals were obtained at 37 °C, which was perceived as a disadvantage of the method, because many membrane proteins are unstable at high temperatures. In 2005, the group of J.U. Bowie reported the first structure of BR obtained using crystals grown from bicelles at room temperature, which resulted in untwinned crystals with *C2221* symmetry [60]. However, the crystals also contained a monomeric non-physiological form of BR.

Nevertheless, crystallization from bicelles provided structures of several BR mutants at resolutions up to 1.6 Å [61–66]. It was also used to crystallize and solve 3D structures of xanthorhodopsin (XR) at 1.9 Å in 2008 [67] and proteorhodopsin from *Med12* (*Med12*BPR) at 2.31 Å in 2013 in its hexameric form [14]. In 2016, the highest resolution (1.43 Å) structure of BR crystallized in bicelles was deposited in the Protein Data Bank (PDB ID: 4Y9H); however, to the best of our knowledge, this structure has not yet been published.

Despite the success of LCP crystallization, crystallization from bicelles is still currently being used for structural studies of microbial rhodopsins. In 2016, two high-resolution structures of BR obtained using serial femtosecond crystallography at XFEL were reported at 2.1 and 2.35 Å [68]. In 2019, the structure of *Gloeobacter* rhodopsin (GR) was obtained at 2 Å [69]. Finally, a 2.33 Å structure of the light-driven chloride pump from *Mastigocladopsis repens* (*Mr*HR) was published in 2020 [70].

1.5 Membrane Fusion Method

Another relatively popular approach for crystallization of microbial rhodopsins is the membrane fusion (or vesicle-fusion) method [71, 72]. This method was first introduced in 1998 by K. Takeda et al. [72] with the demonstration of BR crystals in the novel *P622* symmetry diffracting to 3.5 Å [71] and subsequently improved to 2.5 Å. The approach has been used actively by the group of T. Kouyama to obtain numerous structures of BR in different states [73–75] and its functional mutants [76]. It was also used for the crystallization of archaerhodopsin-1 and -2 (AR-1 and AR-2) at 3.4 and 2.5 Å in 2003 and 2006 [77] and deltarhodopsin and cruxrhodopsin-3 in 2013 and 2014, respectively [78, 79]. Again with the membrane fusion method the resolution of AR-2 structures was improved to 2.1 Å in 2008 [80] and finally to 1.8 Å in 2014 [81]. In 2015, a new type of crystal packing was reported for the light-driven chloride pump from *Halobacterium salinarum* (*Hs*HR) using the membrane fusion approach [82]. Later in 2016, the same group determined the structure of the yellow form of *Hs*HR with a deprotonated retinal Schiff base at 2.6 Å [83]. T. Kouyama and colleagues also performed structural studies of another chloride-pumping rhodopsin from *Natronomonas pharaonis* (*Np*HR) using the membrane fusion technique. The first 2.0 Å resolution structure of the ground state of *Np*HR was reported in 2010 [84]. In 2011, azide- and nitrate-bound *Np*HR structures at 1.9 and 2.2 Å, respectively, became available. Later in 2015, using the same protocols for *Np*HR crystallization, the group provided the first insights into the structures of the intermediate states of light-driven chloride-pumping rhodopsins [85].

The membrane fusion approach includes two major steps. First, rhodopsin-rich membranes (e.g., purple membranes in the case of BR) are incubated with a low concentration of detergent, below its critical micellar concentration (CMC). The incubation results in the formation of small lipid vesicles with incorporated target proteins. Second, the suspension of such vesicles is dehydrated, typically using the vapor diffusion method. The dehydration of the system leads to the fusion of the vesicles into type I crystals of a high quality.

The method has several advantages. For instance, it allows to crystallize membrane proteins without their complete removal from their native membranes, which could be detrimental. As mentioned above, in the case of BR, it produced type I protein crystals of P*622* symmetry, free from merohedral twinning. In contrast, in meso crystallization of BR often results in crystals of P*63* symmetry with a high twin fraction, which is a critical problem for the data interpretation [86]. On the other hand, the membrane fusion method requires using protein-rich lipid membranes, which are not always readily available. Also, the highest resolution diffraction obtained from crystals grown using this approach is 1.8 Å and only for one model, while the mean resolution is around 2.3 Å. Such resolution is insufficient for localization of water molecules in the protein and resolving multiple side chain conformations, important for rhodopsin studies. It might be a reason why this method has not been widely used in structural studies of microbial rhodopsins.

2 Materials

2.1 Precipitants

For initial crystallization screening, it is convenient to use commercial sparse matrix screen kits in 96-well plates that can be purchased from Hampton Research, Molecular Dimensions, Rigaku Reagents, Jena Bioscience, Qiagen, and other vendors. On average, however, about 30% of commercial screen conditions are not compatible with LCP crystallization, transforming LCP into a lamellar or hexagonal phase or completely dissolving the lipid [87]. Such conditions can be diluted two times with water to increase their compatibility. Additionally, more focused grid or sparse matrix screens can be prepared following results of prescreening assays such as LCP-FRAP [88] or LCP-Tm [89].

2.2 Lipids

The most frequently used host lipid for LCP crystallization is monoolein (MO, 1-Oleoyl-*rac*-glycerol). It can be ordered from Sigma or Nu-Chek Prep. Apart from MO, several other host lipids have been used for the crystallization of membrane proteins and microbial rhodopsins, in particular. Thus, monoundecenoin (MU, Nu-Chek Prep) was used for the crystallization of the light-driven inward proton pump xenorhodopsin *Ns*XeR [35]; monopalmitolein (MP, Nu-Chek Prep) was used for the crystallization of the natural light-gated anion channelrhodopsin *Gt*ACR1 [38]. Additive lipids can also be ordered from Sigma or Avanti Polar Lipids. Chloroform (HPLC grade) for lipid preparation can be ordered from Sigma.

2.3 Equipment for LCP Preparation

1. Two 100 μl gas-tight removable needle syringes without a needle (Hamilton, cat.# 7656-01).
2. One 25 μl (Hamilton, cat.# 7654-01) or 50 μl (Hamilton, cat.# 7655-01) gas-tight syringe with a flat tip (point style 3) removable needle (gauge 26s) (Hamilton, cat.# 7768-01).
3. Syringe coupler can be made using two gauge 22 removable needles (Hamilton, cat.# 7770-02) and two removable needle (RN) nuts (Hamilton, cat.# 30902) as described in ref. 90. Alternatively, a similar coupler can be purchased from Formulatrix—SKU 209526 or from Rigaku–SKU 1006905.
4. Short LCP dispensing needle (Hamilton, cat.# 7804-03, gauge 26, point style 3, custom length 10 mm).

2.4 Crystallization Tubes and Plates

1. 96-well glass sandwich plates (Paul Marienfeld GmbH, cat# 0890003; or Molecular Dimensions, cat# MD11-55). These plates are suitable for nanovolume robotic crystallization.
2. IMISX (in meso in situ serial crystallography) Crystallization Plate Kit [91] (Mitegen, SKU: M-IMISX-B-01). These plates are useful for growing crystals for in situ crystallography.
3. 0.5- or 1.5-ml Eppendorf tubes for large-scale crystallization. 200 μl PCR tubes are suitable for initial trial of crystallization scaling. Tubes can be ordered from Sigma.

2.5 Robotic Systems for Crystallization

The following commercial robotic systems for LCP crystallization use similar principles and are available:

1. NT8-LCP from Formulatrix.
2. Gryphon LCP from Art Robbins Instruments.
3. Mosquito LCP from SPT Labtech.
4. ProCrys Meso from Zinsser-Analytic.

3 Methods

3.1 Protein Reconstitution in LCP

The first step of LCP crystallization involves reconstitution of the target membrane protein in the lipid bilayer of LCP. The formation of LCP and protein reconstitution occur spontaneously upon mechanical mixing of the host lipid (typically MO) with an aqueous buffer containing the target protein under proper conditions. While there are several methods of protein reconstitution in LCP, we describe here the most common procedure using a twin-syringe lipid mixer [16]:

1. Prepare the host lipid for making LCP. MO or other lipids that form LCP can be ordered from Sigma, Nu-Chek Prep, or Avanti Polar Lipids. After receiving, aliquot the lipid into small volumes (an average lipid volume used for 1 crystallization setup +20%) in 0.5-ml plastic tubes. The tubes should be sealed using parafilm and stored at −20 °C or −80 °C. Use the following protocol to supplement the host lipid with an additive lipid such as cholesterol, if needed:
   (a) Weigh a small amount (few mg) of the additive lipid in a small (1–2 ml) amber glass vial with a Teflon-lined cap.
   (b) Add an appropriate amount of MO (or other LCP host lipid) to obtain required concentration of the additive lipid (*see* **Note 1**).
   (c) Dissolve the lipids in ~200–400 μl of chloroform.
   (d) Evaporate bulk of the solvent using a gentle stream of dry, filtered nitrogen, keeping the vial warm (~37 °C) to prevent lipid from freezing.
   (e) Remove remaining traces of chloroform under a vacuum (<100 mTorr) for at least 4 h (preferably overnight).
   (f) Aliquot in 0.5-ml tubes as described in step 1 and store at −20 °C or lower temperature until used.
2. Melt lipid at 42 °C. The melting temperature of MO is 37 °C. Other host lipids can melt at higher or lower temperatures. Adjust the incubation temperature to a few degrees above the lipid melting temperature. The incubation time should be no longer than necessary to melt the lipid, in order to avoid potential degradation.
3. Attach a syringe coupler to a 100-μl gas-tight syringe.
4. Remove the plunger from the syringe and transfer molten lipid into the syringe barrel through the plunger end using an adjustable volume pipette. After taking the molten lipid from incubator, it remains liquid at room temperature for a few minutes before it solidifies, allowing for transferring it with a pipette (Fig. 1a).
5. Insert the plunger back and slowly move the lipid up in the syringe to remove any trapped air bubbles. Stop when the lipid reaches the end of the coupler needle (Fig. 1a).

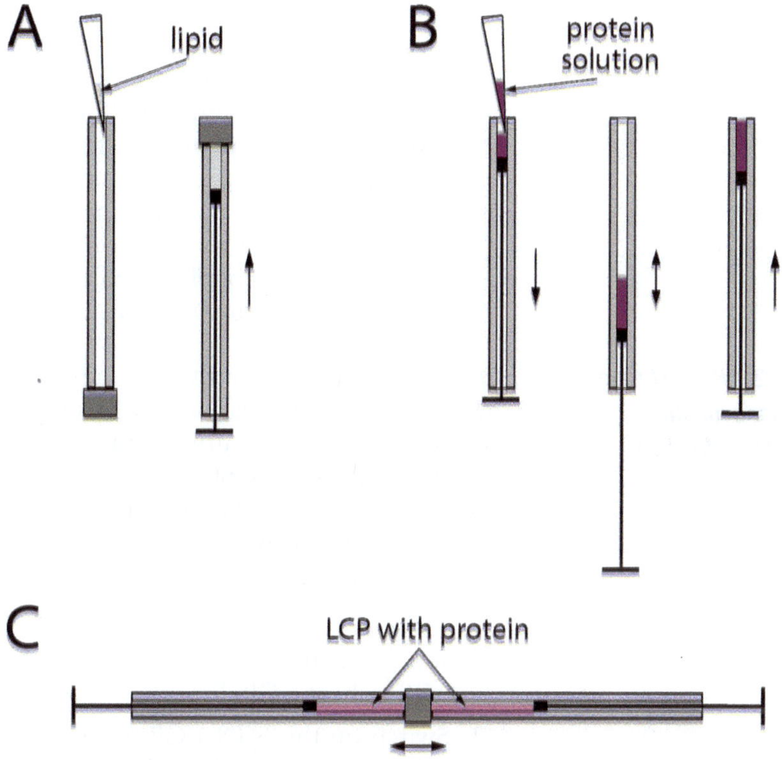

Fig. 1 LCP preparation using a syringe mixer. (**a**) Loading the host lipid (MO) into a Hamilton syringe. (**b**) Loading the protein solution into another Hamilton syringe. (**c**) Connecting the two Hamilton syringes and mixing the lipid with protein to produce LCP

6. Determine the lipid volume based on the syringe volume graduation scale, taking in account that the coupler holds an additional ~2 μl.
7. Use a pipette or 25- or 50-μl syringe with a flat tipped 26 s-gauge needle to transfer an appropriate amount of the protein solution into the second 100-μl gas-tight syringe to achieve 40% v/v of protein solution in the final mixture (e.g., for 30 μl of lipid use 20 μl of protein solution) (Fig. 1b). Avoid trapping air bubbles during the syringe loading (*see* **Note 2**).
8. Move the protein solution with the plunger up as far as possible to minimize the air trapping after assembling the syringe mixer (Fig. 1b).
9. Attach the syringe with protein solution to the open end of the coupler mounted on the syringe with lipid (Fig. 1c).
10. Move the protein solution and the lipid through the coupler's inner needle back and forth from one syringe to another by pushing alternatively on the corresponding plungers until the lipid mesophase in the syringe mixer becomes homogeneous

and transparent (Fig. 1c). This sequence of motions will mechanically mix the lipid with the protein solution through the action of shearing forces forming inside the narrow coupler needle. Upon mixing, an LCP will form spontaneously, and the protein will become reconstituted into the lipid bilayer of the LCP. Complete mixing requires a few hundred passages and typically takes less than 5 min (*see* **Note 3**). The syringe mixer can warm up due to a friction between the lipidic mesophase and the syringe coupler needle. To avoid heating up the sample, limit the mixing rate to ~1 stroke/s.

3.2 High-Throughput Nanovolume Crystallization

Once the protein-laden LCP is ready, one can proceed to setting up nanovolume crystallization in 96-well glass-sandwich plates. Nanovolume crystallization utilizes a robotic system for precise dispensing of small boluses of gel-like LCP (as low as 25 nl, typically 50–100 nl) and corresponding precipitant solutions (typically 400–800 nl). We will describe the procedure for the use of an NT8-LCP robotic system (Formulatrix, USA) as an example.

1. Move all protein-laden LCP into one syringe. Detach the second syringe by unscrewing the coupler. Replace the coupler with a short LCP dispensing needle.
2. Switch on the NT8-LCP robot. Start the corresponding software and wait until the relative humidity inside the machine has reached at least 90%.
3. Open the cover and place the syringe into the holder. Start the calibration procedure using the software. Save the syringe position.
4. Adjust the syringe-dispensing motor position so that it touches the end of the plunger. Push the sample slightly with the motor to prime the needle. Use a clean lint-free tissue to remove the excess of LCP from the tip of the needle.
5. Open the first precipitant kit in a 96-well plate and place it in the corresponding slot inside the NT-8 robot. Do not leave the kit opened for a long time since it might lead to the evaporation of the solutions and consequently to the uncontrolled changes in the crystallization conditions.
6. Load a predefined script for setting up LCP crystallization. The main script parameters are: the LCP drop volume (50–150 nl), the precipitant drop volume (400–800 nl), and the number of pre-dispense drops ($\geq$3). Launch the script.
7. While the plate is in preparation, take a cover glass out of the box and place near the robot. It is useful to inspect the cover glass and blow off any dust using oil-free compressed air.

8. Once the plate is ready, quickly open the NT8-LCP robot cover, take out the plate and cover it with the glass slide from **step** 7.
9. Use a tissue paper or a soft rubber brayer roller to firmly press down on the cover glass to attach it to the plate. Then use a 1000-μl pipette nozzle tip or a wooden toothpick to press on the cover glass around each well to ensure a tight seal.
10. Store the plate at 18–22 °C in the dark.

3.3 Large-Volume Crystallization

As mentioned above, the search for initial crystallization conditions and subsequent optimization are most commonly performed using the high-throughput nanovolume crystallization. When the proper conditions are established, crystallization setup can be scaled up and further optimized to obtain crystals suitable for different types of structural investigations of microbial rhodopsins. Here one could distinguish two specific tasks: (1) obtaining large protein crystals for collecting ultrahigh-resolution X-ray crystallography data or neutron diffraction data to visualize protons and (2) obtaining densely packed protein microcrystals in a sufficient LCP volume (>100 μl) for serial crystallography at room temperature, including time-resolved serial femtosecond and millisecond crystallography at X-ray free-electron lasers (XFELs) and synchrotron sources. Below we present protocols and approaches to achieve each of these goals.

3.3.1 Scaling Up LCP Crystallization for Growing Large Crystals

Growing large crystals (>100 μm) in LCP requires scaling up crystallization volume to allow for (1) more space for the crystal growth and (2) more protein available to form a large crystal. In principle, large-volume crystallization setups could be realized in different types of containers, such as tubes, capillaries, syringes, etc.; however, the more simple and secure system is a 200-μl tube, typically used for setting up polymerize chain reaction (PCR tube).

1. Reconstitute the target protein in LCP as described in Subheading 3.1.
2. Dispense several (typically 8–12) microliters of the protein-laden LCP into a PCR tube. The tube should be at room temperature (18–22 °C). Close the tube immediately after dispensing to avoid phase dehydration. Spin the tube at 4000 × *g* for 10 s.
3. Alternatively, the target protein can be reconstituted in LCP directly inside a PCR tube as described in ref. 92 (Fig. 2). In brief:
   (a) Melt MO at 42 °C. Transfer ~4–6 mg of MO in a PCR tube with a pipette. If MO is smeared on the tube walls, spin it down for 10 min at 13,000 × *g* at room temperature (Fig. 2a).

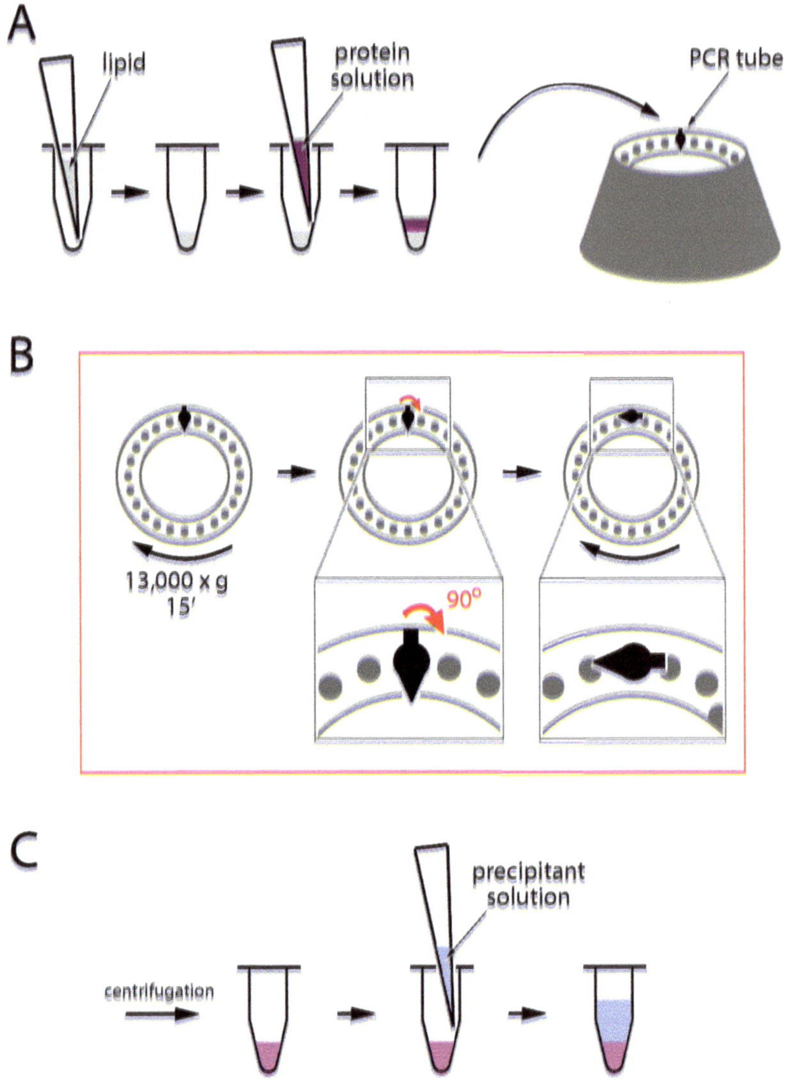

Fig. 2 LCP preparation and crystallization setup in PCR tubes. (**a**) Loading the host lipid and the protein solution into a PCR tube. (**b**) Mixing the lipid and protein by centrifugation. (**c**) Adding a precipitant solution to the protein-laden LCP

(b) To obtain an isotropic lipidic phase, keep MO at 42 °C for an additional 20 min. Then let the lipid to cool down to room temperature. The molten lipid must remain transparent. It may be important for obtaining a homogeneous LCP, which may be important for the quality of the grown crystals. If lipid is not transparent, the melting procedure should be repeated.

(c) Add an equal volume of protein solution (at a protein concentration of ~10 mg/ml) to achieve a 1:1 volume ratio with the host lipid (Fig. 2a).

(d) The formation of LCP is achieved by the following centrifugation procedure (Fig. 2b). Spin the PCR tube with the sample at 22 °C and 13,000 × *g* using a fixed-angle rotor for 15 min. Rotate the tube inside the rotor by 90° and spin again (Fig. 2b). Repeat this spinning procedure four times to obtain a homogenous mixture.

(e) Incubate the sample for 24 h in the dark at 18–22 °C to allow for the protein to equilibrate and become more homogeneously distributed in LCP, which may be important for the quality of the grown crystals.

4. Prepare precipitant solutions. It is useful to make at least 10 ml of each to be able to use the same solutions in several crystallization trials for better reproducibility. Prepare several solutions by varying the following parameters (±20% of salt or PEG concentrations, pH, additives) around the optimal conditions found previously using nanovolume crystallization (*see* **Note 4**).
5. Aspirate the precipitant (about twice the volume of LCP in the tube) using a pipette.
6. Open the PCR tube and quickly dispense the precipitant solution onto the phase. It is better to dispense the precipitant steadily and on the tube wall first (Fig. 2c).
7. Close the PCR tube. Make sure that the precipitant is in contact with LCP. Do not shake the tube.
8. Put the tube into an incubator at 18–22 °C in the dark.

3.3.2 Large-Volume LCP Crystallization for Serial Crystallography

To prepare microcrystals for serial crystallography, the precipitant conditions should be optimized to induce a high nucleation rate uniformly inside the LCP bolus. In the case of microbial rhodopsins, this is most commonly achieved by increasing the concentration of the precipitant agent (salt or PEG) that leads to the formation of protein microcrystals in LCP at a high density. Such optimization should be done using high-throughput nanovolume crystallization in 96-well plates prior to scaling up LCP volume to prepare samples for serial crystallography. When the crystallization conditions are optimized, the scaling up procedure described in Subheading 3.3.1 could be used. However, in some cases, it is essential to preserve the same geometry of the crystallization setup as that used in glass-sandwich plates. This could be achieved by using one of the two procedures described below.

Procedure 1 (crystallization in syringes [93, 94]):

1. Reconstitute the target protein in LCP using a lipid syringe mixer as described in Subheading 3.1.

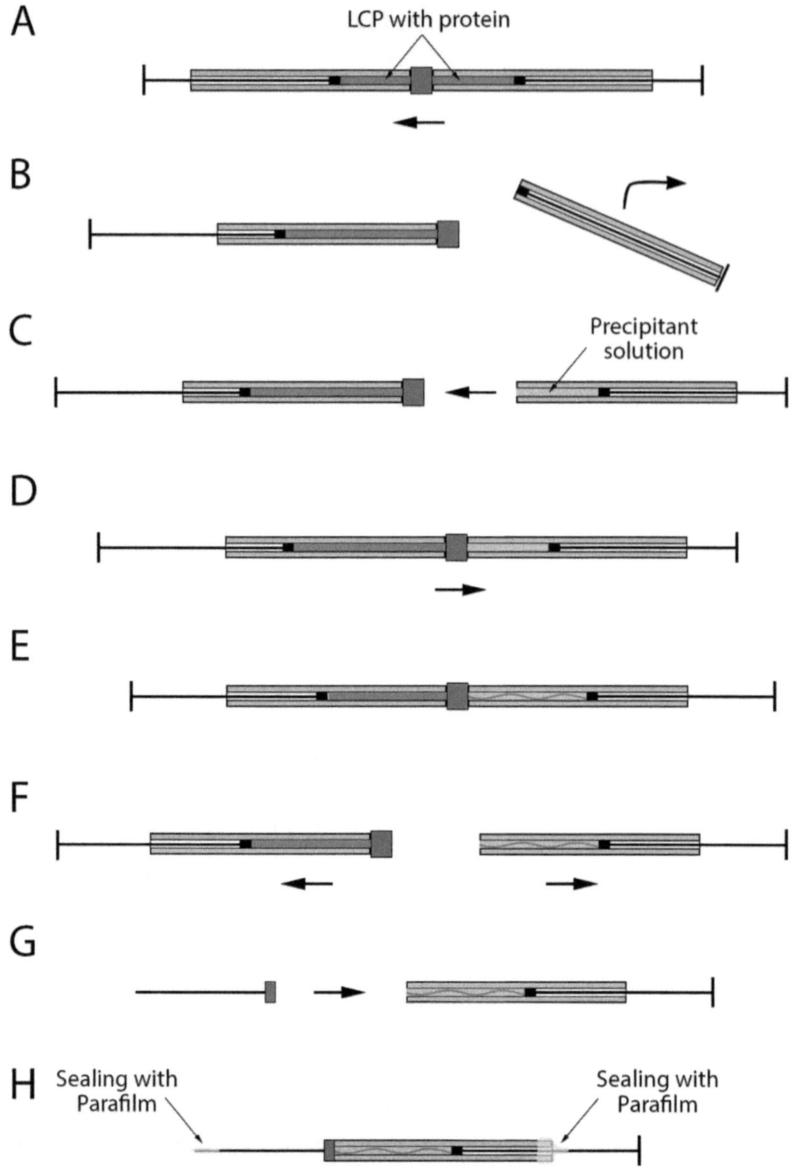

Fig. 3 Crystallization setup in Hamilton syringes for serial crystallography

2. Move all the protein-containing LCP into one syringe of the syringe mixer (Fig. 3a) and detach the empty syringe (Fig. 3b). Keep the coupler attached to the syringe with the LCP sample.
3. Fill a clean 100-μl gas-tight syringe with 50–60 μl of a precipitant solution. Try to avoid trapping air bubbles (*see* **Note 2**).
4. Attach the syringe with the precipitant to the coupler with the LCP-filled syringe. Tighten the connection (Fig. 3c).
5. Inject the protein-laded LCP into the precipitant solution by gently pressing on the plunger (Fig. 3d). The gel-like LCP will come out as a snake-like thread inside the precipitant (*see* **Note 5**). Keep injecting until LCP reaches the tip of the plunger (Fig. 3e).

6. Detach the syringe with the precipitant and loaded LCP (Fig. 3f) and quickly connect it to a needle gauge 26s (Fig. 3g). Tighten the connection. Wrap the syringe from both ends and the needle tip carefully with parafilm (Fig. 3h), and place the syringe in a zip-locked plastic bag to avoid dehydration of the sample. Also, cover the coupler tip with parafilm to avoid LCP dehydration while you are preparing the next syringe with precipitants.
7. Put the syringe into an incubator at 18–22 °C in the dark.
8. Repeat **steps 3**–7 using different precipitant solutions.

Procedure 2 (crystallization in tubes):

1. Reconstitute the target protein in LCP as described in Subheading 3.1.
2. Move all the protein-containing LCP into one syringe (Fig. 4a) and detach the empty syringe along with the coupler and the Teflon ferrule (Fig. 4b). The Hamilton syringe with the LCP sample should have nothing attached to its tip.

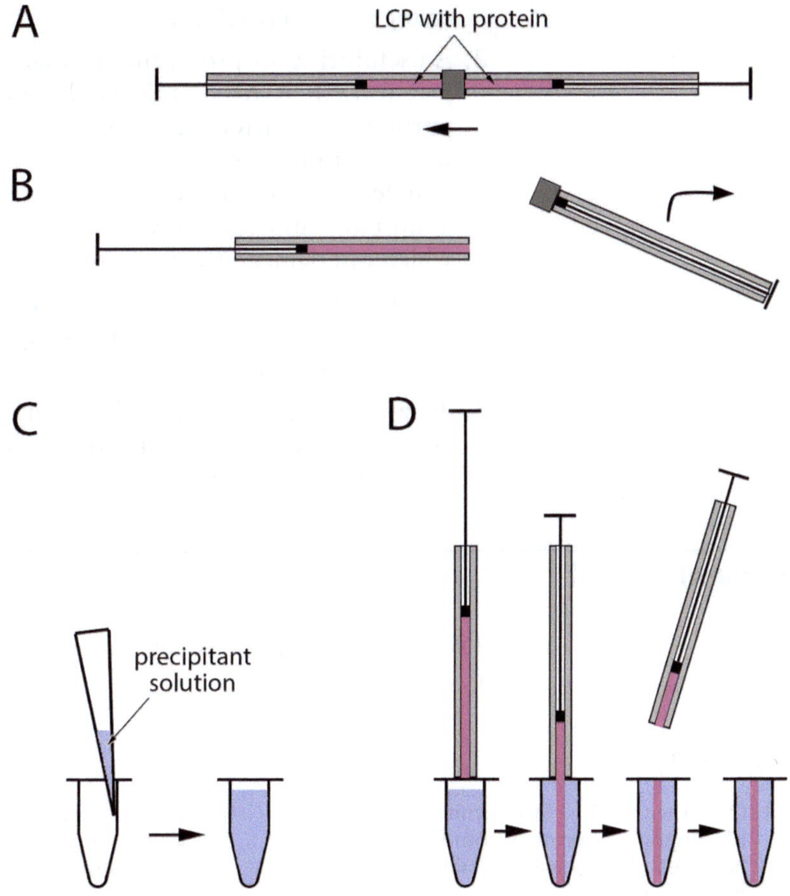

Fig. 4 Crystallization setup in tubes for serial crystallography

3. Take a 1.5-ml Eppendorf tube and fill it with 1.4 ml of a precipitant solution (Fig. 4c).
4. Quickly inject the LCP sample into the Eppendorf tube by pressing the plunger. The phase should come out of the syringe as a thick cylinder (Fig. 4d).
5. Inject until the LCP cylinder reaches the bottom of the tube. Cut the phase using a scalpel and close the tube (Fig. 4d).
6. Put the tube into an incubator at 18–22 °C in the dark.
7. Repeat **steps 2** and **4–6** with different precipitant solutions.

4 Notes

1. Do not exceed 100–200 mg of the total lipid mixture per vial; otherwise, a complete removal of solvent will be difficult. Use several vials or a larger volume vial for making stocks in excess of 200 mg of mixed lipids.
2. To remove air bubbles trapped in solution try moving the plunger slowly down and then quickly up while holding the syringe vertically (Fig. 1b).
3. A useful trick for promoting formation a homogeneous transparent LCP is to slightly cool down the syringe mixer by putting it for a few seconds on ice or in a refrigerator, and then continue mixing. Be careful to avoid overcooling the sample, however, as below 17 °C the monoolein-based LCP is unstable and can convert into a lamellar crystalline phase, which can damage the protein.
4. If you store precipitant stock solutions in a refrigerator (e.g., in +4 °C or −20 °C), warm them up to 18–22 °C and shake before dispensing.
5. To avoid LCP sticking to the syringe barrel during injection, keep syringes vertically with the syringe containing the precipitant solution on top.

Acknowledgment

This work was supported by Russian Science Foundation (RSF) Project 21-64-00018.

References

1. Oesterhelt D, Stoeckenius W (1971) Rhodopsin-like protein from the purple membrane of Halobacterium halobium. Nat New Biol 233:149–152
2. Michel H, Oesterhelt D (1980) Three-dimensional crystals of membrane proteins: bacteriorhodopsin. PNAS 77:1283–1285

3. Henderson R, Unwin PNT (1975) Three-dimensional model of purple membrane obtained by electron microscopy. Nature 257: 28–32
4. Grigorieff N, Ceska TA, Downing KH et al (1996) Electron-crystallographic refinement of the structure of bacteriorhodopsin. J Mol Biol 259:393–421
5. Tsygannik IN, Baldwin JM (1987) Three-dimensional structure of deoxycholate-treated purple membrane at 6 Å resolution and molecular averaging of three crystal forms of bacteriorhodopsin. Eur Biophys J 14:263–272
6. Henderson R, Baldwin JM, Ceska TA et al (1990) Model for the structure of bacteriorhodopsin based on high-resolution electron cryo-microscopy. J Mol Biol 213:899–929
7. Pebay-Peyroula E (1997) X-ray structure of bacteriorhodopsin at 2.5 Angstroms from microcrystals grown in lipidic cubic phases. Science 277:1676–1681
8. Landau EM, Rosenbusch JP (1996) Lipidic cubic phases: a novel concept for the crystallization of membrane proteins. PNAS 93: 14532–14535
9. Hasegawa N, Jonotsuka H, Miki K et al (2018) X-ray structure analysis of bacteriorhodopsin at 1.3 Å resolution. Sci Rep 8:13123
10. Bada Juarez JF, Judge PJ, Adam S et al (2021) Structures of the archaerhodopsin-3 transporter reveal that disordering of internal water networks underpins receptor sensitization. Nat Commun 12:629
11. Essen L-O, Siegert R, Lehmann WD et al (1998) Lipid patches in membrane protein oligomers: crystal structure of the bacteriorhodopsin-lipid complex. PNAS 95: 11673–11678
12. Schertler GFX, Bartunik HD, Michel H et al (1993) Orthorhombic crystal form of bacteriorhodopsin nucleated on benzamidine diffracting to 3·6 Å resolution. J Mol Biol 234: 156–164
13. Wang N, Wang M, Gao Y et al (2012) Crystallization and preliminary X-ray crystallographic analysis of a blue-light-absorbing proteorhodopsin. Acta Cryst F 68:281–283
14. Ran T, Ozorowski G, Gao Y et al (2013) Cross-protomer interaction with the photoactive site in oligomeric proteorhodopsin complexes. Acta Cryst D 69:1965–1980
15. Cherezov V (2011) Lipidic cubic phase technologies for membrane protein structural studies. Curr Opin Struct Biol 21:559–566
16. Caffrey M, Cherezov V (2009) Crystallizing membrane proteins using lipidic mesophases. Nat Protoc 4:706–731
17. Cherezov V, Clogston J, Misquitta Y et al (2002) Membrane protein crystallization in meso: lipid type-tailoring of the cubic phase. Biophys J 83:3393–3407
18. Cherezov V, Clogston J, Papiz MZ et al (2006) Room to move: crystallizing membrane proteins in swollen lipidic mesophases. J Mol Biol 357:1605–1618
19. Luecke H, Schobert B, Richter HT et al (1999) Structure of bacteriorhodopsin at 1.55 Å resolution. J Mol Biol 291:899–911
20. Lanyi JK, Schobert B (2002) crystallographic structure of the retinal and the protein after deprotonation of the Schiff base: the switch in the bacteriorhodopsin photocycle. J Mol Biol 321:727–737
21. Schobert B, Cupp-Vickery J, Hornak V et al (2002) Crystallographic structure of the K intermediate of bacteriorhodopsin: conservation of free energy after photoisomerization of the retinal. J Mol Biol 321:715–726
22. Kolbe M (2000) Structure of the light-driven chloride pump halorhodopsin at 1.8 Å resolution. Science 288:1390–1396
23. Luecke H, Schobert B, Lanyi JK et al (2001) Crystal structure of sensory rhodopsin II at 2.4 angstroms: insights into color tuning and transducer interaction. Science 293: 1499–1503
24. Royant A, Nollert P, Edman K et al (2001) X-ray structure of sensory rhodopsin II at 2.1-Å resolution. PNAS 98:10131–10136
25. Gordeliy VI, Labahn J, Moukhametzianov R et al (2002) Molecular basis of transmembrane signalling by sensory rhodopsin II-transducer complex. Nature 419:484–487
26. Vogeley L (2004) Anabaena sensory rhodopsin: a photochromic color sensor at 2.0 A. Science 306:1390–1393
27. Wada T, Shimono K, Kikukawa T et al (2011) Crystal structure of the eukaryotic light-driven proton-pumping rhodopsin, acetabularia rhodopsin II, from marine alga. J Mol Biol 411: 986–998
28. Kato HE, Zhang F, Yizhar O et al (2012) Crystal structure of the channelrhodopsin light-gated cation channel. Nature 482: 369–374
29. Gushchin I, Chervakov P, Kuzmichev P et al (2013) Structural insights into the proton pumping by unusual proteorhodopsin from nonmarine bacteria. PNAS 110:12631–12636
30. Shevchenko V, Gushchin I, Polovinkin V et al (2014) Crystal structure of Escherichia coli-expressed Haloarcula marismortui bacteriorhodopsin I in the trimeric form. PLoS One 9: e112873

31. Furuse M, Tamogami J, Hosaka T et al (2015) Structural basis for the slow photocycle and late proton release in Acetabularia rhodopsin I from the marine plant Acetabularia acetabulum. Acta Cryst D 71:2203–2216
32. Kim K, Kwon S-K, Jun S-H et al (2016) Crystal structure and functional characterization of a light-driven chloride pump having an NTQ motif. Nat Commun 7:12677
33. Hosaka T, Yoshizawa S, Nakajima Y et al (2016) Structural mechanism for light-driven transport by a new type of chloride ion pump, nonlabens marinus rhodopsin-3. J Biol Chem 291:17488–17495
34. Broecker J, Eger BT, Ernst OP (2017) Crystallogenesis of membrane proteins mediated by polymer-bounded lipid nanodiscs. Structure 25:384–392
35. Shevchenko V, Mager T, Kovalev K et al (2017) Inward H+ pump xenorhodopsin: mechanism and alternative optogenetic approach. Sci Adv 3:e1603187
36. Volkov O, Kovalev K, Polovinkin V et al (2017) Structural insights into ion conduction by channelrhodopsin 2. Science 358:eaan8862
37. Oda K, Vierock J, Oishi S et al (2018) Crystal structure of the red light-activated channelrhodopsin Chrimson. Nat Commun 9:3949
38. Kim YS, Kato HE, Yamashita K et al (2018) Crystal structure of the natural anion-conducting channelrhodopsin GtACR1. Nature 561:343–348
39. Li H, Huang C-Y, Govorunova EG et al (2019) Crystal structure of a natural light-gated anion channelrhodopsin. eLife 8:e41741
40. Needham DM, Yoshizawa S, Hosaka T et al (2019) A distinct lineage of giant viruses brings a rhodopsin photosystem to unicellular marine predators. PNAS 116:20574–20583
41. Bratanov D, Kovalev K, Machtens J-P et al (2019) Unique structure and function of viral rhodopsins. Nat Commun 10:4939
42. Fudim R, Szczepek M, Vierock J et al (2019) Design of a light-gated proton channel based on the crystal structure of *Coccomyxa* rhodopsin. Sci Signal 12:eaav4203
43. Shihoya W, Inoue K, Singh M et al (2019) Crystal structure of heliorhodopsin. Nature 574:132–136
44. Kovalev K, Volkov D, Astashkin R et al (2020) High-resolution structural insights into the heliorhodopsin family. PNAS 117:4131–4141
45. Zabelskii D, Alekseev A, Kovalev K et al (2020) Viral rhodopsins 1 are an unique family of light-gated cation channels. Nat Commun 11: 5707
46. Yun J-H, Park J-H, Jin Z et al (2020) Structure-based functional modification study of a cyanobacterial chloride pump for transporting multiple anions. J Mol Biol 432: 5273–5286
47. Hayashi T, Yasuda S, Suzuki K et al (2020) How does a microbial rhodopsin RxR realize its exceptionally high thermostability with the proton-pumping function being retained? J Phys Chem B 124:990–1000
48. Kovalev K, Astashkin R, Gushchin I et al (2020) Molecular mechanism of light-driven sodium pumping. Nat Commun 11:2137
49. Ikuta T, Shihoya W, Sugiura M et al (2020) Structural insights into the mechanism of rhodopsin phosphodiesterase. Nat Commun 11: 5605
50. Higuchi A, Shihoya W, Konno M et al (2021) Crystal structure of schizorhodopsin reveals mechanism of inward proton pumping. PNAS 118:e2016328118
51. Lu Y, Zhou XE, Gao X et al (2020) Crystal structure of heliorhodopsin 48C12. Cell Res 30:88–90
52. Kovalev K, Polovinkin V, Gushchin I et al (2019) Structure and mechanisms of sodium-pumping KR2 rhodopsin. Sci Adv 5:eaav2671
53. Gushchin I, Shevchenko V, Polovinkin V et al (2015) Crystal structure of a light-driven sodium pump. Nat Struct Mol Biol 22: 390–396
54. Kato HE, Inoue K, Abe-Yoshizumi R et al (2015) Structural basis for Na+ transport mechanism by a light-driven Na+ pump. Nature 521:48–53
55. Skopintsev P, Ehrenberg D, Weinert T et al (2020) Femtosecond-to-millisecond structural changes in a light-driven sodium pump. Nature 583:314–318
56. Sanders CR, Prosser RS (1998) Bicelles: a model membrane system for all seasons? Structure 6:1227–1234
57. Ujwal R, Bowie JU (2011) Crystallizing membrane proteins using lipidic bicelles. Methods 55:337–341
58. Poulos S, Morgan JLW, Zimmer J et al (2015) Chapter nineteen—bicelles coming of age: an empirical approach to bicelle crystallization. In: Shukla AK (ed) Methods in enzymology. Academic, pp 393–416
59. Faham S, Bowie JU (2002) Bicelle crystallization: a new method for crystallizing membrane proteins yields a monomeric bacteriorhodopsin structure. J Mol Biol 316:1–6
60. Faham S, Boulting GL, Massey EA et al (2005) Crystallization of bacteriorhodopsin from

bicelle formulations at room temperature. Prot Sci 14:836–840

61. Cao Z, Bowie JU (2012) Shifting hydrogen bonds may produce flexible transmembrane helices. PNAS 109:8121–8126
62. Joh NH, Min A, Faham S et al (2008) Modest stabilization by most hydrogen-bonded side-chain interactions in membrane proteins. Nature 453:1266–1270
63. Yohannan S, Yang D, Faham S et al (2004) Proline substitutions are not easily accommodated in a membrane protein. J Mol Biol 341: 1–6
64. Yohannan S, Faham S, Yang D et al (2004) The evolution of transmembrane helix kinks and the structural diversity of G protein-coupled receptors. PNAS 101:959–963
65. Joh NH, Oberai A, Yang D et al (2009) Similar energetic contributions of packing in the core of membrane and water-soluble proteins. J Am Chem Soc 131:10846–10847
66. Yohannan S, Faham S, Yang D et al (2004) A C $_{\alpha}$ –H···O hydrogen bond in a membrane protein is not stabilizing. J Am Chem Soc 126: 2284–2285
67. Luecke H, Schobert B, Stagno J et al (2008) Crystallographic structure of xanthorhodopsin, the light-driven proton pump with a dual chromophore. PNAS 105:16561–16565
68. Nakane T, Hanashima S, Suzuki M et al (2016) Membrane protein structure determination by SAD, SIR, or SIRAS phasing in serial femtosecond crystallography using an iododetergent. PNAS 113:13039–13044
69. Morizumi T, Ou W-L, Van Eps N et al (2019) X-ray crystallographic structure and oligomerization of gloeobacter rhodopsin. Sci Rep 9: 11283
70. Besaw JE, Ou W-L, Morizumi T et al (2020) The crystal structures of a chloride-pumping microbial rhodopsin and its proton-pumping mutant illuminate proton transfer determinants. J Biol Chem 295:14793–14804
71. Sato H, Takeda K, Tani K et al (1999) Specific lipid–protein interactions in a novel honeycomb lattice structure of bacteriorhodopsin. Acta Cryst D 55:1251–1256
72. Takeda K, Sato H, Hino T et al (1998) A novel three-dimensional crystal of bacteriorhodopsin obtained by successive fusion of the vesicular assemblies. J Mol Biol 283:463–474
73. Yamamoto M, Hayakawa N, Murakami M et al (2009) Crystal structures of different substates of bacteriorhodopsin's M intermediate at various pH levels. J Mol Biol 393:559–573
74. Matsui Y, Sakai K, Murakami M et al (2002) Specific damage induced by X-ray radiation and structural changes in the primary photoreaction of bacteriorhodopsin. J Mol Biol 324: 469–481
75. Okumura H, Murakami M, Kouyama T (2005) Crystal structures of acid blue and alkaline purple forms of bacteriorhodopsin. J Mol Biol 351:481–495
76. Zhang J, Yamazaki Y, Hikake M et al (2012) Crystal structure of the O intermediate of the Leu93→Ala mutant of bacteriorhodopsin. Proteins 80:2384–2396
77. Enami N, Yoshimura K, Murakami M et al (2006) Crystal structures of archaerhodopsin-1 and -2: common structural motif in archaeal light-driven proton pumps. J Mol Biol 358: 675–685
78. Zhang J, Mizuno K, Murata Y et al (2013) Crystal structure of deltarhodopsin-3 from Haloterrigena thermotolerans. Proteins 81: 1585–1592
79. Chan SK, Kitajima-Ihara T, Fujii R et al (2014) Crystal structure of cruxrhodopsin-3 from Haloarcula vallismortis. PLoS One 9:e108362
80. Yoshimura K, Kouyama T (2008) Structural role of bacterioruberin in the trimeric structure of archaerhodopsin-2. J Mol Biol 375: 1267–1281
81. Kouyama T, Fujii R, Kanada S et al (2014) Structure of archaerhodopsin-2 at 1.8 Å resolution. Acta Cryst D 70:2692–2701
82. Schreiner M, Schlesinger R, Heberle J et al (2015) Structure of Halorhodopsin from Halobacterium salinarum in a new crystal form that imposes little restraint on the E–F loop. J Struct Biol 190:373–378
83. Schreiner M, Schlesinger R, Heberle J et al (2016) Crystal structure of Halobacterium salinarum halorhodopsin with a partially depopulated primary chloride-binding site. Acta Cryst F 72:692–699
84. Kouyama T, Kanada S, Takeguchi Y et al (2010) Crystal structure of the light-driven chloride pump halorhodopsin from Natronomonas pharaonis. J Mol Biol 396:564–579
85. Kouyama T, Kawaguchi H, Nakanishi T et al (2015) Crystal structures of the L1, L2, N, and O states of pharaonis Halorhodopsin. Biophys J 108:2680–2690
86. Borshchevskiy V, Efremov R, Moiseeva E et al (2010) Overcoming merohedral twinning in crystals of bacteriorhodopsin grown in lipidic mesophase. Acta Cryst D 66:26–32
87. Cherezov V, Fersi H, Caffrey M (2001) Crystallization screens: compatibility with the lipidic cubic phase for in meso crystallization of membrane proteins. Biophys J 81:225–242

88. Cherezov V, Liu J, Griffith M et al (2008) LCP-FRAP assay for pre-screening membrane proteins for in meso crystallization. Cryst Growth Des 8:4307–4315
89. Liu W, Hanson MA, Stevens RC et al (2010) LCP-Tm: an assay to measure and understand stability of membrane proteins in a membrane environment. Biophys J 98:1539–1548
90. Cheng A, Hummel B, Qiu H et al (1998) A simple mechanical mixer for small viscous lipid-containing samples. Chem Phys Lipids 95: 11–21
91. Huang C-Y, Olieric V, Ma P et al (2015) In meso in situ serial X-ray crystallography of soluble and membrane proteins. Acta Cryst D 71: 1238–1256
92. Gordeliy VI, Schlesinger R, Efremov R et al (2003) Crystallization in lipidic cubic phases. In: Selinsky BS (ed) Membrane protein protocols: expression, purification, and characterization. Humana Press, Totowa, NJ, pp 305–316
93. Ishchenko A, Cherezov V, Liu W (2016) Preparation and delivery of protein microcrystals in lipidic cubic phase for serial femtosecond crystallography. J Vis Exp 115:e54463
94. Liu W, Ishchenko A, Cherezov V (2014) Preparation of microcrystals in lipidic cubic phase for serial femtosecond crystallography. Nat Protoc 9:2123–2134

Chapter 7

Crystallographic Studies of Rhodopsins: Structure and Dynamics

Marie Luise Grünbein, Gabriela Nass Kovacs, Marco Kloos, Alexander Gorel, R. Bruce Doak, Robert L. Shoeman, Thomas R. M. Barends, and Ilme Schlichting

Abstract

Crystal structures have provided detailed insight in the architecture of rhodopsin photoreceptors. Of particular interest are the protein–chromophore interactions that govern the light-induced retinal isomerization and ultimately induce the large structural changes important for the various biological functions of the family. The reaction intermediates occurring along the rhodopsin photocycle have vastly differing lifetimes, from hundreds of femtoseconds to milliseconds. Detailed insight at high spatial and temporal resolution can be obtained by time-resolved crystallography using pump-probe approaches at X-ray free-electron lasers. Alternatively, cryotrapping approaches can be used. Both the approaches are described, including illumination and sample delivery. The importance of appropriate photoexcitation avoiding multiphoton absorption is stressed.

Key words X-ray free-electron laser, XFEL, Serial femtosecond crystallography, SFX, Serial crystallography, Nanocrystallography, Time-resolved crystallography, Pump probe experiment, Photoexcitation, Rhodopsin, Retinal

1 Introduction

Rhodopsins constitute an important class of photoreceptor proteins: the energy contained in an absorbed visible-light photon (150–300 kJ/mol) drives isomerization of retinal about a double bond, a ubiquitous and fundamental reaction in photochemistry (for a recent review, *see* [1]). This ultrafast reaction then triggers the structural changes important for the specific physiological function of the protein: animal rhodopsins function as G-protein-coupled receptors involved in visual and nonvisual phototransduction (visual rhodopsin [2] and, e.g., extra-retinal pinopsin [3]), in the maintenance of the circadian clock (blue-light-sensitive melanopsin [4]) and as photoisomerases [5]. In the microbial rhodopsins found

Valentin Gordeliy (ed.), *Rhodopsin: Methods and Protocols*, Methods in Molecular Biology, vol. 2501,
https://doi.org/10.1007/978-1-0716-2329-9_7,

in Archaea, Eubacteria, fungi and algae, the functions include light-energy conversion (e.g., by the proton pump bacteriorhodopsin (bR) [6]), chloride pumping (halorhodopsin (HR) [7, 8]), and signaling (e.g., sensory rhodopsin I (SRI) [9] and II (SRII) [10, 11] as well as channelrhodopsins (ChR) [12–14]).

Compared to the isomerization of retinal in solution [15], rhodopsins show a very high specificity of the isomerizing bond and a four- to fivefold higher quantum yield. Generally, in microbial rhodopsins, all-*trans*-retinal isomerizes to the 13-*cis* form with 100% isomerization selectivity and high quantum yields (0.64 for bR). In animal rhodopsins, such as bovine rhodopsin (Rho), 11-cis retinal isomerizes to all-trans-retinal with 100% selectivity and a quantum yield of 0.67. However, there are also microbial and animal rhodopsins that catalyze photoisomerization in the opposite direction [1]. Clearly, a comprehensive understanding of the protein–chromophore interactions that govern retinal isomerization is desired as well as detailed insight into factors that determine the bond specificity and the quantum yield of the isomerization reaction. This is important from both a basic science point of view and for rationalizing the design of better optogenetic tools. Important information is provided by high-resolution crystal structures of well-defined functional states along the reaction coordinate, which also serve as starting point of computational analysis. These structures comprise the stable dark state as well as light-induced intermediates of vastly differing lifetimes (sub-picoseconds to milliseconds; *see* [1] and footnote [1]). To capture these states, diffraction (or spectroscopic) data are collected at a short time delay after triggering the photoreaction in the rhodopsin sample, an approach dubbed pump-probe. Increasing the time delay allows following the entire reaction as it unfolds. In this series, the lifetime of the shortest-lived reaction intermediate to be captured sets an upper limit on the allowable duration of both optical illumination and probe exposure times. The latter is no longer an issue for crystallography and X-ray solution scattering experiments with the advent of X-ray free-electron lasers (XFELs) delivering femtosecond (fs) highly intense X-ray pulses. These allow the collection of a diffraction pattern with a single shot, in the course of which the sample is destroyed, an approach referred to as "diffraction before destruction" [16, 17]. This affords essentially radiation damage-free data collection using very small (typically micron-sized) crystals at ambient temperature. The "downside" of this approach is that each X-ray exposure requires a new crystal, yielding a "still"

[1] The photocycle intermediates and their nomenclature in microbial rhodopsins were originally established in bR: K and O are red-shifted intermediates, L, M, and N are blue-shifted. Animal rhodopsins have similar intermediates called Batho, Lumi, Meta I, and Meta II. Additional transient species exist in both systems. The I (Photo) intermediate, an excited state, is formed within 200 fs, the K (Batho) intermediate within a few ps, Lumi in ns, L and then M (Meta I) in µs, N (Meta II) in ms, followed by O.

diffraction image. A full data set, consisting of integrated diffraction intensities, is assembled from thousands of such snapshots of randomly oriented crystals; this approach is referred to as serial femtosecond crystallography (SFX). Since the method [18] and SFX data analysis [19] were reviewed recently in this series, they will not be expanded on here.

In SFX, fresh crystals need to be delivered into the X-ray beam for each shot, ideally with a similar repetition rate as the XFEL pulses. There are various ways to do this [20]; the method of choice depends among other things on how the microcrystals are grown (in lipidic cubic phase (LCP), sponge phase, or liquid phase) and how much sample is available (*see* **Note 1** for crystal related issues and *see* **Note 2** for sample delivery into the X-ray beam). If relatively large amounts of liquid or sponge phase-grown microcrystals are available, liquid microjet injection using gas dynamic virtual nozzles (GDVN) [21] is a possible option for data collection. Crystals grown in LCP can be delivered into the XFEL beam in sheet-on-sheet chips [22] or by high viscosity extrusion (HVE) [23, 24]. Both the approaches are also well suited if the amount of sample is small; crystals that do not grow naturally in LCP can be embedded [25] in a variety of viscous media [26] for HVE injection.

High-throughput serial data collection, including rapid sample exchange, was developed at XFELs out of necessity. However, sample exchange using chips [27, 28] or HVE injection [24, 29] is also compatible with efficient sample use for data collection at synchrotron sources, an approach called serial synchrotron crystallography (SSX) and reviewed recently in this series [30]. Microfocusing beamlines, in particular at the new diffraction-limited storage rings, provide high enough flux for data collection on the micro-to-millisecond time scale when using suitable detectors. This allows structure determination of long-lived intermediates using diffraction data acquired by SSX [31, 32].

XFELs have opened a new area in time-resolved crystallography, particularly by making the ultra-fast time scale accessible for pump-probe experiments, thus allowing to capture the early events in photoreactions following photon absorption. However, other methods exist to determine structures of light-triggered reaction intermediates, most importantly cryo-trapping approaches [33]. These have provided a great deal of structural insight into all photocycle intermediates (K to O), in particular for bR [34]. Using appropriate photoexcitation conditions is absolutely critical for the observation of the desired reaction intermediate. In case of intense illumination, more than one photon is absorbed by retinal, resulting in various side reactions. This applies not only to excited state absorption when using fs lasers [35, 36] but also to continuous illumination, in particular at cryogenic temperature, and for all intermediates [33]. Thus for all studies on reaction intermediates, it is essential to use small enough crystals that can be photoexcited efficiently and quasi uniformly, avoiding multiphoton absorption. Crystallographers, in contrast, typically want to use large crystals since they generally diffract to higher resolution. This nexus, which is at the heart of single photon excitation, is one of the biggest challenges in performing mechanistically relevant pump-probe experiments [35, 37].

As of this moment (September 2020), there are many hundreds of hits when the Protein Data Bank is queried for "rhodopsin." The tools are now in hand to extend the information obtained from the dark resting states to insights from reaction intermediate structures. The pioneering experiments on bR [38–40] and, somewhat later, on the sodium pump KR2 [41] have demonstrated that time-resolved SFX (TR-SFX) can be used to follow structural changes with up to fs time resolution in rhodopsins. As these experiments were performed early on in the development of TR-SFX techniques, it may have been an understandable approach to maximize signal under multiphoton conditions. Now, however, it is time to learn from both the successes and the failures of the initial studies [38–46] mentioned in this review, and to properly perform future studies on the diverse family of rhodopsin photoreceptors.

2 Minimizing Effects of Radiation Damage

Rhodopsin crystals are prone to X-ray-induced radiation damage [47–49]: with increasing irradiation, changes occur around the bound retinal, e.g., involving the water molecules close to the Schiff base linking retinal to Lys216 in bacteriorhodopsin (bR) [47, 48], and following X-ray-induced decarboxylation of the conserved aspartate (Asp85), large structural changes ensue [47, 48]. It is thus essential to collect the diffraction data in a way that reduces radiation damage. This can be done by either "outrunning" the ionization-induced structural changes via SFX at XFELs [17] or by distributing the dose, i.e., the absorbed energy/kg matter, by collecting the data in wedges or serially (see below). The choice of method depends on the size and number of the crystals available. The easiest way to determine the dose is via the program RADDOSE (http://www.raddo.se/) [50, 51].

The tolerable X-ray dose, however, cannot be calculated but needs to be determined experimentally not only for each system, but it may also depend on a specific functional state. The tolerable dose is typically defined as the dose at which the magnitude of the radiation-induced change of a certain parameter is 50% of its maximal value. This parameter can be resolution [52], specific features in the electron density (decarboxylations [53, 54], S-S bond cleavage [53, 54], ...) or, in the case of chromophore containing proteins, certain spectroscopic signatures [47, 49, 55]. In the latter case, radiation-induced changes can be followed by performing spectral measurements (UV-Vis, fluorescence, IR, Raman, X-ray emission, ...) on (often cryocooled) crystals as they are being irradiated with X-rays. Setups for performing such microspectrophotometric studies are available at a number of synchrotron beamlines (e.g., Swiss Light Source (PX 10) [56], ESRF (several beamlines [57]), SSRL (9-2), ...), often in close collaboration with the beamline scientists. In case of rhodopsins, it is easiest to measure the optical absorption spectrum of retinal and follow its changes with X-ray irradiation [47, 49]. Due to the high optical density of rhodopsin crystals, it is important to perform the spectroscopic measurements either using thin crystals or at thin edges of thicker crystals. Otherwise, important absorption bands are saturated and X-ray-induced changes may be undetectable. At the very least, comparison with published spectra, in particular of reaction intermediates, becomes complicated.

3 Structures of Light-Induced Reaction Intermediates

3.1 General Considerations

The magnitude of the light-induced structural changes in rhodopsins is expected to scale with reaction time. Bond length alterations (BLA) in the polyene chain of retinal take place during the first 200 fs [58], followed by retinal twisting in the K- (batho-) intermediate ultimately inducing large structural changes in later intermediates. Analyzing small structural changes is challenging, in particular in case of low occupancy states. For this reason, and to increase the signal-to-noise ratio of features in electron density maps, crystallographers desire strong light-induced signals. Thus, the general tendency is to excite as many molecules in the crystal as possible. To this end, the number of retinal molecules in a given crystal is calculated. Unfortunately, however, this commonly used approach is often wrong in several aspects.

Retinal is a strongly absorbing chromophore. This, together with the high protein concentration in crystals, renders rhodopsin crystals optically dense (point in case: bacteriorhodopsin crystals have a molar extinction coefficient $\varepsilon_{532\mathrm{nm}} = 45{,}600\ \mathrm{M}^{-1}\ \mathrm{cm}^{-1}$, bR retinal concentration 27 mM). Light penetration (transmission) is described by Lambert Beer's law. In case of optically thick samples and high light intensities, multiple photons are likely absorbed by each molecule at the front of the sample (crystal), and since the light is strongly attenuated upon passing through the sample, the photoexcitation is lower at the rear (the $1/e$ penetration depth in bR crystals is 3.5 μm at 532 nm.). The latter results in an undesired decrease of the light-induced difference signal $F_{\mathrm{obs}}^{\mathrm{dark}} - F_{\mathrm{obs}}^{\mathrm{light}}$. The net result is lower than the expected yield of the light signal, typically prompting the experimenter to increase the laser energy. While this may boost the magnitude of the difference signal, it also increases the number of absorbed photons/chromophore, in particular also at the front of the sample, leading to a strongly increased likelihood for multiphoton absorption.

Multiphoton absorption is to be avoided at all cost because the natural, biologically relevant reaction takes place in the linear single-photon absorption regime. Multiphoton excitation can amplify nuclear motion (see Supplement of [42]), lead to additional radical intermediates [40, 44], and open non-productive higher excited state relaxation channels, decreasing the single-photon reaction yield [36, 40, 59].

Appropriate photoexcitation conditions avoiding such artefacts must be established experimentally, by performing a power titration to determine the power density range where an increase in pump laser energy scales linearly with light-induced signal. This requires a full time-resolved spectroscopic characterization of the photoreaction both in solution and crystals because the kinetics of the photocycle, including the yield of various intermediates can differ between the two, including on the ultrafast time-scale [45, 60]. Since high occupancy of the intermediates facilitates accurate structural interpretation, it is important to identify the photoexcitation wavelength that gives the highest yield of the desired species. The optimal excitation wavelength balances yield with penetration depth, and thus allowable crystal thickness. The crystal thickness should not exceed the penetration depth of the photoexcitation light. Should it not be possible to grow small well-diffracting crystals, one should consider using milling or fractionating approaches to break large crystals. Both crystals grown in LCP and crystals grown in a liquid phase can be reduced in size by flowing the crystal containing medium through sets of stainless-steel filters using an HPLC pump.

Since a detailed protocol for preparing time-resolved pump-probe experiments has been published recently (see Supplement in ref. 37), it will not be repeated here. A Python script and Excel analysis sheet for calculating the number of photons/chromophore, including the depth-dependent photon-absorption regimes inside crystals, is available at https://github.com/MGruenbein/pump-probe-experiments [37].

A common justification for using high pump laser intensity is that a large fraction of the incident pump beam scatters off of the jet employed to deliver the crystals and is therefore lost before reaching the interaction region [39, 61]. However, the actual losses were either never measured at all [61] or misinterpreted by not taking into account the strong divergence of light beams passing through cylindrical jets [39]. In fact, absorption by common jet media is low ($<2\%$), as is reflection at jet and crystal interface ($<10\%$ at each interface). In contrast, refraction at the jet interfaces changes the incident power density distribution significantly, leading to inhomogeneous illumination inside the jet and increasing excitation intensity in large regions of the interior jet volume (on average a 10–60% increase is observed depending on jet type (large diameter viscous jet vs. small diameter GDVN jet) and relative alignment of pump and probe optical axes) [37].

In general, the photoexcitation power density should not exceed 100 GW/cm^2 (see e.g. ref. 36). This is particularly an issue for ultrafast pump probe experiments.

3.2 Steady State Illumination: Pre-illumination

In some microbial rhodopsins, the protein-bound all-*trans* retinal is not stable in the dark; it can isomerize thermally to the 13-*cis*,15-*syn* form which is referred to as dark-adaption. This thermally stable form differs from the metastable 13-*cis*,15-*anti* retinal that is formed during the functional photocycle upon photoexcitation. Moreover, upon photoexcitation, only all-*trans* retinal undergoes a functional (productive) photocycle inducing the structural changes required for the biological function. Thus, for experiments analyzing the photocycle or specific intermediates, it is desirable to maximize the all-*trans* retinal form and thus the fraction of the productive reaction path while minimizing overlaying features acting as noise. This can be done by "light-adaption," converting 13-*cis*,15-*syn* to all-*trans* retinal, resulting in a high steady state occupancy of the functional dark state form. It is of note that in some sensory rhodopsins such as *Anabaena* Sensory Rhodopsin (ASR), the directions of dark and light adaption are reversed. Light adaption yield can be assessed by HPLC analysis [40, 62, 63] of the illuminated sample.

Light adaption of crystalline microbial rhodopsins occurs when looking at the crystals under a white-light microscope [62, 64, 65]. During the mounting process, the crystals are often rotated, resulting in illumination from several sides and ensuring that also in case of thick crystals large fractions thereof are light-adapted. All subsequent crystal manipulations, including data collection at ambient temperature, need to be done before the light-adapted state has converted back to the dark-adapted form. In bacteriorhodopsin, the light-adapted state decays with a time constant of ~20 min. This time frame is no issue when cryocooling the crystals for diffraction data acquisition. However, for data collection approaches using crystals kept at ambient temperature, the decay time of the light-adapted state becomes an issue if it is faster than data acquisition. This is often the case when the pre-illuminated crystals are delivered into the X-ray beam with chips or injectors. To ensure a constant high level of light-adaptation during data acquisition, it is then necessary to perform light-adaptation in situ. However, when conducting time-resolved pump-probe experiments, constant pre-illumination is problematic since not only the light-adapted retinal form accumulates but also intermediates that are formed faster than they decay and hence accumulate (e.g., the M-intermediate). Thus, the pre-illumination protocol must contain both illumination and dark phases to ensure light-adaption as well as return to the ground state, respectively. Such a setup for crystal delivery by high viscosity extrusion injection is described in detail in the supplement of ref. 40. The much simpler approach of illuminating the syringes with the crystal-containing LCP before loading it into the injector [38, 39] has two issues: (1) light penetration through the thick LCP "sausage" is low since crystals shield each other, preventing efficient illumination and thus light-adaptation, and (2) filling and mounting the injector take time. This temporal overhead and the typical data collection times far exceed the lifetime of light adaption. Thus, due to the decay of the light adapted state, the sample composition changes during data collection. This change needs to be reflected in both the dark state data used as a reference and the pump-probe data, or systematic errors are introduced.

3.3 Data Collection Strategies for Reaction Intermediates

The structure of a reaction intermediate of interest can only be determined if the intermediate is formed more rapidly than it decays, accumulates to a reasonable occupancy, and if the structure differs significantly from that of other states. The limits of what can be detected depend on resolution and data quality.

3.3.1 Trapping

The photocycle intermediates of rhodopsins have very different energy barriers and thus lifetimes. The primary photoproduct is formed at liquid nitrogen temperatures with high quantum yield, but the forward reaction is energetically blocked (up to 130 K). This allows structure determination of the K-intermediate (by illuminating cryo-cooled crystals during X-ray data collection, generating a photostationary state [47, 66, 67]). Illumination at higher but still cryogenic temperature allows trapping of the later intermediates in the photocycle (L [68, 69], M [70]). Reducing the temperature for data collection [69] not only "freezes in" the intermediate but also slows radiation damage. The structure of the M intermediate was also determined using room temperature illumination followed by cryogenic trapping [71, 72]. In addition to temperature trapping, mutagenesis has been used to stabilize various intermediates.

Prolonged intense illumination results in side reactions and thus mixed states [33], since all intermediates can absorb further photons, complicating analysis. Since temperature and experimental parameters such as pH can affect the yield of intermediate formation and the distribution (occupancies) of intermediates as known from ambient temperature solution spectroscopy, it is important to characterize the trapped state spectroscopically [33]. Moreover, online spectroscopic examination is also important to monitor radiation damage [47].

3.3.2 Time-Resolved Crystallography

Time-resolved experiments rely on well-defined, known starting conditions of the reaction, an appropriate fast and efficient trigger and suitable detection. In case of pump-probe experiments on photoreceptor proteins, the reaction is started by a light flash, resulting in photoexcitation of the chromophore. As already mentioned above, appropriate photoexcitation conditions are essential (*see* **Note 3**) for the functional relevance of the experiment since the natural photoreaction occurs in the single-photon regime. Spectroscopic studies are performed in this regime. Therefore, a comparison of these results with those obtained by TR-SFX is only valid if both investigate the same reaction (ideally the one induced by the absorption of a single photon). Importantly, this constraint applies not only to comparison with published data but also to comparative spectroscopic investigations on the photoreaction in crystals and solutions to establish whether or not the reaction is similar in both cases. Extrapolation of these results to validate TR-SFX experiments performed at much higher photoexcitation power [41] is highly questionable at best (*see* [35]).

In general, the starting point of the photoreaction is the stable (possibly light-adapted, see above) dark state of the photoreceptor. Ensuring that a pump-probe cycle actually starts from the dark state can be challenging when collecting many data points from a sample: when (1) exposing the same sample spot, it is important to wait until any exposed sample returns to the ground (dark) state, when (2) translating the sample to a new spot (for example, in case of irreversible reactions), it is critical to ensure that this new measurement spot was not exposed to stray light. The latter is an issue when using slow viscous jets or chips for sample delivery. In the latter case, apertures need to be used and translations between subsequent shots must be large enough to ensure that unexposed sample is available for each shot. Crystal-containing high viscosity jets, obtained by mixing crystals into a viscous carrier medium or naturally by growing crystals in LCP jets, are notoriously inhomogeneous, with often very non-uniform crystal concentrations. This results in unequal flow characteristics, with often strongly varying jet speeds [26]. The latter can be smoothened by adding, for example, pluronic [26] or paraffin oil [39].

When performing pump-probe experiments on a moving target such as a free-flowing jet, jet speed, the spatial extent of the pump excitation region and the repetition rate of the experiment have to be matched. After initiating a reaction in a given segment of the jet, the very same segment must be located in the X-ray interaction region at the time the X-ray pulse arrives; otherwise, the ground state is probed instead of the evolving reaction. Moreover, the excited sample segment must have left the interaction region before the next pump-probe event such that each pump-probe event starts with the same, clean conditions in the ground state (the same applies when measuring light–dark sequences). This constrains jet velocity for a desired time delay to a defined regime: (1) if the jet is too fast, the desired segment has already traveled beyond the interaction point before the X-ray probe pulse arrives, and (2) if it is too slow, it has not yet cleared the interaction region in time for the subsequent (pump-)probe exposure, or may have not yet even arrived in the interaction region if the point of reaction initiation is upstream from the X-ray interaction region [20].

For successful pump-probe measurements, the jet speed must therefore be known accurately, and, together with other experimental parameters, adjusted where necessary and if possible to the repetition rate of the experiment as well as the extent of the sample segment in which the pump laser initiates the reaction. This is particularly important for pump-probe measurements on viscous jets, where jet speed is often not a constant value, but may vary significantly over time due to sample inhomogeneity affecting jet flow. Both unavoidable inhomogeneities resulting from the mixing of different substances for the final matrix and the crystals themselves disturb jet flow, leading to changes in jet shape due to changing surface tension and changing propensity to adhere to the sample capillary tip. Moreover, the pressure required to transport the sample through the sample capillary depends on sample viscosity as well as crystal size and density. Changes in these parameters that naturally occur in any sample of microcrystals embedded in a viscous matrix lead to variations in sample flow rate and thus variations in jet speed due to the lag time required for the hydraulic system driving sample flow to re-adjust the driving pressure. It is thus not sufficient to assume a constant velocity derived from flow rates and the inner diameter of the sample capillary [39].

Instead, it is critical that jet speed is measured in situ during the experiment when working with crystals embedded in a viscous material: jet speed can vary wildly, and sometimes jets even come to a complete, albeit temporary, stop. Moreover, jet behavior is highly dependent on sample consistency which often varies from sample batch to sample batch. Measuring jet speed of similar samples before the beam time is a good preparation for the experiment [26, 73] but like calculating jet speed based on sample flow rate and diameter of the injection insufficient for knowing the accurate jet speed during the TR-SFX/TR-SSX experiment. The jet speed can be derived manually by tracking features carried in the crystal-carrying jet over time. This allows excluding data in which jet speed variation was too large for reliable pump excitation conditions [40]. At XFELs, this method can be improved by making use of X-ray pulse-induced changes in the consistency of the viscous matrix in a localized region around the site of impact, appearing as a dark stripe on the jet. Consecutive X-ray pulses intersecting the moving jet, thus leave behind several of such stripes, appearing as dark "ladder steps" in the brighter jet, with the step size depending on both jet speed and X-ray repetition rate. The distance between these marks of impact is easily obtained from jet images and provides, together with the known X-ray pulse repetition rate, immediate feedback on jet speed.

In conclusion, time-resolved pump-probe experiments comprise a number of experimental details that have to be understood and continuously checked during the experiment in order to probe the desired state and nothing else. In a first step, suitable excitation conditions have to be established that avoid multiphoton conditions. Then, during the experiment, pump excitation region, pump-probe repetition rate, and jet speed have to be coordinated. Instead of collecting light and dark data sets separately, it is highly advisable to collect light–dark sequences or modifications thereof. This not only ensures that systematic errors show up in both data sets (*e.g.*, decay of the light-adapted state, different properties of different crystal batches, . . .) but also allows checking for systematic and conceptional errors concerning the illumination protocol.

3.4 Data Analysis for Time-Resolved Crystallographic Studies

The occupancies of intermediate states populated in pump-probe experiments are generally low. Even in the case of unity quantum yield, at most 50% of the molecules will be in the excited state (saturation state, equilibrium of stimulated emission, and absorbance) in a two-level system. In the linear regime, one should not expect more than 30% excited molecules. Given the small light-induced difference signal, sufficient data needs to be collected to reach a sufficiently high signal-to-noise ratio. However, with this data collection strategy, it takes a long time until a decent SNR is reached. When several time points need to be collected, beam time becomes a precious commodity, and online feedback from a dedicated data analysis team is essential to make the most efficient use of any beam time available, as will be set out below.

At XFELs, even more so than at synchrotrons, beam times are highly dynamic, with beam parameters, injection parameters, and samples constantly changing. Therefore, a close eye must be kept on the data, by processing in real time or at least as quickly as possible after collection of a (partial) data set. Highly beneficial are automated scripts that plot I/sigma, unit cell parameters, N (Z), Wilson plots, etc. in an easy-to-view format. This will allow the quick detection of any problems such as poor sample batches, sample-to-detector distance changes, etc. Also, as XFELs are still highly experimental one should be very vigilant even about aspects that at synchrotrons would never be questioned. For instance, there is a good chance that even something like the correct detector distance is actually not exactly known at an XFEL but can only be estimated.

These uncertainties and variabilities do not only apply to the beam line as such but also to the pump laser and the samples. As pump laser setups may accidentally be bumped and sample batches may vary, the photolysis yield should be constantly monitored by calculating $F^{light} - F^{dark}$ maps as often as possible. Thus, those processing data during the beam time share a huge responsibility in ensuring such unforeseen occurrences do not result in a wasted beam time.

For the same reasons, one should regularly calculate difference maps between partial data sets that should be the same, but may not be. Due to drifts in laser, XFEL, and sample properties and differences between sample batches, one may have to deal with different parts of the data differently. During one of our experiments, we found that two sample batches differed considerably in unit cell, necessitating careful matching of light data sets with the most isomorphous dark data sets [40].

Online data processing and evaluation are also crucial to planning the course of the beam time. For these purposes, one can assume that signal strengths scale with the square root of the number of images. Thus, collecting twice as many images will increase a peak height in a map by a factor of approximately 1.4. When interpreting maps and refined structures, one should be aware that the signal-to-noise level of SFX data is inherently much lower than what is usual in synchrotron-based crystallography. Thus, model bias is an even bigger danger. To get a feeling for the quality of diffraction intensities, simulated annealing omit maps should be calculated [74]. Simple omit maps can also be used to evaluate signal strength; for instance, if a map after molecular replacement shows clear water molecules that were not in the model, then the data are good enough to see a small molecule.

When calculating difference maps, Q-weighting [75] may increase the signal-to-noise level. This appears to be project-specific; it may work for a particular system, or it may not. Many smaller light-induced structural changes may not be immediately apparent from $F^{light} - F^{dark}$ maps but only become visible after refinement. As full occupancy of a light-activated state cannot typically be attained for most systems, including rhodopsins, such refinement must necessarily either be ensemble refinement [40] or refinement against extrapolated structure factors. Here the occupancy of the activated state is estimated, and the value is used to extrapolate the observed $F^{light} - F^{dark}$ differences to full occupancy using the following formula:

$$F_{ext} = \frac{F_{obs}^{light} - F_{obs}^{dark}}{n} + F_{obs}^{dark}$$

where n is the estimated occupancy [76, 77].

While this offers a way to estimate the structure factor amplitudes of an imaginary state with 100% occupancy, it also magnifies the measurement errors of the intensities, which for SFX are typically already much higher than is usual in synchrotron-based crystallography. Thus, even more care should be applied when interpreting maps and refined structures. Here, too, omit maps can help estimate the information content of the extrapolated structure factors amplitudes.

Whatever refinement strategy is used, if structural differences are only apparent from refined structures and not from maps, the changes are likely to be small and may be on the order of the typical coordinate errors encountered in protein crystallography. To distinguish such apparent changes from noise, resampling methods such as bootstrapping or jackknifing can be used to obtain error estimates for the atomic positions [41].

In conclusion, those responsible for data processing and interpretation should interact closely with the experimentalists and provide constant feedback during the beam time about the rate at which data is coming in, as well as its quality. To this end, they should be prepared to do any aspect of data processing and interpretation as quickly and efficiently as possible during the actual beam time, and they should be available for this task round the clock as long as data is being collected. After the beam time, this close interaction should be maintained as the experimentalists likely have important information on data quality (such as on runs during which illumination conditions may not have been optimal, etc.)

4 Notes

Note 1

- If feasible, test the sample at an XFEL during a preceding screening beamtime to identify the best sample in terms of shape (if multiple) and size. For a time-resolved experiment, the smallest crystals yielding sufficient resolution should be chosen. For a static experiment, in case of HVE injection, larger crystals will require a larger nozzle diameter. This does not necessarily result in higher sample consumption.
- If a test at an XFEL is not feasible, test some of the larger microcrystals at a synchrotron beamline. Identify the space group and ideally collect a reference data set.
- Any changes of purification or crystallization conditions, e.g., during up-scaling should always be followed by a diffraction test and space group check. Even small changes can result in different space groups. Due to the small size of the microcrystals, this may not be apparent from the shape of the crystals.
- Estimate the amount of crystallization-ready sample required to cover all necessary/feasible measurements. Typically, a significantly larger amount of sample needs to be available than what

will be consumed in the experiment as even a very reproducible crystallization setup can yield a batch-dependent distribution of crystal sizes. Consequently, only a part of the sample may be useful, in particular for pump probe experiments.

- There are two strategies for chip data collection: (1) indiscriminate rastering of the chip area, (2) identifying individual crystals on the chip, collect data only at these positions. In general, the two approaches differ in crystal density, with (1) profiting from high density. LCP-grown crystals can be concentrated by centrifugation as described in detail in the supplement of ref. 40. The SOS chip [22] is uniquely suited for LCP-grown crystals using strategy (1) at XFELs.

  Crystals can also be grown on the solid target, either silicon chips for soluble proteins [78] or between X-ray transparent membranes when using LCP for crystallization [79, 80]. The latter approach is well suited using approach (1) for data collection at synchrotrons [81].

Note 2

- The choice of sample delivery method needs to reflect the requirements of the experiment, both scientific and technical ones, and of the sample. The former includes the scientific goal—(1) obtain static or also time-resolved structures: (2) if only static—at RT or cryogenic temperatures, (3) if time-resolved, is it light- or ligand-triggered, (4) if light-induced time-resolved, is it on an ultrafast or μs and slower timescale. Among the technical aspects are following questions—(5) is the experiment performed at atmospheric pressure or in vacuum, (6) can the sample delivery method match the given repetition rate of the XFEL, (7) can the facility physically accommodate this method and the entire set-up that is required, and (8) is low X-ray background crucial either due to likely weak diffraction or, e.g., because highly accurate data for phasing is needed. The determining sample questions encompass—(9) what is the size, homogeneity, and quantity available of the sample, (10) what is the crystallization condition (in solution—composition of the mother liquor, or in LCP), (11) is the sample light sensitive or has another particular sensitivity. This incomplete list of questions should provide a sneak peek into the diverse parameters important for sample delivery methods, and for a high-level classification, see ref. 20.
- Not all HVE media tolerate vacuum conditions or low temperatures. Some mixtures of lipids used to form a lipid cubic phase for protein crystallization are not stable under these conditions and form strongly diffracting lamellar phases. Some media freeze quickly when injected into vacuum, leading to pronounced ice

rings in the scattering pattern and potential damage to the crystal. Use additives and cryoprotectants. Test the influence of potential compounds on crystal quality ahead of time at a synchrotron.

- Injectors always need to be regularly and thoroughly tested in the home laboratory for leaks and other potential defects due to, e.g., damaged sealing rings. Make sure enough capillaries/nozzles of dimensions needed/likely needed for the experiment are prepared. Always have necessary consumables available during the experiment. Never plan to perform an experiment with an untested injector, you risk severe sample and time loss.
- Which injection conditions (sample flow rate, gas pressure) work need to be tested before the experiment, first on (1) same medium without crystals, then (2) final test on medium with crystals. In particular, it needs to be tested which injection conditions work for starting a jet reliably, without it curling up on the nozzle tip. These starting conditions may differ from the (minimum) conditions for smooth flow once the jet is running. If so, start the jet with the starting conditions and gradually change flow rate and gas pressure toward the tested and desired conditions for data collection.
- The sample needs to be as homogeneous as possible to allow smooth sample injections. This requires that any mixing (e.g., crystal into viscous material, or mixing two viscous materials for better fluid dynamic properties) is done efficiently. Moreover, mixing needs to be done with great care such that no air bubbles are introduced. Air bubbles cause the jet to fragment. Thereafter, the jet sometimes curls back on the nozzle after the bubble, complicating data collection due to irregular flow properties (speed and location) or due to the presence of multiple jet streams or loops.
- Important for time-resolved experiments: Jet speed needs to be measured during the experiment. It is not sufficient to measure jet speed for similar injection conditions at home since every HVE sample runs differently (different crystal concentration, homogeneity after mixing, ...). Even more so, it is unacceptable to only calculate jet speed based on flow rate and diameter of nozzle since the diameter of the extruded jet may vary, and inhomogeneities in the sample lead to variations in jet speed even when nominally driving the sample at constant flow rate, because the hydraulic system needs time to equilibrate required pressures. To measure jet speed, decrease X-ray intensity at XFELs, such that the viscous stream in not cleaved but instead bubbles are visible. Use these for tracking. At synchrotrons use inhomogeneities (e.g., crystals) in the jet. Measure the displacement of the tracked feature with time.

- Interaction of the jet with the X-ray pulse changes the consistency of LCP and likely also other matrices in a localized region around the site of impact, appearing as a dark stripe on the jet if suitable jet imaging conditions are established. Consecutive X-ray pulses intersecting the moving jet thus leave behind multiple such stripes, appearing as a "ladder steps" in the brighter jet with the step size depending on both jet speed and X-ray repetition rate. The distance between these marks of impact is easily obtained from jet images and provides, together with the known X-ray pulse repetition rate, immediate feedback on jet speed. X-ray power needs to be sufficiently low to prevent severing of the HVE jet.
- A catcher below HVE jet improves flow behavior since the jet can attach to the catcher wall, which reduces or eliminates wild jet oscillations. This and a modest He focusing gas stream produce the straightest jets.
- Triboelectric charging can occur with certain combinations of injectors, specific injection media (chemical composition), and injection flow rates. The effect is larger at higher flow rates and can make it difficult to achieve a stable jet. The charged jet often whips around after extrusion and can even curl back onto the injector, making data collection difficult or impossible. Cleaning the outside of the injection nozzle with an alcohol-saturated wiper sometimes reduces this phenomena.
- Ideally, sample loading jigs are employed to enable sample loading quickly and in a manner that prevents air being introduced into the injector. The common injection systems are designed to be hydraulic systems, both for the displacement/pumping liquid (or the plunger of a syringe system) and for the sample. Air or other gases are poor media for hydraulic systems since they are compressible and greatly distort/slow the response behavior of the system.

Note 3

- Spectroscopic power titrations should be performed before the time-resolved SFX experiment to establish safe excitation regimes. This is critical to avoid measuring multiphoton response.
- The peak fluence and peak power density at the center of the beam are often not calculated correctly. A Gaussian beam is described by the intensity distribution $I_0 \cdot \exp(-2r^2/w^2)$ at the focus position. Here, I_0 is the intensity at the center of the beam, w the beam's waist radius (distance to beam center where intensity falls to $1/e^2$ of the maximum value) and r the distance to the beam center. The peak fluence F is given by $F = 2E/(\pi w^2)$ and likewise the peak power P by $P = 2E/(\pi w^2 \tau)$ where τ is the pulse

length. The factor of 2 is often forgotten, resulting in an incorrect value.

- A Gaussian intensity distribution also means that experiments employing the specific pump-probe sequence "pump–probe–probe," aiming to collect data of two or more time delays following one pump pulse, use different excitation intensities for both time delays. For easier interpretation of the results, this should be avoided.
- Translationally offsetting the pump laser focus from the X-ray focus [39, 40] is tempting when using viscous jets for sample delivery. This decreases the time needed to flow any illuminated material out the interaction zone by the time the next pump and/or probe arrives. However, it is often difficult to control (reproduce) this offset reliably during the sometimes very long data collection times. Small variations can result in large changes of effective laser power. Therefore, this setup should be avoided.
- Scattering by the jet does not reduce pump laser intensity significantly [37]. This has been falsely claimed by some publications [39]. However, a thorough investigation showed that pump laser intensity indeed may be increased inside the jet due to refraction on the jet interface [37]. Measuring the intensity change within the jet cannot be (easily) done by measuring transmission through the jet because the focusing effect of the cylindrically shaped jet leads to a large divergence of the beam after passage through the jet such that it is extremely difficult to capture all transmitted light.
- Irreproducible and numerous scattering centers in the jet, as for example intrinsically contained nanoparticles in the commonly employed grease matrix [25] or inhomogeneities introduced by mixing (e.g., gas or liquid bubbles) [46], change the incident intensity in an undefined manner by redirecting light from the original beam path. These changes cannot be quantified by transmission measurements since a downstream detector generally does not capture all scattering angles, such that the measured intensity loss is much larger and not representable of the intensity change within the medium itself. Introduction of scattering centers in undefined quantities and size ranges therefore must be avoided.

References

1. Ernst OP, Lodowski DT, Elstner M, Hegemann P, Brown LS, Kandori H (2014) Microbial and animal rhodopsins: structures, functions, and molecular mechanisms. Chem Rev 114:126–163. https://doi.org/10.1021/cr4003769
2. Palczewski K (2006) G protein-coupled receptor rhodopsin. Annu Rev Biochem 75: 743–767. https://doi.org/10.1146/annurev.biochem.75.103004.142743
3. Shichida Y, Matsuyama T (2009) Evolution of opsins and phototransduction. Philos Trans R

Soc B 364:2881–2895. https://doi.org/10.1098/rstb.2009.0051

4. Provencio I, Rodriguez IR, Jiang G, Hayes WP, Moreira EF, Rollag MD (2000) A novel human opsin in the inner retina. J Neurosci 20: 600–605. https://doi.org/10.1523/jneurosci.20-02-00600.2000
5. Terakita A (2005) The opsins. Genome Biol 6: 213. https://doi.org/10.1186/gb-2005-6-3-213
6. Oesterhelt D, Stoeckenius W (1971) Rhodopsin-like protein from purple membrane of *Halobacterium halobium*. Nat New Biol 233:149–152
7. Matsuno-Yagi A, Mukohata Y (1977) 2 Possible roles of bacteriorhodopsin - comparative-study of strains of *Halobacterium halobium* differing in pigmentation. Biochem Biophys Res Commun 78(1):237–243
8. Schobert B, Lanyi JK (1982) Halorhodopsin is a light-driven chloride pump. J Biol Chem 257: 10306–10313
9. Bogomolni RA, Spudich JL (1982) Identification of a 3rd rhodopsin-like pigment in phototactic *Halobacterium halobium*. Proc Natl Acad Sci U S A 79:6250–6254. https://doi.org/10.1073/pnas.79.20.6250
10. Tomioka H, Takahashi T, Kamo N, Kobatake Y (1986) Flash spectrophotometric identification of a fourth rhodopsin-like pigment in *Halobacterium halobium*. Biochem Biophys Res Commun 139:389–395. https://doi.org/10.1016/S0006-291X(86)80003-1
11. Spudich EN, Sundberg SA, Manor D, Spudich JL (1986) Properties of a second sensory receptor protein in *Halobacterium halobium* phototaxis. Proteins 1:239–246. https://doi.org/10.1002/prot.340010306
12. Nagel G, Ollig D, Fuhrmann M, Kateriya S, Mustl AM, Bamberg E, Hegemann P (2002) Channelrhodopsin-1: a light-gated proton channel in green algae. Science 296: 2395–2398. https://doi.org/10.1126/science.1072068
13. Nagel G, Szellas T, Huhn W, Kateriya S, Adeishvili N, Berthold P, Ollig D, Hegemann P, Bamberg E (2003) Channelrhodopsin-2, a directly light-gated cation-selective membrane channel. Proc Natl Acad Sci U S A 100:13940–13945. https://doi.org/10.1073/pnas.1936192100
14. Govorunova EG, Sineshchekov OA, Janz R, Liu XQ, Spudich JL (2015) Natural light-gated anion channels: a family of microbial rhodopsins for advanced optogenetics. Science 349:647–650. https://doi.org/10.1126/science.aaa7484
15. Song L, El-Sayed MA, Lanyi JK (1993) Protein catalysis of the retinal subpicosecond photoisomerization in the primary process of bacteriorhodopsin photosynthesis. Science 261: 891–894
16. Neutze R, Wouts R, van der Spoel D, Weckert E, Hajdu J (2000) Potential for biomolecular imaging with femtosecond X-ray pulses. Nature 406:752–757. https://doi.org/10.1038/35021099
17. Chapman HN, Caleman C, Timneanu N (2014) Diffraction before destruction. Philos Trans R Soc B 369:20130313. https://doi.org/10.1098/rstb.2013.0313
18. Chapman HN (2017) Structure determination using X-ray free-electron laser pulses. Methods Mol Biol 1607:295–324
19. White TA (2017) Processing of XFEL data. Methods Mol Biol 1607:325–347
20. Grunbein ML, Nass Kovacs G (2019) Sample delivery for serial crystallography at free-electron lasers and synchrotrons. Acta Crystallogr D 75:178–191. https://doi.org/10.1107/S205979831801567X
21. Weierstall U, Spence JCH, Doak RB (2012) Injector for scattering measurements on fully solvated biospecies. Rev Sci Instrum 83: 035108. https://doi.org/10.1063/1.3693040
22. Doak RB, Kovacs GN, Gorel A, Foucar L, Barends TRM, Grunbein ML, Hilpert M, Kloos M, Roome CM, Shoeman RL, Stricker M, Tono K, You D, Ueda K, Sherrell DA, Owen RL, Schlichting I (2018) Crystallography on a chip - without the chip: sheet-on-sheet sandwich. Acta Crystallogr D 74: 1000–1007. https://doi.org/10.1107/s2059798318011634
23. Weierstall U, James D, Wang C, White TA, Wang D, Liu W, Spence JCH, Doak RB, Nelson G, Fromme P, Fromme R, Grotjohann I, Kupitz C, Zatsepin NA, Liu H, Basu S, Wacker D, Han GW, Katritch V, Boutet S, Messerschmidt M, Williams GJ, Koglin JE, Seibert MM, Klinker M, Gati C, Shoeman RL, Barty A, Chapman HN, Kirian RA, Beyerlein KR, Stevens RC, Li D, Shah STA, Howe N, Caffrey M, Cherezov V (2014) Lipidic cubic phase injector facilitates membrane protein serial femtosecond crystallography. Nat Commun 5:3309. https://doi.org/10.1038/ncomms4309
24. Botha S, Nass K, Barends TRM, Kabsch W, Latz B, Dworkowski F, Foucar L, Panepucci E, Wang MT, Shoeman RL, Schlichting I, Doak RB (2015) Room-temperature serial crystallography at synchrotron X-ray sources using slowly flowing free-

standing high-viscosity microstreams. Acta Crystallogr D 71:387–397

25. Sugahara M, Mizohata E, Nango E, Suzuki M, Tanaka T, Masudala T, Tanaka R, Shimamura T, Tanaka Y, Suno C, Ihara K, Pan D, Kakinouchi K, Sugiyama S, Murata M, Inoue T, Tono K, Song C, Park J, Kameshima T, Hatsui T, Joti Y, Yabashi M, Iwata S (2015) Grease matrix as a versatile carrier of proteins for serial crystallography. Nat Methods 12:61–63. https://doi.org/10.1038/nmeth.3172
26. Kovacsova G, Grunbein ML, Kloos M, Barends TRM, Schlesinger R, Heberle J, Kabsch W, Shoeman RL, Doak RB, Schlichting I (2017) Viscous hydrophilic injection matrices for serial crystallography. IUCrJ 4:400–410. https://doi.org/10.1107/S2052252517005140
27. Zarrine-Afsar A, Barends TR, Muller C, Fuchs MR, Lomb L, Schlichting I, Miller RJ (2012) Crystallography on a chip. Acta Crystallogr D Biol Crystallogr 68:321–323. https://doi.org/10.1107/S0907444911055296
28. Owen RL, Axford D, Sherrell DA, Kuo A, Ernst OP, Schulz EC, Miller RJ, Mueller-Werkmeister HM (2017) Low-dose fixed-target serial synchrotron crystallography. Acta Crystallogr D Struct Biol 73:373–378. https://doi.org/10.1107/S2059798317002996
29. Nogly P, James D, Wang DJ, White TA, Zatsepin N, Shilova A, Nelson G, Liu HG, Johansson L, Heymann M, Jaeger K, Metz M, Wickstrand C, Wu WT, Bath P, Berntsen P, Oberthuer D, Panneels V, Cherezov V, Chapman H, Schertler G, Neutze R, Spence J, Moraes I, Burghammer M, Standfuss J, Weierstall U (2015) Lipidic cubic phase serial millisecond crystallography using synchrotron radiation. IUCrJ 2:168–176
30. Diederichs K, Wang M (2017) Serial synchrotron X-ray crystallography (SSX). Methods Mol Biol 1607:239–272
31. Mehrabi P, Schulz EC, Dsouza R, Muller-Werkmeister HM, Tellkamp F, Miller RJD, Pai EF (2019) Time-resolved crystallography reveals allosteric communication aligned with molecular breathing. Science 365:1167–1170. https://doi.org/10.1126/science.aaw9904
32. Weinert T, Skopintsev P, James D, Dworkowski F, Panepucci E, Kekilli D, Furrer A, Brunle S, Mous S, Ozerov D, Nogly P, Wang M, Standfuss J (2019) Proton uptake mechanism in bacteriorhodopsin captured by serial synchrotron crystallography. Science 365:61–65. https://doi.org/10.1126/science.aaw8634
33. Balashov SP, Ebrey TG (2001) Trapping and spectroscopic identification of the photointermediates of bacteriorhodopsin at low temperatures. Photochem Photobiol 73:453–462. https://doi.org/10.1562/0031-8655(2001)073<0453:tasiot>2.0.co;2
34. Wickstrand C, Dods R, Royant A, Neutze R (2015) Bacteriorhodopsin: would the real structural intermediates please stand up? Biochim Biophys Acta 1850:536–553
35. Miller RJD, Pare-Labrosse O, Sarracini A, Besaw JE (2020) Three-dimensional view of ultrafast dynamics in photoexcited bacteriorhodopsin in the multiphoton regime and biological relevance. Nat Commun 11:1240. https://doi.org/10.1038/s41467-020-14971-0
36. Prokhorenko VI, Halpin A, Johnson PJM, Miller RJD, Brown LS (2011) Coherent control of the isomerization of retinal in bacteriorhodopsin in the high intensity regime. J Chem Phys 134:085105
37. Grunbein ML, Stricker M, Nass Kovacs G, Kloos M, Doak RB, Shoeman RL, Reinstein J, Lecler S, Haacke S, Schlichting I (2020) Illumination guidelines for ultrafast pump-probe experiments by serial femtosecond crystallography. Nat Methods 17:681–684. https://doi.org/10.1038/s41592-020-0847-3
38. Nango E, Royant A, Kubo M, Nakane T, Wickstrand C, Kimura T, Tanaka T, Tono K, Song C, Tanaka R, Arima T, Yamashita A, Kobayashi J, Hosaka T, Mizohata E, Nogly P, Sugahara M, Nam D, Nomura T, Shimamura T, Im D, Fujiwara T, Yamanaka Y, Jeon B, Nishizawa T, Oda K, Fukuda M, Andersson R, Bath P, Dods R, Davidsson J, Matsuoka S, Kawatake S, Murata M, Nureki O, Owada S, Kameshima T, Hatsui T, Joti Y, Schertler G, Yabashi M, Bondar AN, Standfuss J, Neutze R, Iwata S (2016) A three-dimensional movie of structural changes in bacteriorhodopsin. Science 354:1552–1557
39. Nogly P, Weinert T, James D, Carbajo S, Ozerov D, Furrer A, Gashi D, Borin V, Skopintsev P, Jaeger K, Nass K, Bath P, Bosman R, Koglin J, Seaberg M, Lane T, Kekilli D, Brunle S, Tanaka T, Wu WT, Milne C, White T, Barty A, Weierstall U, Panneels V, Nango E, Iwata S, Hunter M, Schapiro I, Schertler G, Neutze R, Standfuss J (2018) Retinal isomerization in bacteriorhodopsin captured by a femtosecond x-ray laser. Science 361:eaat0094
40. Kovacs GN, Colletier J-P, Grünbein ML, Yang Y, Stensitzki T, Batyuk A, Carbajo S, Doak RB, Ehrenberg D, Foucar L, Gasper, R, Gorel, A, Hilpert, M, Kloos, M, Koglin, JE,

Reinstein, J, Roome, CM, Schlesinger, R, Seaberg, M, Shoeman, RL, Stricker, M, Boutet, S, Haacke, S, Heberle, J, Heyne, K, Domratcheva, T, Barends, TRM, Schlichting, I (2019) Three-dimensional view of ultrafast dynamics in photoexcited bacteriorhodopsin. Nat Commun 10:1–17

41. Skopintsev P, Ehrenberg D, Weinert T, James D, Kar RK, Johnson PJM, Ozerov D, Furrer A, Martiel I, Dworkowski F, Nass K, Knopp G, Cirelli C, Arrell C, Gashi D, Mous S, Wranik M, Gruhl T, Kekilli D, Brunle S, Deupi X, Schertler GFX, Benoit RM, Panneels V, Nogly P, Schapiro I, Milne C, Heberle J, Standfuss J (2020) Femtosecond-to-millisecond structural changes in a light-driven sodium pump. Nature 583:314–318. https://doi.org/10.1038/s41586-020-2307-8

42. Barends TRM, Foucar L, Ardevol A, Nass K, Aquila A, Botha S, Doak RB, Falahati K, Hartmann E, Hilpert M, Heinz M, Hoffmann MC, Kofinger J, Koglin JE, Kovacsova G, Liang M, Milathianaki D, Lemke HT, Reinstein J, Roome CM, Shoeman RL, Williams GJ, Burghardt I, Hummer G, Boutet S, Schlichting I (2015) Direct observation of ultrafast collective motions in CO myoglobin upon ligand dissociation. Science 350: 445–450

43. Pande K, Hutchison CDM, Groenhof G, Aquila A, Robinson JS, Tenboer J, Basu S, Boutet S, DePonte DP, Liang MN, White TA, Zatsepin NA, Yefanov O, Morozov D, Oberthuer D, Gati C, Subramanian G, James D, Zhao Y, Koralek J, Brayshaw J, Kupitz C, Conrad C, Roy-Chowdhury S, Coe JD, Metz M, Xavier PL, Grant TD, Koglin JE, Ketawala G, Fromme R, Srajer V, Henning R, Spence JCH, Ourmazd A, Schwander P, Weierstall U, Frank M, Fromme P, Barty A, Chapman HN, Moffat K, van Thor JJ, Schmidt M (2016) Femtosecond structural dynamics drives the trans/cis isomerization in photoactive yellow protein. Science 352:725–729

44. Coquelle N, Sliwa M, Woodhouse J, Schiro G, Adam V, Aquila A, Barends TRM, Boutet S, Byrdin M, Carbajo S, De la Mora E, Doak RB, Feliks M, Fieschi F, Foucar L, Guillon V, Hilpert M, Hunter MS, Jakobs S, Koglin JE, Kovacsova G, Lane TJ, Levy B, Liang MN, Nass K, Ridard J, Robinson JS, Roome CM, Ruckebusch C, Seaberg M, Thepaut M, Cammarata M, Demachy I, Field M, Shoeman RL, Bourgeois D, Colletier JP, Schlichting I, Weik M (2018) Chromophore twisting in the excited state of a photoswitchable fluorescent protein captured by time-resolved serial femtosecond crystallography. Nat Chem 10:31–37

45. Woodhouse J, Nass Kovacs G, Coquelle N, Uriarte LM, Adam V, Barends TRM, Byrdin M, de la Mora E, Bruce Doak R, Feliks M, Field M, Fieschi F, Guillon V, Jakobs S, Joti Y, Macheboeuf P, Motomura K, Nass K, Owada S, Roome CM, Ruckebusch C, Schiro G, Shoeman RL, Thepaut M, Togashi T, Tono K, Yabashi M, Cammarata M, Foucar L, Bourgeois D, Sliwa M, Colletier JP, Schlichting I, Weik M (2020) Photoswitching mechanism of a fluorescent protein revealed by time-resolved crystallography and transient absorption spectroscopy. Nat Commun 11:741. https://doi.org/10.1038/s41467-020-14537-0

46. Claesson E, Wahlgren WY, Takala H, Pandey S, Castillon L, Kuznetsova V, Henry L, Panman M, Carrillo M, Kübel J, Nanekar R, Isaksson L, Nimmrich A, Cellini A, Morozov D, Maj M, Kurttila M, Bosman R, Nango E, Tanaka R, Tanaka T, Fangjia L, Iwata S, Owada S, Moffat K, Groenhof G, Stojković EA, Ihalainen JA, Schmidt M, Westenhoff S (2020) The primary structural photoresponse of phytochrome proteins captured by a femtosecond X-ray laser. elife 9: e53514. https://doi.org/10.7554/eLife.53514

47. Matsui Y, Sakai K, Murakami M, Shiro Y, Adachi S, Okumura H, Kouyama T (2002) Specific damage induced by X-ray radiation and structural changes in the primary photoreaction of bacteriorhodopsin. J Mol Biol 324: 469–481. https://doi.org/10.1016/s0022-2836(02)01110-5

48. Borshchevskiy VI, Round ES, Popov AN, Bueldt G, Gordeliy VI (2011) X-ray-radiation-induced changes in bacteriorhodopsin structure. J Mol Biol 409:813–825. https://doi.org/10.1016/j.jmb.2011.04.038

49. Borshchevskiy V, Round E, Erofeev I, Weik M, Ishchenko A, Gushchin I, Mishin A, Willbold D, Buldt G, Gordeliy V (2014) Low-dose X-ray radiation induces structural alterations in proteins. Acta Crystallogr D Biol Crystallogr 70:2675–2685. https://doi.org/10.1107/S1399004714017295

50. Zeldin OB, Gerstel M, Garman EF (2013) RADDOSE-3D: time- and space-resolved modelling of dose in macromolecular crystallography. J Appl Crystallogr 46:1225–1230. https://doi.org/10.1107/S0021889813011461

51. Bury CS, Brooks-Bartlett JC, Walsh SP, Garman EF (2018) Estimate your dose: RADDOSE-3D. Protein Sci 27:217–228. https://doi.org/10.1002/pro.3302

52. Owen RL, Rudino-Pinera E, Garman EF (2006) Experimental determination of the radiation dose limit for cryocooled protein crystals. Proc Natl Acad Sci U S A 103: 4912–4917. https://doi.org/10.1073/pnas.0600973103
53. Burmeister WP (2000) Structural changes in a cryo-cooled protein crystal owing to radiation damage. Acta Crystallogr D Biol Crystallogr 56:328–341. https://doi.org/10.1107/s0907444999016261
54. Weik M, Ravelli RB, Kryger G, McSweeney S, Raves ML, Harel M, Gros P, Silman I, Kroon J, Sussman JL (2000) Specific chemical and structural damage to proteins produced by synchrotron radiation. Proc Natl Acad Sci U S A 97: 623–628. https://doi.org/10.1073/pnas.97.2.623
55. Beitlich T, Kuhnel K, Schulze-Briese C, Shoeman RL, Schlichting I (2007) Cryoradiolytic reduction of crystalline heme proteins: analysis by UV-Vis spectroscopy and X-ray crystallography. J Synchrotron Radiat 14:11–23. https://doi.org/10.1107/S0909049506049806
56. Pompidor G, Dworkowski FS, Thominet V, Schulze-Briese C, Fuchs MR (2013) A new on-axis micro-spectrophotometer for combining Raman, fluorescence and UV/Vis absorption spectroscopy with macromolecular crystallography at the Swiss Light Source. J Synchrotron Radiat 20:765–776. https://doi.org/10.1107/S0909049513016063
57. von Stetten D, Giraud T, Carpentier P, Sever F, Terrien M, Dobias F, Juers DH, Flot D, Mueller-Dieckmann C, Leonard GA, de Sanctis D, Royant A (2015) In crystallo optical spectroscopy (icOS) as a complementary tool on the macromolecular crystallography beamlines of the ESRF. Acta Crystallogr D Biol Crystallogr 71:15–26. https://doi.org/10.1107/S139900471401517X
58. Song L, El-Sayed MA (1998) Primary step in bacteriorhodopsin photosynthesis: bond stretch rather than angle twist of its retinal excited-state structure. J Am Chem Soc 120: 8889–8890
59. Hutchison CDM, Kaucikas M, Tenboer J, Kupitz C, Moffat K, Schmidt M, van Thor JJ (2016) Photocycle populations with femtosecond excitation of crystalline photoactive yellow protein. Chem Phys Lett 654:63–71. https://doi.org/10.1016/j.cplett.2016.04.087
60. Konold PE, Arik E, Weissenborn J, Arents JC, Hellingwerf KJ, van Stokkum IHM, Kennis JTM, Groot ML (2020) Confinement in crystal lattice alters entire photocycle pathway of the Photoactive Yellow Protein. Nat Commun 11:4248. https://doi.org/10.1038/s41467-020-18065-9
61. Arnlund D, Johansson LC, Wickstrand C, Barty A, Williams GJ, Malmerberg E, Davidsson J, Milathianaki D, DePonte DP, Shoeman RL, Wang D, James D, Katona G, Westenhoff S, White TA, Aquila A, Bari S, Berntsen P, Bogan M, van Driel TB, Doak RB, Kjaer KS, Frank M, Fromme R, Grotjohann I, Henning R, Hunter MS, Kirian RA, Kosheleva I, Kupitz C, Liang M, Martin AV, Nielsen MM, Messerschmidt M, Seibert MM, Sjohamn J, Stellato F, Weierstall U, Zatsepin NA, Spence JC, Fromme P, Schlichting I, Boutet S, Groenhof G, Chapman HN, Neutze R (2014) Visualizing a protein quake with time-resolved X-ray scattering at a free-electron laser. Nat Methods 11:923–926
62. Scherrer P, Mathew MK, Sperling W, Stoeckenius W (1989) Retinal isomer ratio in dark-adapted purple membrane and bacteriorhodopsin monomers. Biochemistry 28:829–834
63. Muders V, Kerruth S, Lorenz-Fonfria VA, Bamann C, Heberle J, Schlesinger R (2014) Resonance Raman and FTIR spectroscopic characterization of the closed and open states of channelrhodopsin-1. FEBS Lett 588: 2301–2306. https://doi.org/10.1016/j.febslet.2014.05.019
64. Wand A, Friedman N, Sheves M, Ruhman S (2012) Ultrafast photochemistry of light-adapted and dark-adapted bacteriorhodopsin: effects of the initial retinal configuration. J Phys Chem B 116:10444–10452. https://doi.org/10.1021/jp2125284
65. Stoeckenius W, Lozier RH, Bogomolni RA (1979) Bacteriorhodopsin and the purple membrane of halobacteria. Biochim Biophys Acta - Rev Bioenerg 505:215–278. https://doi.org/10.1016/0304-4173(79)90006-5
66. Schobert B, Cupp-Vickery J, Hornak V, Smith S, Lanyi J (2002) Crystallographic structure of the K intermediate of bacteriorhodopsin: conservation of free energy after photoisomerization of the retinal. J Mol Biol 321:715–726. https://doi.org/10.1016/s0022-2836(02)00681-2
67. Edman K, Nollert P, Royant A, Belrhali H, Pebay-Peyroula E, Hajdu J, Neutze R, Landau EM (1999) High-resolution X-ray structure of an early intermediate in the bacteriorhodopsin photocycle. Nature 401:822–826. https://doi.org/10.1038/44623
68. Edman K, Royant A, Larsson G, Jacobson F, Taylor T, van der Spoel D, Landau EM, Pebay-Peyroula E, Neutze R (2004) Deformation of helix C in the low temperature L-intermediate of bacteriorhodopsin. J Biol Chem 279:

2147–2158. https://doi.org/10.1074/jbc.M300709200

69. Lanyi JK, Schobert B (2007) Structural changes in the L photointermediate of bacteriorhodopsin. J Mol Biol 365:1379–1392. https://doi.org/10.1016/j.jmb.2006.11.016
70. Facciotti MT, Rouhani S, Burkard FT, Betancourt FM, Downing KH, Rose RB, McDermott G, Glaeser RM (2001) Structure of an early intermediate in the M-state phase of the bacteriorhodopsin photocycle. Biophys J 81(6):3442–3455. https://doi.org/10.1016/S0006-3495(01)75976-0
71. Schobert B, Brown LS, Lanyi JK (2003) Crystallographic structures of the M and N intermediates of bacteriorhodopsin: assembly of a hydrogen-bonded chain of water molecules between Asp-96 and the retinal schiff base. J Mol Biol 330:553–570. https://doi.org/10.1016/S0022-2836(03)00576-X
72. Yamamoto M, Hayakawa N, Murakami M, Kouyama T (2009) Crystal structures of different substrates of bacteriorhodopsin's M intermediate at various pH levels. J Mol Biol 393:559–573. https://doi.org/10.1016/j.jmb.2009.08.047
73. James D, Weinert T, Skopintsev P, Furrer A, Gashi D, Tanaka T, Nango E, Nogly P, Standfuss J (2019) Improving high viscosity extrusion of microcrystals for time-resolved serial femtosecond crystallography at X-ray lasers. J Vis Exp (144):e59087. https://doi.org/10.3791/59087
74. Nass K, Foucar L, Barends TRM, Hartmann E, Botha S, Shoeman RL, Doak RB, Alonso-Mori R, Aquila A, Bajt S, Barty A, Bean R, Beyerlein KR, Bublitz M, Drachmann N, Gregersen J, Joensson HO, Kabsch W, Kassemeyer S, Koglin JE, Krumrey M, Mattle D, Messerschmidt M, Nissen P, Reinhard L, Sitsel O, Sokaras D, Williams GJ, Hau-Riege S, Timneanu N, Caleman C, Chapman HN, Boutet S, Schlichting I (2015) Indications of radiation damage in ferredoxin microcrystals using high-intensity X-FEL beams. J Synchrotron Radiat 22:225–238. https://doi.org/10.1107/s1600577515002349
75. Ursby T, Bourgeois D (1997) Improved estimation of structure-factor difference amplitudes from poorly accurate data. Acta Crystallogr A 53:564–575
76. Genick UK (2007) Structure-factor extrapolation using the scalar approximation: theory, applications and limitations. Acta Crystallogr D 63(Pt 10):1029–1041
77. Genick UK, Borgstahl GE, Ng K, Ren Z, Pradervand C, Burke PM, Srajer V, Teng TY, Schildkamp W, McRee DE, Moffat K, Getzoff ED (1997) Structure of a protein photocycle intermediate by millisecond time-resolved crystallography. Science 275(5305):1471–1475
78. Lieske J, Cerv M, Kreida S, Komadina D, Fischer J, Barthelmess M, Fischer P, Pakendorf T, Yefanov O, Mariani V, Seine T, Ross BH, Crosas E, Lorbeer O, Burkhardt A, Lane TJ, Guenther S, Bergtholdt J, Schoen S, Tornroth-Horsefield S, Chapman HN, Meents A (2019) On-chip crystallization for serial crystallography experiments and on-chip ligand-binding studies. IUCrJ 6(Pt 4):714–728. https://doi.org/10.1107/S2052252519007395
79. Huang CY, Olieric V, Ma P, Panepucci E, Diederichs K, Wang M, Caffrey M (2015) In meso in situ serial X-ray crystallography of soluble and membrane proteins. Acta Crystallogr D Biol Crystallogr 71(Pt 6):1238–1256. https://doi.org/10.1107/S1399004715005210
80. Khvostichenko DS, Schieferstein JM, Pawate AS, Laible PD, Kenis PJ (2014) X-ray transparent microfluidic chip for mesophase-based crystallization of membrane proteins and on-chip structure determination. Cryst Growth Des 14(10):4886–4890. https://doi.org/10.1021/cg5011488
81. Huang CY, Olieric V, Caffrey M, Wang M (2020) In meso in situ serial X-ray crystallography (IMISX): a protocol for membrane protein structure determination at the Swiss light source. Methods Mol Biol 2127:293–319. https://doi.org/10.1007/978-1-0716-0373-4_20

Chapter 8

Time-Resolved UV-VIS Spectroscopy of Microbial Rhodopsins

Dmytro Soloviov, Valentin Borshchevskiy, and Igor Chizhov

Abstract

Absorption of light quanta by microbial rhodopsins (or more generally by retinal proteins) leads to conversion of the light energy to the generation of transmembrane anion or cation gradients, optically gated channels, or signal states in photoreception. All these processes are accompanied by series of reaction steps with half-times ranging from femtoseconds to seconds or longer (photocycles). The number of these steps and their kinetic and spectral properties are the essential experimental information required for determination of the mechanism of light energy conversion in these proteins. Here we describe experiments and data analysis providing this information.

Key words Laser flash photolysis, Time-resolved spectroscopy, Retinal proteins, Bacteriorhodopsin, Photocycle, Kinetics

1 Introduction

Microbial rhodopsins undergo cyclic photoinduced transitions (photocycles) including structural and chemical changes. The result may be active ion-translocation through a membrane, channel opening or closing, or generation of a signal. A photocycle starts with the absorption of a photon of visible light leading to isomerization of the retinal chromophore of the protein and includes several intermediate states—their particular number depends on the specific microbial rhodopsin.

Structural changes in the vicinity of retina shift the retinal absorption maximum in the range 350–650 nm. A typical photocycle lasts for 10–1000 ms, and then the protein returns to its ground state. The analysis of the photocycle of microbial rhodopsins and their mutants under different conditions has been and remains a valuable tool for studying their functions and accessing their potential as optogenetic tools. In the case of microbial rhodopsins, flash photolysis is a technique [1] that allows tracking

Valentin Gordeliy (ed.), *Rhodopsin: Methods and Protocols*, Methods in Molecular Biology, vol. 2501,
https://doi.org/10.1007/978-1-0716-2329-9_8,

spectral changes in a time-resolved manner. A nanosecond-long laser pulse starts the protein photocycle. A continuous lamp, which covers a broad visual spectrum passed through a monochromator and then with much lower intensity through the sample, is used to determine time-resolved spectral changes of the protein.

In this chapter, we describe the construction of a flash photolysis setup and a data treatment algorithm that has already been used in numerous transient spectroscopy studies of retinal proteins [2–6].

2 Laser Flash Photolysis Experiments: Protein Kinetics from Nanoseconds to Seconds

2.1 Optical Layout of the Experimental System and Major Electronic Components

A photograph of our Laser Flash Photolysis Spectrometer (LFPS) that we have dubbed "the Cycloscope" is shown in Fig. 1. This setup is located at the Research Center for Molecular Mechanisms of Aging and Age-related Diseases, Moscow Institute of Physics and Technology, Dolgoprudny, Russia. Several prototypes of the LFPS were developed and described by Chizhov and coworkers [1, 7–12]. One of the major motivations for development of these particular LFPS systems was the investigation of the first microbial retinal protein bacteriorhodopsin (bR) from *Halobacterium salinarium* discovered in the early 1970s [13]. Here we describe the method for performing experiments and data analysis using bR as an example. One should note, however, that over the last 6 years, we have investigated almost ten newly discovered retinal proteins or their modifications and mutants (e.g., *see* refs. 2–6, 14). Figure 2 depicts the major spectroscopic and electronic components of the Cycloscope and the optical layout of the setup. All measurements and data analysis are performed using our custom-built computer program that allows full control during data acquisition. The Cycloscope's optics, electronics, and software were designed in such way that proteins with different spectro-temporal properties can be investigated.

A Brilliant BRILL/IR-10 Nd:YAG Q-switched laser (1) with Rainbow OPO (Optical Parametric Oscillator; Quantel, France) provides pulses of 4 ns duration at wavelengths that can be chosen from 420 to 680 nm and energy up to 40 mJ/pulse with maximum repetition rate of 10 Hz. The sample (300 μl volume, optical density OD_{max} = 0.5–1.0, 5 × 5 mm pathlength spectroscopic quartz cuvette (Hellma GmbH & Co, Germany)) is placed in a thermostated housing (7) between two synchronized stepper-motor-controlled monochromators (MSH150, LOT-Quantum Design, Germany). The probing light (Xe-arc lamp (4) L2174-01, 75 W, Hamamatsu Photonics, Japan) passes the first monochromator (5) and the sample and arrives after the second monochromator (5) at a PMT (photomultiplier tube) detector

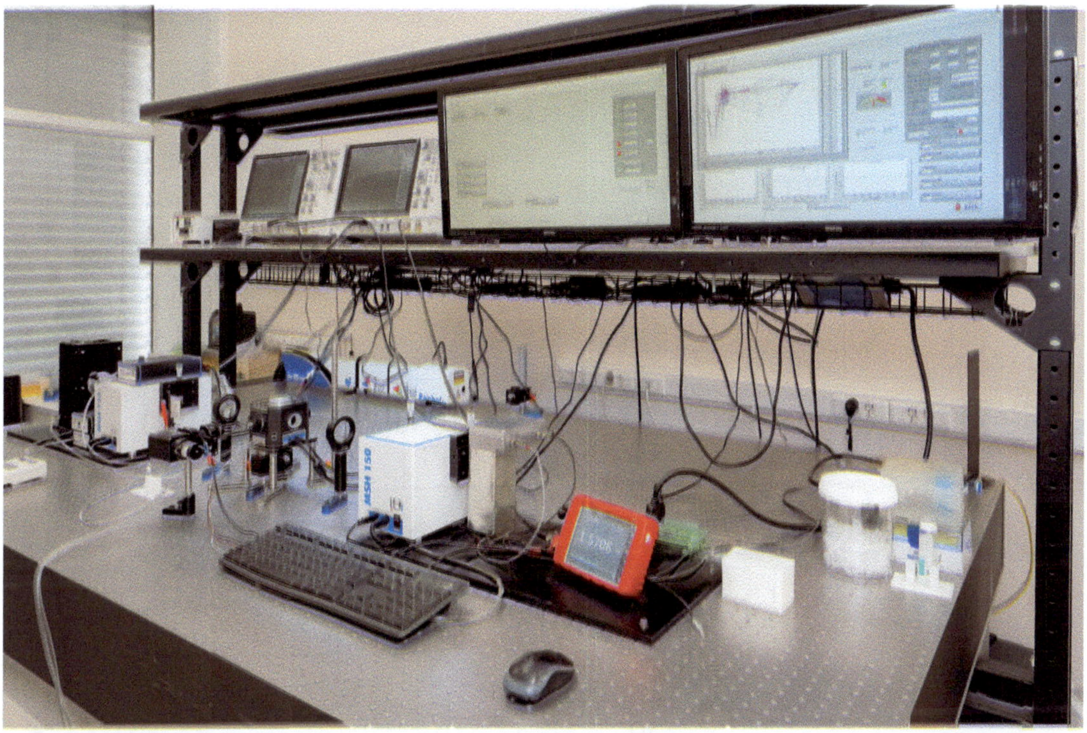

Fig. 1 Photograph of the Cycloscope laser flash photolysis spectrometer built at the Research Center for Molecular Mechanisms of Aging and Age-related Diseases, Moscow Institute of Physics and Technology, Dolgoprudny, Russia

(R12829, Hamamatsu Photonics, Japan). Two digital oscilloscopes (10) (Keysight DSO-X 4022A, Keysight, USA) with 16 kilobytes of buffer memory per channel are used to record the traces of transient transmission changes in two overlapping time windows. The maximal digitizing rate is 10 ns per data point.

2.2 Data Acquisition

Figure 3a shows the transient light intensity changes at 570 nm after laser pulse excitation of bR. The two traces are recorded by the two oscilloscopes in two different time windows (ns – μs: fast oscilloscope, black line, and ms – s: slow oscilloscope, red line). The traces in Fig. 3a are the result of averaging the signal following a series of 25 laser pulses. Thus, the signal-to-noise ratio is improved fivefold. This number is optimal for precise kinetic analysis of the resulting data. Signals from the two oscilloscopes are further transformed to differential absorption changes (ΔOD) at each time point i after the laser pulse using the formula 1.

$$\Delta\mathrm{OD}_i = -\lg\left(\frac{V_i}{V_0}\right) \quad (1)$$

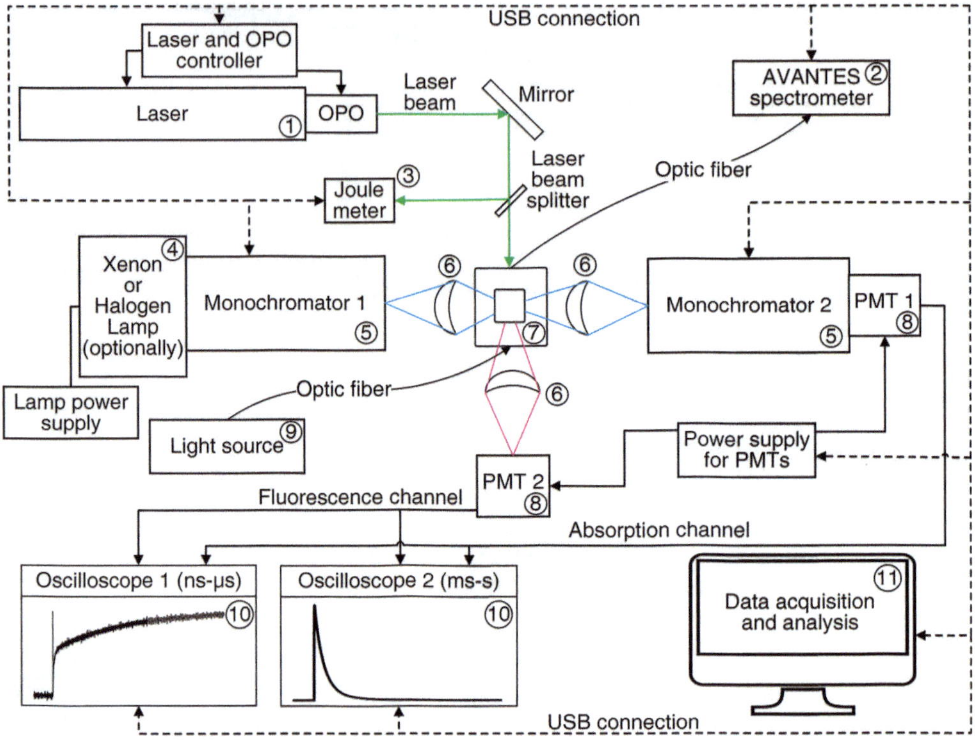

Fig. 2 Major optical layout and electronic components of the Cycloscope: (1) Nd-YAG Q-switched pump laser Brilliant BRILL/IR-10 (10 Hz, 360 mJ, with second and third harmonic generators) with Rainbow OPO 420–680 nm with up to 40 mJ (Quantel, France). (2) Fiber-optic Spectrometer AvaSpec-ULS2048L (Avantes, The Netherlands). (3) Laser Joule Meter ES111C with control unit PM200 (Thorlabs, USA). (4) MLS-75-150 light source (LOT-Quantum Design, Germany) with L2174-01 75 W xenon arc lamp (Hamamatsu Photonics, Japan). Optionally, 100 W halogen light source LSA132 (LOT-Quantum Design, Germany). (5) Monochromators MSH150 (LOT-Quantum Design, Germany). (6) Quartz lenses (focal length 90 mm, diameter 50 mm) (QIOPTIQ, UK). (7) Temperature-controlled cuvette holder CUV-UV/VIS-TC-ABS/FL (temperature range from -20 °C to $+80$ °C, accuracy ± 0.1 degree) (Avantes, The Netherlands) with thermostat Ministat 125 (Huber, Germany). The laser beam excites the sample from the bottom of the quartz cuvette (5 mm optical pathlength, Hellma, Germany). (8) Photomultiplier tubes R12829 with power supply module C7169, power supply DAP socket C12597-01 (Hamamatsu Photonics, Japan) and homemade current to voltage converter (20 mV/μA, min rise time 5 ns; two operational amplifiers MAX4227, Maxim Integrated, USA). (9) Deuterium and halogen light source AvaLight-DHc (Avantes, The Netherlands). (10) Digital Oscilloscopes Keysight DSO-X 4022A (two channels, 200 MHz bandwidth, Keysight, USA). (11) Data acquisition and analysis PC (Windows 10 Pro, LabView-based data acquisition program, MEXFIT program for global multiexponential fitting of experimental kinetic data, Origin 2020 Pro for graphic representation of results and spectral analysis). The PC controls all experimental equipment via USB interfaces. Additionally, it is connected to the USB Modular Multifunction Data Acquisition device (U2351A 16-Channel A/D converter 250 kSa/s and D/A converter, Keysight, USA). The analogous output of this block allows computer feedback control of the PMT1 and PMT2 detector gains during the wavelength scanning acquisition process

The signal level of the unexcited protein (pre-triggering data points) is used for the determination of V_0 and for calculating the initial noise level. Note that the noise is higher in the fast channel.

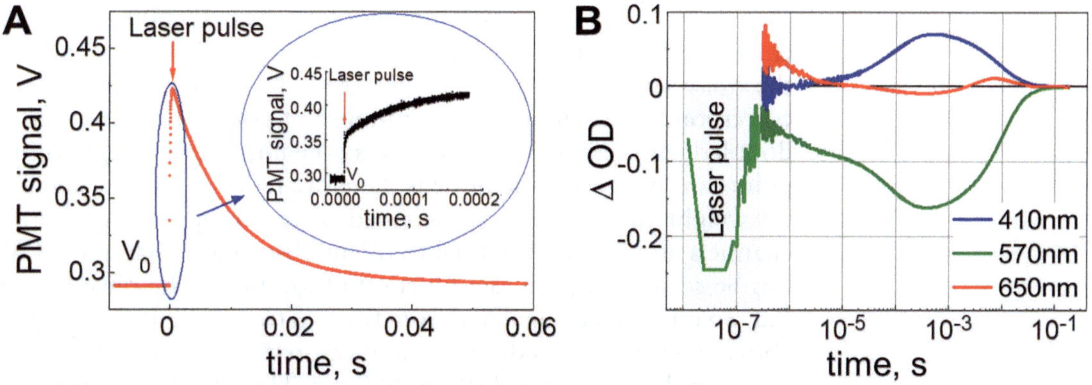

Fig. 3 (**a**) Transient changes in light intensity transmitted through the sample at 570 nm after excitation of bacteriorhodopsin (bR) by nanosecond laser pulse (540 nm, 5 ns, 8 mJ/cm^2). Red points measured by the ms – s oscilloscope; black points—measured by the ns – μs oscilloscope. Each trace is the result of averaging of 25 laser pulses. (**b**) Transient absorption changes of the bR photocycle at three characteristic wavelengths (570, 410, and 650 nm) shown versus the logarithm of time. The traces are calculated from the records of the two linear-timescale oscilloscopes as exemplified in (**a**) at 570 nm wavelength

This channel has a rise time of about 5 ns, whereas the slow channel of the PMT's current-to-voltage converter is adjusted to a respond time of about 1 μs. The reciprocal of noise is assigned as the initial statistical data weights of the ΔOD transients. These values are different for the fast and slow channels and, additionally, they depend on the probing wavelength. The weights are included in global multiexponential nonlinear least-square data fitting. Moreover, the noise of the data is further reduced by quasi-logarithmic data compression. This algorithm on each channel reduces the transient ΔOD data from the initial equally spaced ~16,000 points to ~400 points evenly distributed over a log time scale giving ~50 points per time decade. The algorithm of the data compression is illustrated in the following chart:

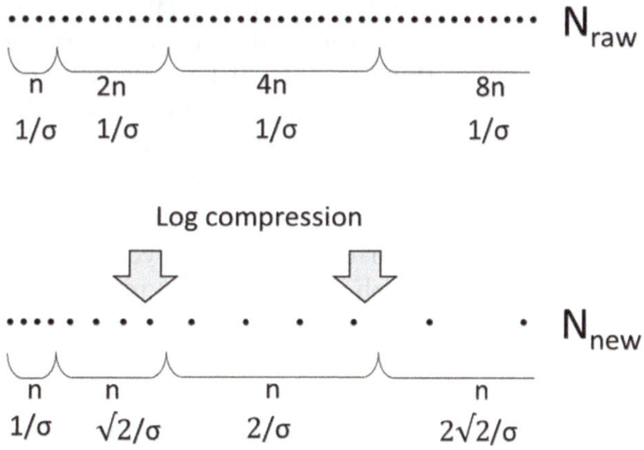

where N_{raw} is the initial number of points (~16,000 points per channel), N_{new} is the new number of points (~400 points), and σ is the noise level of the baseline. Accordingly, in each domain with correspondent doubling of the number of points, the time step is doubled with respect to the previous domain, and the neighboring points are averaged. The noise is therefore reduced and the statistical weight of the point is increased. After compression of both channels and joining the final transient absorption changes, ΔOD can be seen in Fig. 3b as a function of log time. The olive line in Fig. 3b is the trace calculated from the two traces of Fig. 3a with the above-described procedure. It was measured at 570 nm. Two more wavelengths are shown: 410 and 650 nm. These three characteristic wavelengths show the kinetics and spectral features of the bR photocycle. The time domain highlighted by gray shows the laser pulse profile detected with the fast amplifier channel of the PMT. This profile indicates that the actual instrumental response time is about 100 ns. Note that the true laser pulse duration is only 4 ns. Thus, in order to exclude the laser pulse from the analysis, we truncated the points earlier than 300 ns as it is shown at 410 and 650 nm in Fig. 3b. The whole experimental data set including the three traces in Fig. 3b is shown in Fig. 4a. This is the transient differential absorption kinetics of the bacteriorhodopsin photocycle collected on the Cycloscope. The data were obtained by sequentially scanning the wavelength from 330 to 730 nm in 10-nm steps (41 wavelengths). At each wavelength, the photocycle traces induced by 25 laser pulses (about 1 pulse/s) were averaged. The photocycle kinetics are shown in the time domain from 500 ns to 200 ms, e.g., over almost six decades in time. The red positive peak around 400 nm wavelength and few hundred microseconds of time is characteristic for the formation of the so-called M_{410} intermediate of the bR photocycle [13]. This state of bR occurs after the light-induced transient de-protonation of the retinal chromophore. The blue negative peak around 570 nm and at the same time is the depletion of the ground absorption state of bR. The amplitude of this peak of −0.17 (*see* also Fig. 3b) corresponds to ~30% of the ground state depletion under these experimental conditions. This latter value will be determined in the data analysis part below. The small positive peak around 650 nm at early times is the K_{600} intermediate,and the peak at a few milliseconds is the O_{630} intermediate of the photocycle (compare also with Fig. 3b; [1, 13]). The bR photocycle shown in Fig. 4a was measured at 20 °C. The same measurements were performed at other temperatures from 0 to 50 °C (10 °C steps, six temperature points). The temperature dependence of the bR photocycle kinetics is illustrated in Fig. 4b. Six transient absorption traces are shown at 650 nm. The slowest trace corresponds to 0 °C and the fastest to 50 °C. It is obvious that the photocycle kinetics gradually accelerate with rising

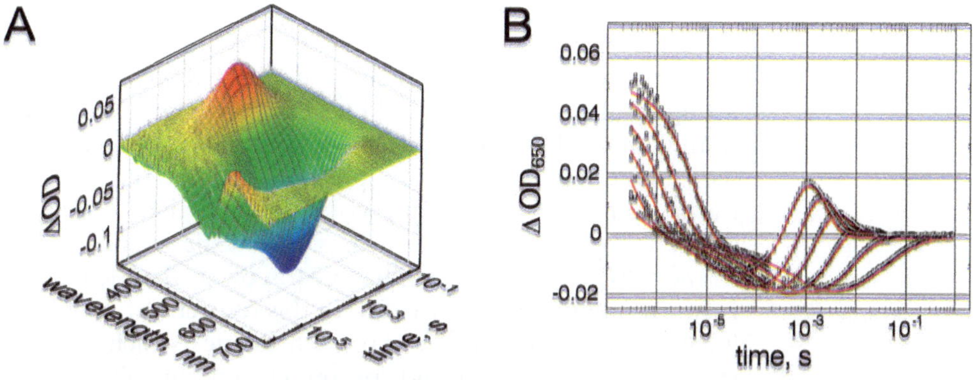

Fig. 4 (**a**) Experimentally measured differential transient absorption changes of the bR photocycle at 20 °C. Each temporal trace contains ~400 points which were obtained after quasi-logarithmic data compression of the initial 32,000 linear time points measured in the two linear time windows (16,000 points from each oscilloscope). The data were acquired at 41 wavelengths from 330 to 730 nm with 10-nm steps. Similar data sets were obtained at other temperatures (from 0 to 50 °C with 10-degree steps). (**b**) An example of the global eight-exponential fit of the data measured at 650 nm at six temperatures (black: data, red: fit)

temperature. The experimental data (black points) are shown together with eight exponential fit of the data (red solid lines).

2.3 Global Multi-exponential Nonlinear Least-Square Fitting of the Data

The data set for each temperature was independently analyzed using the global multi-exponential nonlinear least-squares fitting program MEXFIT [15]. This program was written in the late 1980s in the Fortran programming language and uses the Levenberg–Marquardt algorithm adapted for global fitting of the kinetics by a sum of exponential terms. The meaning of global fitting in our case is the obvious postulate that the reaction rate constants do not depend on the probing wavelengths but that their correspondent spectral amplitudes do. Depending on the size of the data set and available computer memory, MEXFIT is able to approximate the experimental data as a sum of up to 10 exponential terms on a set of 1000 wavelengths (or any other kinetic traces, e.g., the transient photovoltage or current records) each containing 1000 time points. This program is available for free on request with permission of the author, K.-H. Mueller. One should note that commercially available data analysis software usually includes build-in fitting functions that are appropriate for kinetics data. It even allows the global options (but for much fewer traces) and the ability to estimate the least-square of weighted residuals (i.e., include in the fit statistical weights of points). But, surprisingly, all these programs assumed that kinetic data do not contain more than three distinct reaction rate constants. That is certainly not the case for bacteriorhodopsin and most of the other photoactive proteins.

As it was shown in [1], the bR photocycle contains at least eight distinct reaction rate constants in the described time and spectral domains and over the wide temperature range. This number increases to nine at high hydrostatic pressure [10]. We show the results of fit by eight exponentials (six red lines in Fig. 4b) for six temperatures at 650 nm. A more detailed representation of the fitting results is published in [1]. Here we represent the data and results on the bR photocycle measured in 2020 using the Cycloscope. One should therefore appreciate very high reproducibility of the determined reaction rate constants and their correspondent spectral amplitudes. That article reports experiments on bR performed 30 years earlier (1990) in Moscow at the General Physics Institute and published in 1996. In this respect, the bR photocycle is a very important example of protein kinetics, which shows almost the same rate constants and spectral changes in spite of the complexity of the reaction mechanism.

The number of rate constants and their spectra were determined according the following procedure. In the MEXFIT program, the number of exponential components applied for fitting was incremented until the standard deviation of weighted residuals did not further improve. Additionally, the randomness of the 3D surface of weighted residuals and the smoothness of the temperature dependence of the derived rate constants and their amplitude spectra were used to determine the number of exponentials.

For instance, it was found that the sensory rhodopsin II from *Natronobacterium pharaonis* contains eight rate constants in photocycle [8], the chloride pump pHR from the same organism six reaction steps [9], and recently published photocycle of xenorhodopsin could be sufficiently approximated by five rate constants [3].

3 Photocycle Modeling

3.1 Model Assumptions: Irreversible Sequential Chain of Reaction Steps with Kinetically Distinct Rate Constants

After establishing the apparent rate constants and their assignment to the internal irreversible transitions of a single chain of relaxation, the amplitude spectra of the exponentials (Fig. 5a, black lines) were transformed to the differential spectra of the corresponding intermediates (Fig. 5a, red lines). An exact mathematical solution of the master equations that describe this kinetic model has been published elsewhere (e.g., *see* ref. 1). Note that one of the first photocycle models of the bR [13] also assumed the single chain of irreversible reaction steps. The authors in [13] introduced for the first time the so-called archetypical spectral intermediates of the bR photocycle and named them as K, L, M, N, and O intermediates. Accordingly, their minimal sequential irreversible kinetic model consists of five reaction steps and corresponding distinct rate constants. Over the years, considerable progress has been made on the accuracy of kinetic measurements and computer-assisted numerical analysis. This led to the discovery that the bR kinetics contains more than five distinct rate constants. Six, seven, and then eight

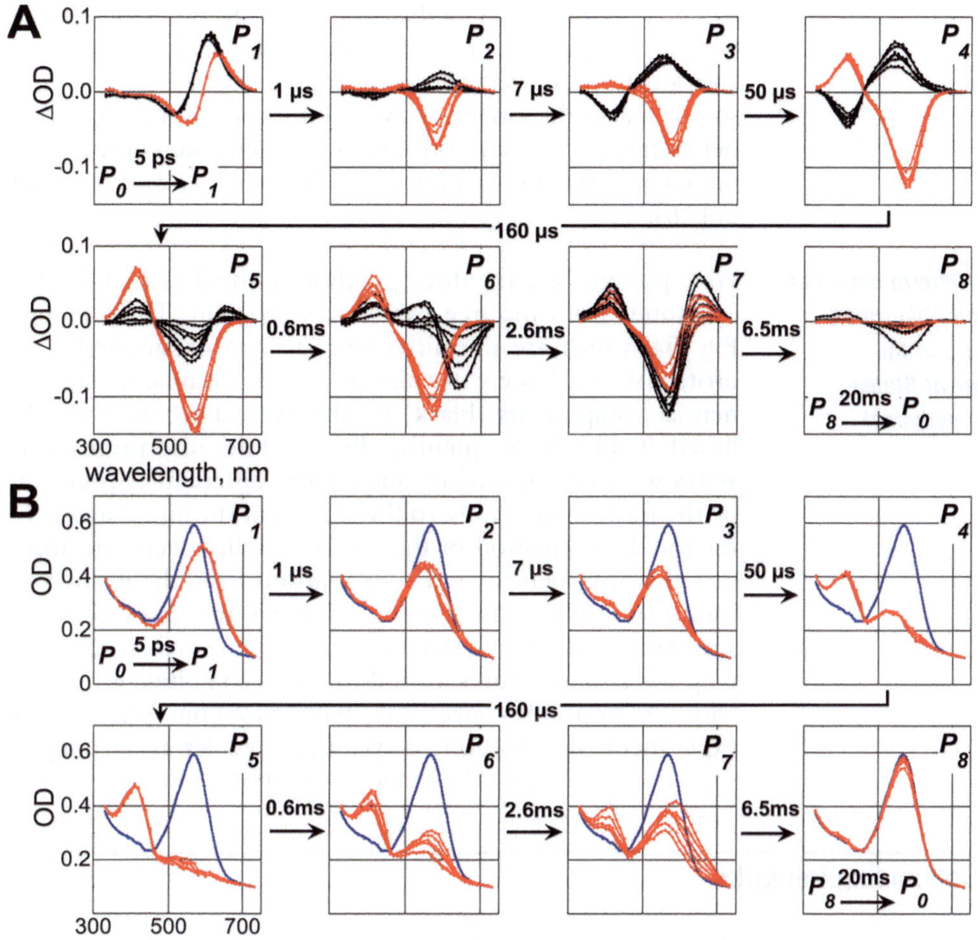

Fig. 5 The bR photocycle: (**a**) Differential spectra of amplitudes of eight exponentials at different temperatures (black), and differential spectra of seven intermediates of the bR photocycle derived from the sequential irreversible scheme of relaxation (red). (**b**) Absolute absorption spectra of eight intermediates of the bR photocycle derived from the sequential irreversible scheme of relaxation at different temperatures. The half-times of reactions corresponds to 20 °C

rate constants have been reported in the literature over the last 45 years. On the other hand, no more than five spectral intermediates were found with more precise measurements. This apparent paradox leads to the enormous modifications of the initially proposed photocycle model [13] of bR. They include parallel or branched photocycles, reversible steps, and spectrally identical intermediates (like L_1, L_2, M_1 and M_2, M_{fast}, M_{slow}, and so on). Discussion on all these variants of the bR photocycle model is beyond the scope of this chapter. We would like only to emphasize that one of the major ambiguities of the discussed variants is the assumption that each kinetically distinct transition of the photocycle should correspond to one of the above-mentioned "pure" spectral intermediates. In [1], it was shown that the apparent paradox can be solved by the concept of spectrally silent transitions without necessity of branched/parallel and/or reversible schemes. Introduction of reversible reaction steps (like K<=>L<=>M, and so on) makes the kinetic analysis almost unsolvable since the number of free parameters considerably exceeds the number of experimentally observable parameters. This problem is deeply evaluated in [16].

3.2 Determination of UV-VIS Absorption Spectra of the Transient States (Intermediates)

We represent here the direct method of the data analysis that does not suffer from the free parameters problem. The red spectra in Fig. 5a, which are the differential spectra of kinetically resolved protein states P_i, were calculated from differential spectra of exponential components (black) by the exact algebraic formula published in [1]. Subsequently, the absolute absorption spectra of states were determined by adding the differential spectra divided by the fraction of cycling to the spectra of the initial states. Criteria for the determination of the fraction cycling were the absence of negative absorbencies and contributions from the ground state to the calculated spectra. Figure 5b shows the absolute absorption spectra of eight kinetic states of bR (red lines) resolved experimentally at six temperatures from 0 to 50 °C combined for comparison with the ground bR spectrum (max. at 570 nm, blue line). In this experiment, the fraction of cycling molecules at all temperatures was about 30%. Very close value was reported in [1].

4 Concluding Remarks

We describe here briefly the method of time-resolved spectroscopy of photoactive proteins using bacteriorhodopsin as an example. In recent years, we have characterized numerous examples of retinal proteins using the Cycloscope laser flash photolysis setup and the method of data analysis described above [2–6, 14]. Experiments can be performed on photoreactions that are not necessarily cyclic. The kinetics of the photoreactions of phytochrome have been reported in [12] using an essentially similar experimental setup with some minor modifications. In [7], an example of kinetic investigation of non-photoactive proteins (actin and myosin) is shown, where the reaction of myosin with its substrate molecules ATP is triggered by laser flash photolysis of the inert precursor called caged ATP. These experiments can be done with a setup similar to the Cycloscope. Recent progress in the creation of new tunable middle infrared quantum cascade lasers opens very attractive new perspectives for the further development of time resolved spectroscopy. Expanding kinetic measurements in the mid-IR range using this new laser technique will reveal more detailed molecular processes that are responsible for the function of proteins.

Acknowledgments

We thank Richard H. Lozier for his continuing interest in our work and assistance in editing. This work is supported by Russian Science Foundation, grant 21-64-00018.

References

1. Chizhov I, Chernavskii DS, Engelhard M et al (1996) Spectrally silent transitions in the bacteriorhodopsin photocycle. Biophys J 71: 2329–2345
2. Bratanov D, Kovalev K, Machtens J et al (2019) Unique structure and function of viral rhodopsins. Nat Commun 10:4939
3. Shevchenko V, Mager T, Kovalev K et al (2017) Inward H+ pump xenorhodopsin: mechanism and alternative optogenetic approach. Sci Adv 3:e1603187
4. Kovalev K, Astashkin R, Gushchin I et al (2020) Molecular mechanism of light-driven sodium pumping. Nat Commun 11:2137
5. Kovalev K, Volkov D, Astashkin R et al (2020) High-resolution structural insights into the heliorhodopsin family. Proc Natl Acad Sci 117:4131–4141
6. Zabelskii D, Alekseev A, Kovalev K et al (2020) Viral channelrhodopsins: calcium-dependent Na+/K+ selective light-gated channels. bioRxiv 2020.02.14.949966
7. Chizhov I (2013) Encyclopedia of biophysics. In: Robers G (ed) Encyclopedia of biophysics. Springer-Verlag, Berlin, pp 765–768
8. Chizhov I, Schmies G, Seidel R et al (1998) The photophobic receptor from natronobacterium pharaonis: temperature and pH dependencies of the photocycle of sensory rhodopsin II. Biophys J 75:999–1009
9. Chizhov I, Engelhard M (2001) Temperature and halide dependence of the photocycle of halorhodopsin from Natronobacterium pharaonis. Biophys J 81:1600–1612
10. Klink BU, Winter R, Engelhard M et al (2002) Pressure dependence of the photocycle kinetics of bacteriorhodopsin. Biophys J 83: 3490–3498
11. Friedrich T, Geibel S, Kalmbach R et al (2002) Proteorhodopsin is a light-driven proton pump with variable vectoriality. J Mol Biol 321: 821–838
12. Chizhov I, Zorn B, Manstein DJ et al (2013) Kinetic and thermodynamic analysis of the light-induced processes in plant and cyanobacterial phytochromes. Biophys J 105: 2210–2220
13. Lozier RH, Bogomolni RA, Stoeckenius W (1975) Bacteriorhodopsin: a light-driven proton pump in halobacterium halobium. Biophys J 15:955–962
14. Bulavin LA, Mikhailov AE, Kuzmichev PK et al (2020) Influence of cholesterol concentration on bacteriorhodopsin photocycle. Ukr J Phys 65(9):778
15. Mueller K-H, Plesser T (1991) Variance reduction by simultaneous multi-exponential analysis of data sets from different experiments. Eur Biophys J 19:231–240
16. Nagle JF (1991) Solving complex photocycle kinetics. Theory and direct method. Biophys J 59:476–487

Chapter 9

Solid-State NMR Spectroscopy on Microbial Rhodopsins

Clara Nassrin Kriebel, Johanna Becker-Baldus, and Clemens Glaubitz

Abstract

Microbial rhodopsins represent the most abundant phototrophic systems known today. A similar molecular architecture with seven transmembrane helices and a retinal cofactor linked to a lysine in helix 7 enables a wide range of functions including ion pumping, light-controlled ion channel gating, or sensing. Deciphering their molecular mechanisms therefore requires a combined consideration of structural, functional, and spectroscopic data in order to identify key factors determining their function. Important insight can be gained by solid-state NMR spectroscopy by which the large homo-oligomeric rhodopsin complexes can be studied directly within lipid bilayers. This chapter describes the methodological background and the necessary sample preparation requirements for the study of photointermediates, for the analysis of protonation states, H-bonding and chromophore conformations, for 3D structure determination, and for probing oligomer interfaces of microbial rhodopsins. The use of data extracted from these NMR experiments is discussed in the context of complementary biophysical methods.

Key words Microbial rhodopsin, Retinal, Structure, Oligomer, Photointermediate, Solid-state NMR, MAS, Dynamic nuclear polarization, PRE, In situ illumination

1 Introduction

1.1 General Background

The discovery of proteorhodopsin as the first bacterial rhodopsin about 20 years ago has triggered a paradigm shift: Until then the extensively studied archaeal bacteriorhodopsin had served as an important "innovation engine" for the development of many biophysical methods for membrane protein research (e.g., cryo-EM, solid-state NMR, X-ray crystallography etc., see e.g. [1]). However, it was also considered an oddity of less evolutionary importance because of the rare occurrence of its extremophile host *H. salinarium*. This view has changed dramatically with rhodopsins found in all phyla of life covering a wide spectrum of functions. They represent actually the most abundant phototrophic systems known to date [2, 3]. At the heart of research efforts in this field lies the question how a rather similar structural fold with seven transmembrane helices and an all-*trans* retinal chromophore bound to a

Valentin Gordeliy (ed.), *Rhodopsin: Methods and Protocols*, Methods in Molecular Biology, vol. 2501,
https://doi.org/10.1007/978-1-0716-2329-9_9,

conserved lysine is able to yield such a diverse array of functions ranging from ion pumps via channels to sensors. Examples are proteorhodopsin (PR) working as light-driven proton pump [4], the light-driven sodium pump KR2 [5], channelrhodopsin-2 (ChR2) functioning as a light-gated ion channel [6], and Anabaena sensory rhodopsin (ASR), a photosensor for chromatic adaptation [7].

To address this question, a deep mechanistic insight on the molecular level needs to be derived coupling functional characterization with optical properties, photocycle kinetics, and 3D structures.

Solid-state NMR offers here unique opportunities since on the one hand, full 3D structure can be determined [8]. On the other hand, also very fine details usually beyond the resolution of X-ray structures such as C-C bond length distortions [9], out-of-plane twists within the retinal chromophore [10], retinal interaction with residues within the binding pocket [11, 12], or the characterization of hydrogen bonds at the Schiff base nitrogen [13] or other residues [14] can be obtained in a site-resolved way. These details are needed to understand for example counter ion interactions or color tuning. Recognizing subtle differences between rhodopsins will have a significant impact for the understanding of their functional mechanism. It became also increasingly clear that these proteins form homo-oligomers, with mainly trimers in case of archaea and predominantly pentamers in case of eubacteria [15]. These large complexes including their lipid bilayer environment can be studied directly using solid-state NMR spectroscopy. In this chapter, we describe briefly the general technical background as well as sample preparation requirements and provide details on how photointermediates can be studied, how 3D structures can be determined, and how oligomerization interfaces can be probed.

1.2 Solid-State NMR Spectroscopy

Most NMR interactions are anisotropic, i.e., orientation-dependent with respect to the magnetic field, which yields broad and complex line shapes. These interactions are averaged to their isotropic value if the molecules are soluble and tumble fast enough, which is however not the case for large, homo-oligomeric, membrane-embedded proteins such as rhodopsins. A solution is offered by magic angle sample spinning (MAS) by which these interactions are averaged through fast macroscopic sample rotation (Fig. 1a). A typical 2D NCA correlation spectrum of the uniformly ^{13}C, ^{15}N-labeled sodium pump KR2 within lipid bilayers is shown in Fig. 1b (left). Spectra are usually recorded via ^{13}C detection and resonance assignments become possible through 3D correlation experiments [16, 18, 19], which is the requirement for further structure and dynamic analysis.

In contrast to solution state NMR, direct detection of protons is challenging since they form a strong, homogeneous dipole–

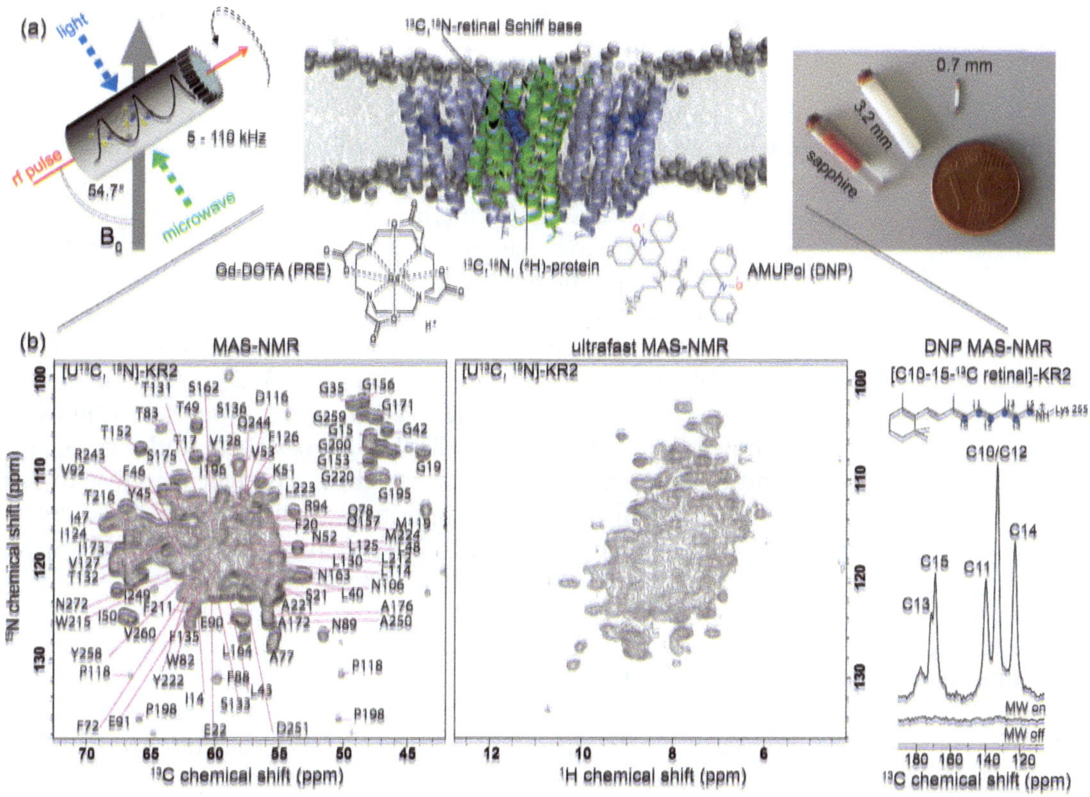

Fig. 1 Solid-state NMR on microbial rhodopsins. (**a**) MAS-NMR optionally coupled with in situ sample illumination for preparing photointermediates and microwave irradiation for DNP (left). Typically, 3.2 mm sapphire rotors are used for irradiation experiments (here filled with PR proteoliposomes). Ultrafast spinning requires even smaller rotors with just 0.7 mm diameter (right). Isotope-labeled microbial rhodopsins such as pentameric PR or KR2 are embedded within lipid bilayers. Paramagnetic compounds can be added for PRE or DNP. (**b**) Example MAS-NMR spectra of the light-driven sodium pump KR2 in DMPC/DMPA lipid bilayers: ^{13}C detected NCA spectrum (left; 850 MHz, 14 kHz MAS, 12 mg KR2, pH 8.5, 280 K); ^{1}H-detected HN spectrum using ultra-fast MAS-NMR (middle; 850 MHz, 110 kHz MAS, 0.2 mg KR2, pH 6.0, 280 K); DNP-enhanced ^{13}C spectrum of C10-15-$^{13}C_6$-retinal within KR2 showing a 50-fold signal enhancement (right; 263 GHz/400 MHz, 8 kHz MAS, 10 mg KR2, AMUPol, 110 K) [16, 17]

dipole coupling network leading to severe line broadenings under conditions of non-isotropic or reduced molecular mobility. Shrinking the outer diameter of the MAS rotors down to 0.7 mm (Fig. 1a, right) enables sample rotation rates of up to 110 kHz leading to improved ^{1}H resolution as demonstrated here for a ^{1}H-detected NH-MAS NMR spectrum of uniformly ^{15}N-labeled KR2 (Fig. 1b, middle). Resonance assignment using fully protonated lipid-embedded proteorhodopsin in such a small rotor has already been achieved [20]. For proton detection at slower MAS rates, additional sample deuteration would be needed [21]. Despite the significant reduction in sample volume, the detection sensitivity of this approach is good due to the high intrinsic sensitivity of protons and because of the high efficiency of MAS probe heads with small radio frequency coils.

A fundamental increase in detection sensitivity is offered by dynamic nuclear polarization (DNP) [22, 23]. A non-Boltzmann nuclear magnetization is created by utilizing stable biradicals such as AMUPol [24] as polarizers and by strong microwave irradiation during the MAS NMR experiment (Fig. 1a, left, middle). The microwave frequency has to match the electron Larmor frequency at the used magnetic field (e.g., 263 GHz/400 MHz at 9 T). As a result, significant signal enhancement can be achieved. Values of 50–80 have been typically reported for rhodopsins enabling a 2500–6400 times faster data acquisition [13, 25]. An example for the signal enhancement of the ^{13}C resonances of retinal within KR2 is shown in Fig. 1b (right) [17]. DNP works best under cryogenic conditions on frozen samples, which is a perfect match with the requirements for trapping of photocycle intermediates. To create such intermediates, in situ photo irradiation within the NMR spectrometer is used (Fig. 1a, left) as described below.

2 General Sample Preparation Requirements

For MAS-NMR, the sample of interest has to be enriched with ^{13}C and ^{15}N. Such isotope enrichment is possible for the retinal and the protein itself. For retinal, numerous labeling schemes involving single, double, multiple, and even uniform ^{13}C labeling have been used all based on in vitro synthesis [9, 11, 26–29]. The preparation of U-^{13}C retinal has even been shown biosynthetically using a prototrophic *E. coli* strain which was transformed with plasmids expressing retinal-producing enzymes [30]. The labeled retinal needs to be incorporated into the opsin of interest. This can either be achieved by adding it directly to the growth medium, which is however material intensive, or by adding it to the opsin in the non-purified, crude [16, 31] membrane fraction, or during solubilization [25]. In cases in which the opsin cannot be stably expressed, a bleaching step can be used at that stage for exchanging unlabeled with labeled retinal [32].

Many of the novel microbial rhodopsins discovered since 2000 can be conveniently expressed in *E. coli*, which offers a multitude of isotope labeling options since *E. coli* can be grown on a minimal medium in which ^{13}C- and ^{15}N-enriched carbon and nitrogen sources (glucose, glycerol, ammonium) can be used.

The type of labeling scheme depends on the experimental aims and the choice of the MAS-NMR approach. Uniform ^{13}C- and ^{15}N-labeled samples are usually used for the structure and dynamic characterization of microbial rhodopsins in their ground states. These experiments are carried out on non-frozen samples at high magnetic fields resulting usually in high spectral resolution so that a large number of site-resolved data can be obtained simultaneously. In addition, ultra-fast MAS-NMR for direct ^{1}H detection often requires deuterated samples in which exchangeable sites have been re-protonated. DNP-enhanced MAS-NMR is performed on frozen samples and often on lower magnetic fields, which results in less well-resolved spectra. Therefore, in order to reduce spectral complexity, often amino-acid-selective labeling is applied (for a recent review on such labeling options in *E. coli*, *see* [33]). Even a perspective for segmental labeling has been demonstrated for proteorhodopsin, which would further reduce spectral complexity [34]. For eukaryotic microbial rhodopsins such as ChR2 and others, ^{13}C as well as ^{15}N isotope labeling options in *P. pastoris* have been successfully demonstrated, which significantly extends the applicability of the approaches shown here [31, 35, 36].

In contrast to bacteriorhodopsin, which can be readily used for MAS-NMR in its native purple membrane, all other microbial rhodopsins are usually purified from the membrane and then reconstituted into lipid bilayers to form proteoliposomes. It is however noteworthy that the possibility to record MAS-NMR spectra of ASR directly within the inner *E. coli* membrane without further purification has been demonstrated [37]. For the NMR measurement, the proteoliposomes/membrane fractions are collected into MAS rotors by a centrifugation step.

Depending on the experimental strategies, further sample modifications might be helpful. For example, the data acquisition time can be shortened by "doping" the samples with paramagnetic compounds such as Gd-DOTA (Fig. 1a), which does not interact with the rhodopsins or the bilayer but reduces the spin-lattice relaxation time of protons efficiently within the whole sample as demonstrated for green PR [38]. For DNP, as mentioned above, stable biradicals have to be added to the sample as polarization source. In case of BR, purple membrane patches are dispersed in "DNP juice" (60% D_8-Glycerol, 30% D_2O, and 10% H_2O (v/v/v) with 10 mM AMUPol) [13], while proteoliposome pellets are best incubated with a top layer of 30% D_8-Glycerol, 60% D_2O, and 10% H_2O (v/v/v) with 20 mM AMUPol, which is then removed before the experiment (*see* [39]).

Labeling of cysteines by stable radicals offers additional options to obtain complementary long-range restraints via paramagnetic relaxation enhancement (PRE) [38, 40] and for site-selective DNP or spectral editing [41].

3 Chromophore and Photocycle Intermediate States

Solid-state NMR has been extensively used for the analysis of photointermediate states of microbial rhodopsins (for an overview, *see* [39, 42]). These studies focus primarily on the active site, i.e.,

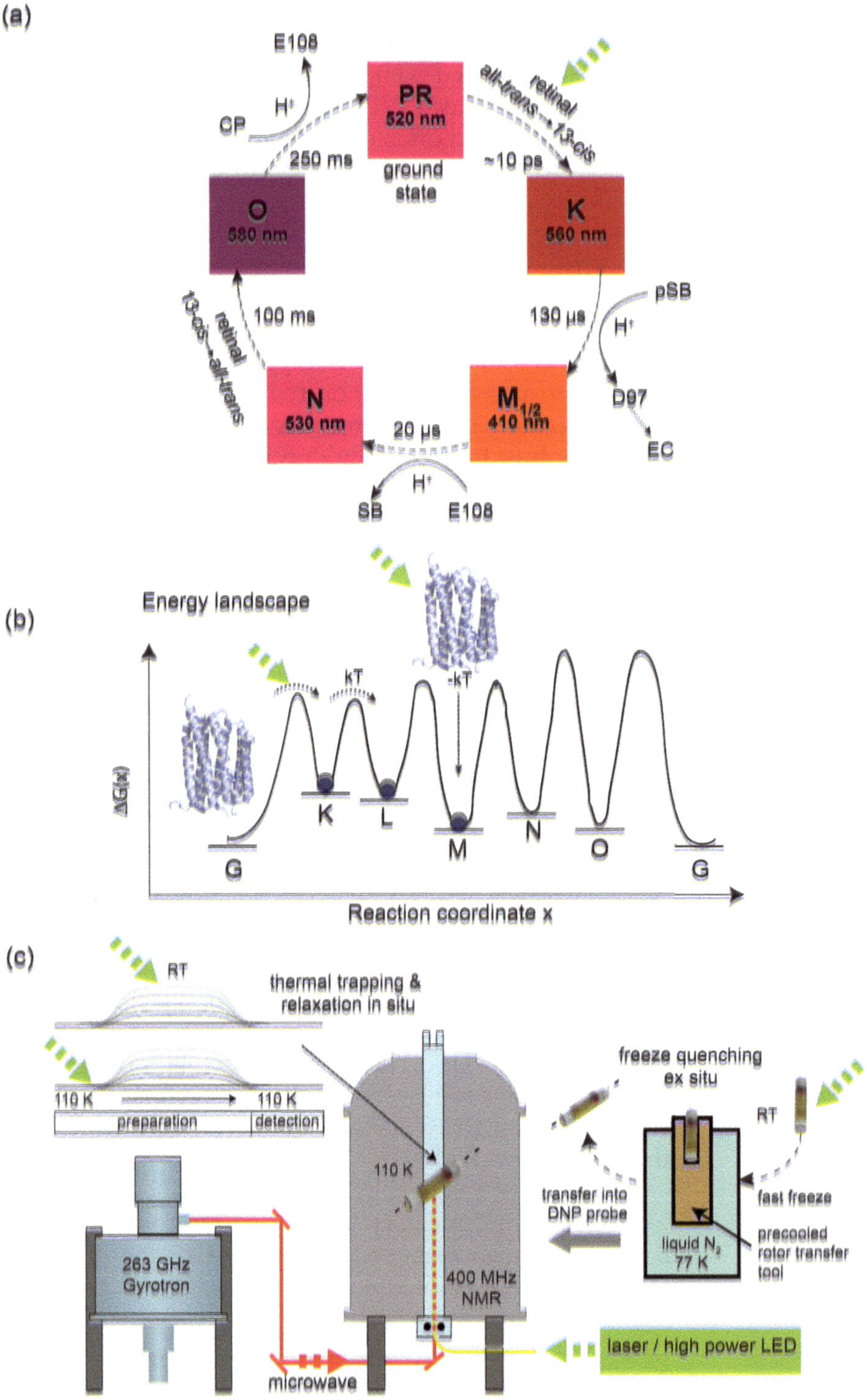

Fig. 2 Catching photointermediate states: (**a**) Schematic photocycle of green PR [43]. Photointermediates are characterized by their absorption maximum and their lifetime. (**b**) They can be trapped by a suitable choice of illumination and cooling conditions. For example, the K-state can be created by irradiating and measuring the sample at low temperature (110 K). Subsequent intermediates could be reached thermally or by illumination at higher temperatures followed by cooling. (**c**) Experimental setup for cryo-trapping of photointermediates for

3.1 Light-Induced Cryo-Trapping and NMR Spectroscopy

the retinal Schiff base (RSB) complex and its counter ion interactions and rely on trapping of photointermediates after photoirradiation under MAS-NMR conditions.

A typical microbial rhodopsin photocycle is shown in Fig. 2a for proteorhodopsin [43]. In the ground state, the chromophore is in the all-*trans*, 15-*anti* conformation. An exception is found in dark-adapted BR with an equilibrium between all-*trans*,15-*anti* and 13-*cis*,15-*syn* retinal [44]. Upon light absorption, the retinal switches from an all-*trans*,15-*anti* to 13-*cis*,15-*anti* conformation, which leads to the K state. This process is fast, and initial evidence exists that also the protein environment in some rhodopsins starts to respond on the same timescale [45]. The absorption maximum is red shifted with respect to the ground state. Next, the protein prepares for the Schiff base deprotonation step which is, in some microbial rhodopsins, associated with the formation of a kinetically distinct L-intermediate. Subsequently, the strongly blue-shifted M-intermediate with the deprotonated Schiff base forms. Its reprotonation usually causes the formation of the N-state and reconversion of the chromophore from 13-*cis* to the all-*trans* creates the red-shifted O-state followed by relaxation back to the ground state.

In order to derive mechanistic models for the multiple functions fulfilled by microbial rhodopsins, it is essential to precisely understand structure, dynamics, and interactions within and between RSB and opsin in all intermediates. Solid-state NMR can provide unique information on the chromophore conformation, chromophore–opsin interactions, and rearrangements within the opsin provided that these intermediates can be trapped.

In some cases, certain intermediates exhibit a much longer half-life than the other states. Such a situation offers the opportunity for *stationary trapping* of these states by continuous light irradiation since the production rate exceeds the relaxation rate. Examples are the extremely long-lived sensory rhodopsin II M-state (for a recent review on solid-state NMR on the SR II photocycle, *see* [42]) and the desensitized state in ChR2 [31]. Another example is the elongation of the lifetime of certain photointermediate states by mutations as shown for the proton donor mutant E108Q in PR which enables enrichment of the M-state [25]. In the latter two examples, the stationary generated states were then trapped using a freeze quench approach. In all other cases, specific schemes for light-induced cryo-trapping have to be applied and have been especially well worked out for bacteriorhodopsin [13]. Exact protocols depend very much on the specific protein characteristics (optical

Fig. 2 (continued) DNP-enhanced solid-state NMR. Using thin films of proteoliposomes within sapphire MAS rotors ensure sufficient light penetration. Lasers or high-power LEDs are suitable light sources. Illumination takes place within the spectrometer for thermal trapping/relaxation approaches or outside when utilizing freeze quenching. DNP-enhanced MAS NMR spectra are usually recorded at low temperature (90–110 K)

absorption properties, intermediate half-life, intermediate populations, thermal equilibria between intermediates), but the general approach can be illustrated based on the schematic energy landscape for the different intermediates in Fig. 2b: The K-state can be generated by illuminating the sample at low temperatures (around 100 K), which prevents its decay. Generally, a mixture between ground and K-state is obtained because absorption maxima overlap and K is converted back. By a subsequent rise in temperature, further intermediates can become accessible through the additional thermal energy ("thermal relaxation"). The intermediates trapped in this way depend very much on the protein-specific energy landscape and the particulars of the applied thermal relaxation protocol. Alternatively, sample illumination with the intermediate-specific wavelength at a suitable temperature can be used ("thermal trapping"). For example, in case of light-adapted BR, the K-state is trapped by green light at 90 K, the L-state by red light at 150 K, the M-state by green light at 210 K, and the N-state by red light at 240 K [13]. The duration of illumination also depends on the sample (optical density) and ranges between minutes and hours. This approach can result in an almost complete elimination of ground state populations in the sample. However, phenomena such as blue-light quenching in PR create an additional pathway back to the ground state [46] or photocycle branching can result in side product usually not observed using the thermal relaxation approach, which has to be considered when designing trapping experiments for a particular rhodopsin.

These trapping schemes can be implemented into the MAS-NMR experiment as shown in Fig. 2c. First of all, sufficient light-penetration into the proteoliposomes within the MAS rotors has to be achieved. Usually, transparent MAS rotors made of sapphire are used (Fig. 1a, right). For in situ illumination, the MAS NMR probe head is modified with a suitable multimode fiber to couple light during the MAS-NMR experiment directly into the sample. Lasers as well as high power LEDs are suitable light sources. Instead of optically dense proteoliposome pellets, thin proteoliposome films spread on the inner surface of the MAS rotors turned out to enable a good trapping efficiency [31].

Highest sensitivity is a crucial factor especially for the detection and analysis of less-populated intermediates. Dynamic nuclear polarization in combination with in situ sample illumination offers here unprecedent possibilities. These experiments are fairly complex as light, microwave, and multiple rf-frequency channels have to be coupled into the sample, but the significant gain in sensitivity enables experiments, which could not be carried out by conventional MAS-NMR.

DNP experiments are carried out usually in the temperature range between 90 and 110 K, which is also a perfect match for cryotrapping of photointermediates. The temperature variation/

illumination schemes described above always include a detection step at the end which is carried out in this temperature range (Fig. 2c). In situ illumination is certainly the preferred and most elegant way for creating photointermediates. There are however cases in which certain intermediates can only be prepared under continuous illumination of non-frozen samples followed by trapping through freeze-quenching. This step cannot easily be performed in situ as the cooling rate within the magnet is not fast enough. Therefore, freeze-quenching outside by dropping the MAS rotor into liquid nitrogen (Fig, 2c) followed by fast transfer into the pre-cooled MAS probe head has to be performed. An improved method for rapid freeze-quenching has recently been demonstrated for DNP-enhanced solid-state NMR applications [47], which might also help to extend the range of accessible photointermediates and to improve trapping efficiency.

3.2 The Retinal-Schiff Base Complex During the Photocycle

The first step in understanding the optical properties of microbial rhodopsins and the link between photocycle and their ion pumping activity is to elucidate the chromophore structure since it transfers the absorbed photon energy into the protein leading to protonation/de-protonation steps which couples back to the chromophore during the photocycle.

It is therefore helpful to summarize briefly key NMR parameters which define isomerization and protonation state of the RSB as shown in Fig. 3a. The isomerization from all-*trans* (1) to 13-*cis* (2) releases the steric interaction between C20 and C15 and creates a new steric contact between C12 and C15 [48, 49]. Therefore, chemical shift changes are observed as C20 becomes more deshielded and C12 more shielded. The C12-C15 or C20-C15 distances can also be used to differentiate both isomers by determining their dipole-dipole couplings. The steric interaction between C14 and Lys-Cε in 15-*syn* retinal (3) leads to an increased shielding of both atoms [50] and shorter distance between them compared to 15-*anti* [51]. In the ground state, the all-*trans* retinal Schiff base is protonated and positively charged (4). The charge is partially delocalized to the C13 atom (5). Deprotonation removes this charge (6) and results in a very large deshielding of the nitrogen by more than 100 ppm [52] and a significant shielding of C13. In microbial rhodopsins, the chromophore is found in its 6-s-*trans* form, whereas in visual rhodopsins, a skewed 6-s-*cis* conformation is found (7). A planar 6-s-*cis* conformation is unfavorable because of a steric clash between C18 and C8. The skewing of the ring weakens the conjugation of the C5 double bond with the π-system resulting in a shielding of C5 [48].

Besides these well-established chemical shift markers, numerous additional parameters can be extracted, which can define the RSB conformation to high precision. The Schiff base ^{15}N chemical shift depends on the C15-N torsion but also to a large extent on the

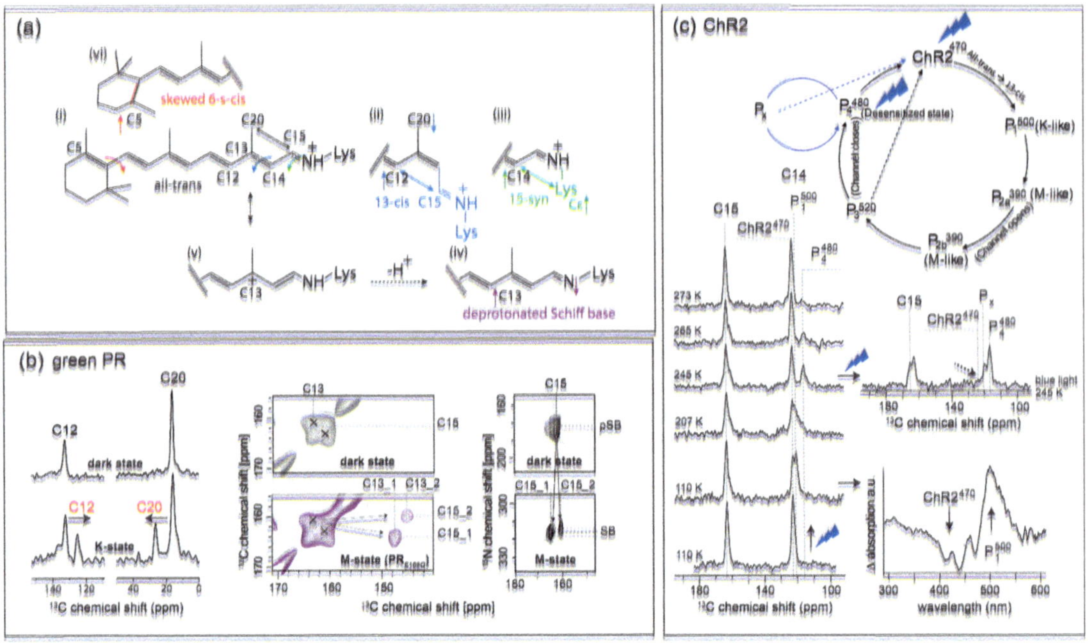

Fig. 3 NMR readouts for the conformation of the retinal Schiff base chromophore and example spectra. (**a**) Key NMR parameters for the conformation of the retinal Schiff base chromophore in rhodopsins. Increased shielding (↑), deshielding (↓), and relevant distances (↔) are indicated (see text for further details). These changes are in the range of 5–15 ppm for shielding/deshielding. Distances change by around 0.8 Å. (**b**) Trapping green PR photointermediates: K-state (left) [11], M-state (middle, right) [25]. (**c**) Trapping ChR2 photointermediates [31]: Schematic photocycle of ChR2 (top). The late desensitized state P_4^{480} shows a strong shielding of ^{13}C14 indicative of a 15-*syn* conformation. The cryo-UV-vis spectrum recorded after illumination at 150 K shows as an additional control ground state depopulation and K-like increase

strength of the H-bond in which it might be involved, which is usually to the counter ion and which again correlates with the optical absorption maximum of the particular rhodopsin. The H-bond character can be further defined by determining the N-H bond length through dipolar distance measurements and by determining the proton chemical shift (see example below). Bond lengths distortion along the polyene chain can be determined by measuring C-C dipole couplings of directly bonded carbons [9, 53], while drilling along the chain is accessible via H-C-C-H torsion angle measurements [25, 54].

The effects of changes in the chromophore conformation are shown in Fig. 3b for green PR in the K- and M-state. Illumination and trapping at 110 K leads to an additional shielding of C12 and a deshielding of C20 [11, 25] indicative of the transition from 13-*anti* to 13-*cis*. Upon M-state trapping, a large additional shielding of C13 is observed, and the signal splits into two populations arising from two trapped M-states. The Schiff base nitrogen resonance shifts from 181 ppm in its protonated form to 312 and 316 ppm upon deprotonation in both M-states [25].

A combined thermal trapping/relaxation approach is illustrated in Fig. 3c for the light-gated ion channel ChR2 [31]. In the dark state, the retinal is solely in the all-*trans*, 15-*anti* form. Upon illumination at 110 K, the K-like state P_1^{500} is formed. Cryo-UV/VIS spectroscopy can be used as an additional control for correct trapping (Fig. 3c). Further thermal relaxation steps show its decay and the gradual rise of the P_4^{480} state, which is the desensitized state after channel closure. C14 is additionally shielded indicative of the formation of a 15-*syn* conformation. Interestingly, illumination of P_4^{480} de-populates the ground state $ChR2^{470}$ and leads to another photoproduct (P_x), which has not been observed by optical spectroscopy.

Initially, retinal extraction experiments followed by HPLC analysis were the standard way to determine the chromophore confirmation in retinal proteins. However, the invasive nature of this method led to unreliable results. In addition, the information on the 15-bond is lost, and it is rarely possible to analyze photo-intermediate states. Vibrational spectroscopy and especially resonance Raman spectroscopy is much more powerful and noninvasive. However, band assignment is not trivial. The vibrational bands are superpositions of many modes, and it is difficult to obtain localized information. In contrast, assignment of NMR chemical shifts is often straightforward, especially with selectively labeled retinal, and provides very localized information. The observed chemical shift changes due to bond-isomerization are large and easy to interpret. Solid-state NMR spectroscopy should therefore be the method of choice to analyze the chromophore conformation.

3.3 Probing-Specific Residues During the Photocycle

The different trapping methods cannot only be used to follow the RSB throughout the photocycle but can be expanded toward monitoring the protein response in a site-resolved manner (by introducing suitable isotope labeling schemes) as illustrated for two specific cases below.

A seminal example for the power and importance of the methods described above is a recent DNP study on the unknown primary proton transfer step in BR [13]. The aim was to discriminate between proton transfer pathways put forward by QM/MM simulations, which either involve a direct transfer from SB to Asp85, a diversion via Asp212, Water 402 to Asp85 or a relay from SB via Thr89 to Asp85 [55]. The issue could be solved by determining the Schiff base ^{15}N and ^{1}H chemical shifts and the ^{15}N-^{1}H distances throughout the photocycle. Suitable trapping conditions have been established for almost all photointermediates in bacteriorhodopsin. In BR, the ^{15}N chemical shifts of the intermediates are well known and can be used to identify the trapped states in the spectra. Proton chemical shifts were determined via 2D ^{15}N-^{1}H HETCOR and distances measured by DIPSHIFT experiments. Surprisingly, the de-protonated nitrogen in the M_0 state shows a through-space correlation to an alcohol proton within H-bond distance (1.45 Å) assigned to Thr89 as proton donor. ^{13}C-^{13}C through-space correlation experiments also establish a closer interaction between Thr89 and the primary proton acceptor Asp85 the L- and M_0-state, i.e., prior and after the proton transfer step. In addition, the distance between the SB nitrogen and Asp85-Cγ is shortened in M_0 while the distance to Asp212-Cγ does not change during the photocycle, which reflects helices C and G moving closer to each other before the proton transfer step and relaxing afterwards. These data clearly show that the proton is indeed relayed to Asp85 via Thr89. It is important to point out that such fine details are difficult to obtain by X-ray crystallography.

In another case, selective DNP based on methyl-labeling was used to monitor changes in the retinal interaction with green PR [11]: The signal of the ^{13}C nuclei in the retinal C20 methyl group could be selectively increased by transferring the DNP-enhanced proton magnetization from the methyl protons (though hetNOE). The created carbon magnetization was then transferred (by spin diffusion) to ^{13}C-labeled residues within the retinal binding pocket in order to probe packing and conformational changes upon light-induced K-state formation. It could be shown that the C20 packing against the opsin is tighter in the K-state compared to the ground state providing experimental evidence for the role of C20 in the first steps of the photocycle. In contrast, the bulky residues in the binding pocket did not react upon the initial retinal isomerization.

Light-induced cryo-trapping of selectively labeled samples also enabled the discovery of photocycle-dependent conformational changes at PR protomer interfaces as highlighted below.

4 3D Structure Determination of Microbial Rhodopsins

Solid-state NMR not only enables hypothesis-driven experiments, as outlined above on photointermediates, but also offers the possibility of complete 3D structure determinations of membrane proteins. Although rather challenging, this approach is attractive despite tremendous progress in X-ray crystallography and cryo-EM, because it is accessible to full oligomeric rhodopsin complexes within lipid bilayers at native-like pH and salt concentrations and under non-freezing conditions. A milestone example has been the structure determination of Anabaena sensory rhodopsin (ASR) [8]. ASR mediates signal transfer via a small soluble cytoplasmic protein between membrane and DNA in order to regulate the expression of proteins involved in photosynthesis and circadian rhythm [7]. In the following for the case of ASR, all necessary steps needed for the 3D structure determination of membrane proteins are outlined as summarized in Fig. 4.

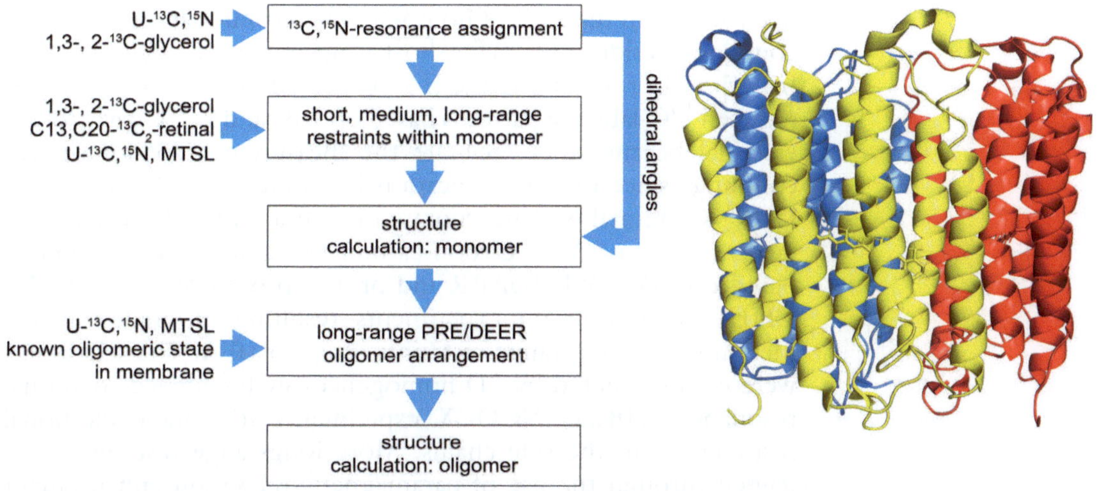

Fig. 4 Determination of the trimeric 3D structures of Anabaena Sensory Rhodopsin (ASR) within lipid bilayers by solid-state NMR [8]. The flowchart represents necessary samples and required experimental steps. The 3D structure shown here was refined by PRE and DEER long-range constraints (5UK6) [40]

4.1 Resonance Assignment

The first prerequisite for a successful NMR structure calculation is a full ^{13}C, ^{15}N NMR resonance assignment. For resolving all ^{13}C and ^{15}N resonances, high magnetic fields (>700 MHz) and a set of multidimensional (2D-4D) correlation experiments are required. Obtaining highly resolved spectra depends on sample homogeneity and intrinsic protein dynamics. Typically, a set of 3D spectra such as NCACX, NCOCX, and CONCA are recorded by which a sequential assignment along the backbone and into the side chains can be achieved. In case of ASR, three samples were prepared. Uniform ^{13}C and ^{15}N labeling allows assigning already a large set of resonances. Remaining spectral overlap was dealt with by using 1,3- and 2-^{13}C glycerol as a carbon source, which results in a reduced ^{13}C-labeling pattern [56]. Often, additional labeling schemes are used to reduce spectral overlap, which include ^{13}C, ^{15}N labeling of only few particular amino acid types (forward labeling) or suppress labeling of certain amino acid types (reverse labeling) or by using specifically labeled amino acids (selective labeling).

4.2 Distance Restraints

The obtained chemical shifts already provide dihedral angle restraints, which define secondary structure regions within the protein [57, 58]. For a 3D structure, a large number of additional internuclear distance restraints are needed. Furthermore, the arrangement of the protomers within the oligomeric complex and the oligomerization interfaces have to be resolved. In addition, in case of microbial rhodopsins, special attention has to be drawn to protein-chromophore contacts which significantly stabilize these seven-helical transmembrane proteins. Therefore, as many short-, medium-, and long-range restraints as possible need to be determined on the basis of homo- and heteronuclear dipole-dipole couplings (residue indices for short range $|i - j| = 2$, medium range $|i - j| = 3$ or 4, long-range $|i - j| \geq 5$). For ^{13}C-^{13}C distances, reduced ^{13}C-labeling patterns based on 1,3- and 2-^{13}C glycerol are essential because they facilitate the measurement of long-range couplings since dipolar truncation is avoided. In addition, C13, C20-$^{13}C_2$-retinal was incorporated to obtain retinal-protein contacts. In case of ASR, carbon-carbon restraints were determined based on 2D PDSD/DARR and proton-proton restraints by 2D CHHC spin diffusion experiments resulting in predominantly medium- and long-range restraints of up to 10 Å. These data sets were complemented by 3D homogeneously broadened rotational resonance (HBR2) NCOCX experiments to obtain additional restraints from the side chains. More long-range restraints were gained through the use of paramagnetic relaxation enhancement (PRE). This experiment requires site-specific protein labeling, often chemically via cysteines, with stable paramagnetic compounds such as nitroxides (MTSL). The dipolar interaction between the unpaired electron in these probes with ^{13}C and ^{15}N nuclei within the protein is long range (up to 15–20 Å) and results in shortened T_1 and T_2 relaxation times. The latter represents itself in a distance-dependent line broadening and hence reduced peak intensity (r^{-6}) [59, 60]. Such long-range constraints are extremely valuable for 3D structure determination in both solution- and solid-state NMR. They have been used here to refine the protomer structure and to define the oligomer interface and the oligomer arrangement (see subsection below) [61].

4.3 Structure Calculation

The experimental data (chemical shifts/dihedral angles, distance restraints) need to be included into a structure calculation algorithm. In case of ASR, CNS 1.21 and ARIA 2.3 were used [62, 63]. First, the structure of the ASR monomer was determined by excluding inter-protomer restraints (previously determined by PRE), which converged into a monomeric, heptahelical structure with a backbone r.m.s.d. of 0.8 Å [8]. A further refinement was carried out using inter-protomer PRE as well as EPR double electron-electron resonance (DEER) restraints, by which the native trimeric structure of ASR was mapped (see subsection below) [40]. Knowledge about the oligomeric order itself was available through a wealth of biochemical data.

4.4 Differences Between the NMR and X-Ray Structures of ASR

The refined solid-state NMR structure shows in general good agreement with the X-ray structure [64] in terms of helical packing around the Schiff base. Conserved residues in the retinal-binding pocket as well as in the conserved cytoplasmic cluster of polar residues, responsible for the ASR photosensory function, are found in the NMR structure at positions very close to those in the crystal structure. There are however three key differences between both structures: (1) The NMR structure in agreement with other data is a lipid-stabilized homo-trimer with helices A, B and D′, E′ facing each other. In crystals, ASR forms undulating layers of dimers, stabilized by hydrophobic interactions between only helices D of two monomers and with antiparallel oriented retinals. (2) A beta-hairpin structure in the periplasmic BC loop present in the NMR structure, a conserved structural feature among microbial rhodopsins, seems disordered in the X-ray structure. Furthermore, the cytoplasmic sites of helices A and E are tilted away from the center of the seven α-helical bundle in the NMR structure, but do not show this tilt in the crystal structure, which could be caused by the different oligomeric states and protein-protein contacts within the crystal. (3) The NMR data show clearly 100% all-*trans* retinal, whereas the crystal structure suggests a mixture between all-*trans* and 13-*cis*.

It is likely that for a detailed understanding of the protein beyond the retinal binding pocket, for example, for phenomena involving the oligomeric interface, the knowledge of the correct protein orientation and interface as given by the solid-state NMR structure is crucial.

4.5 Probing the Conformational Dynamics of Microbial Rhodopsins by Solid-State NMR

A resonance assignment is not only a starting point for a 3D structure determination but provides also the foundation for probing conformational dynamics of microbial rhodopsins within lipid bilayers.

A relatively straightforward approach is offered by H/D exchange experiments. Replacing an exchangeable amide proton by a deuteron will reduce the NMR intensities of the corresponding amide nitrogen in the protein backbone. The deduction in peak intensity depends on the H/D exchange rate, which is affected by the conformational dynamics of the protein. In this way, differences in conformational equilibria under dark and light conditions can be resolved as shown for ASR [65]. Another possibility is to use dipole-dipole couplings as dynamic reporters. The strong heteronuclear N-H or C-H dipole couplings of directly bonded nuclei will be scaled down in the presence of dynamics. This will affect ^{13}C and ^{15}N peak intensities as shown for SR II and proteorhodopsin [66, 67] and enables to discriminate between rigid and mobile regions within the protein (e.g., helices or rigid loops vs. flexible loops). The exact measurement of these dipole couplings also enables to define site-resolved order parameters within the protein as shown for ASR and SR II [67, 68].

Probing alterations in conformational dynamics are especially interesting in systems, which form functional complexes with subunits as for example in case of SR II. Its interaction with the transducer was monitored based on chemical shift perturbations and complemented by the analysis of peak intensities upon complex formation and light activation [67].

5 Oligomer Interactions

Membrane proteins often form homo-oligomers in lipid bilayers. The view on microbial rhodopsins was shaped for decades by the most-studied bacteriorhodopsin, which forms trimers packed in hexagonal, two-dimensional crystalline arrays [69]. With the discovery of many new microbial rhodopsins at the beginning of the century, evidence started to appear that especially eubacterial rhodopsins form larger oligomeric complexes.

In an attempt to prepare 2D crystals of green proteorhodopsin [70], large "donut"-shaped structures could first be observed by cryo-EM, which were then resolved as hexamers by AFM [71]. Interestingly, the protein assembled into a pentameric state in non-crystalline membrane areas in these samples, which was then confirmed by native MS and blue native-PAGE analysis as the predominant form of green PR in DDM and after reconstitution into lipid bilayers [72, 73]. A global picture emerged with a systematic study using a novel high-speed AFM method on a number of microbial rhodopsins which revealed that indeed eubacterial rhodopsins predominantly form pentamers while archaeal rhodopsins exist as trimers within the membrane [15]. However, no unified interaction pattern yielding to such complexes could be identified. These finding raise the question whether the observed homo-oligomeric assemblies are of functional relevance and whether the corresponding oligomer-interfaces could be analyzed in terms of functionally and/or structurally relevant interactions. Such data can indeed be obtained by solid-state NMR as outlined below.

5.1 NMR on Mixed-Labeled Oligomers

The possibility to prepare mixed-labeled complexes offers an attractive opportunity for visualizing cross-protomer contacts by NMR. The basic idea is illustrated in Fig. 5a for the green PR pentamer [72]: The oligomer can be assembled from differently labeled monomers (e.g., using U-^{13}C- and U-^{15}N-PR) resulting in a mixed-labeled pentamer. Interactions across the interface can now be measured by recording through-space ^{13}C-^{15}N dipolar correlation spectra to visualize cross-protomer side chain–side chain contacts. To achieve this goal, suitable conditions have to be found by which the oligomerization can be controlled in a reversible fashion. This is very much protein-specific. For example, the monomeric state of *Gloeobacter* rhodopsin can be controlled by pH [74]. Another possible parameter is the choice of detergent [75]. PR forms mainly pentamers in dodecyl β-maltoside (DDM) and remains in that state after reconstitution into lipid bilayers [72]. However, solubilization, e.g., in octyl glucoside (OG) or Triton X-100 (TX-100), leads mainly to monomers. Therefore, solubilized ^{15}N-PR and ^{13}C-PR monomers could be mixed in a

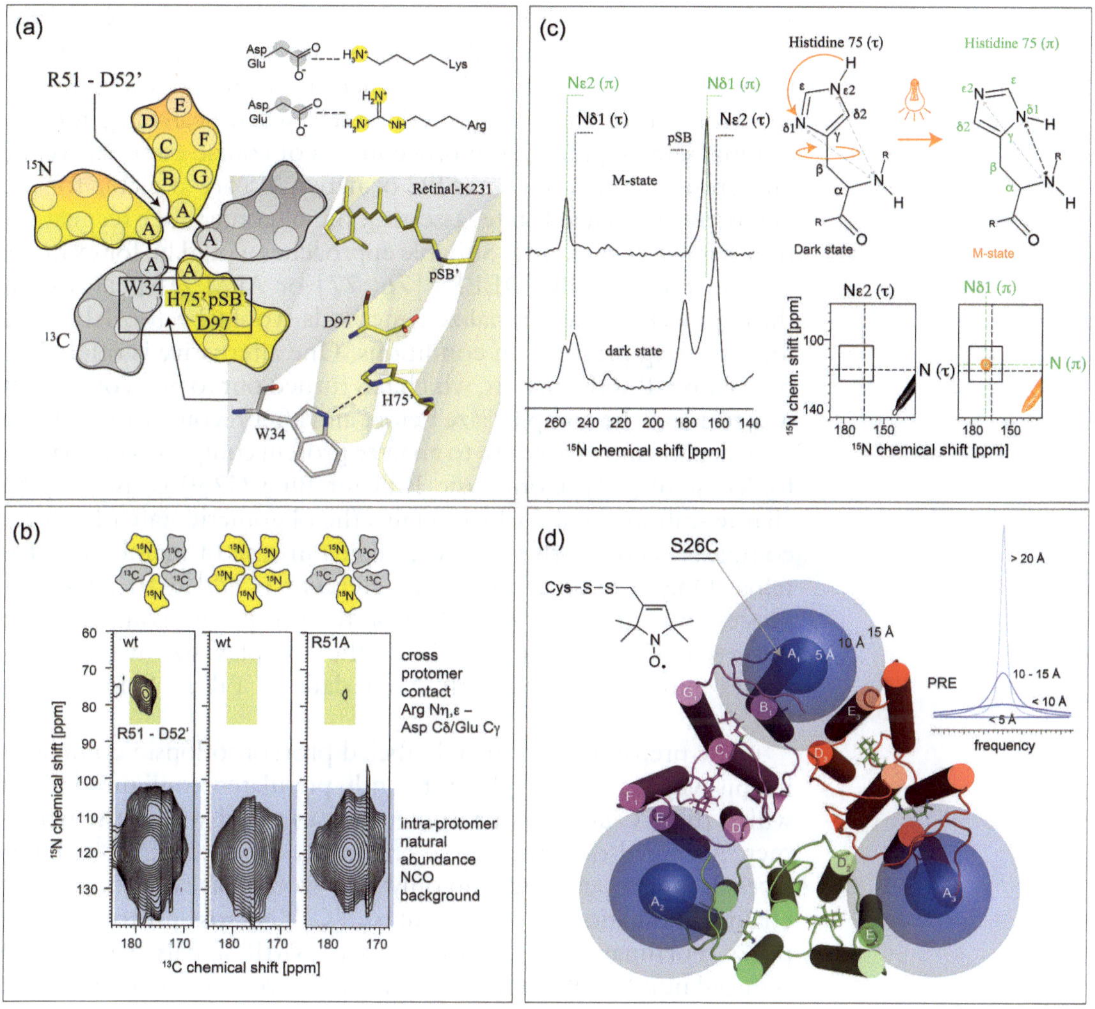

Fig. 5 Probing interactions at the oligomerization interfaces of microbial rhodopsins by solid-state NMR. (**a**) Preparing mixed-labeled PR $^{13}C/^{15}N$ oligomers offers the possibility to identify interactions across the protomer interfaces such as Lys-Asp/Glu or Arg-Asp/Glu salt bridges. Specific interactions between Arg51-Asp52′ and Trp34-His75′ were found by DNP-enhanced solid-state NMR [14, 72]. (**b**) Representative dipolar ^{15}N-^{13}C TEDOR spectrum of mixed-labeled green PR pentamers. The cross peak between $^{15}N\eta\varepsilon$-Arg51 and $C\gamma$-Asp51′ cannot be observed in uniformly ^{15}N-labeled complexes and also disappears upon mutating Arg51 or Asp52 [72]. The NCO cross peak arises from the correlation of ^{15}N- and ^{13}C-enriched sites with the natural abundance background of their neighbors (1.1% ^{13}C, 0.4% ^{15}N). (**c**) Residue His75 at the protomer interface of green PR participates in the photocycle by switching its tautomer in the M-state from τ to π (spectra left). His75 also changes the imidazole ring orientation in the M-state as observed by through-space ^{15}N-^{15}N PDSD spectra. Only in the M-state, a cross peak between an imidazole and the backbone nitrogen can be observed. His75 seems to mediate an interaction between Trp34 and Asp97′ across the interface [14]. (**d**) Probing cross-protomer interfaces by PREs as shown for trimeric ASR. Site-directed spin-labeling of Cys26 with MTSL enhances the nuclear spin relaxation properties depending on r^{-6}, which is observed by a reduced peak height. The spin label creates cross-protomer PREs in helices D and E [61]. The graphical representation is based on the PRE and EPR-refined solid-state NMR structure (PDB: 5UK6) [40]

1:1 ratio followed by reconstitution into lipid bilayers. In the bilayer, the protein assembles again into pentamers, which will then consist of ^{13}C and ^{15}N protomers as illustrated in Fig. 5a.

Monitoring these oligomeric states when searching for the optimal sample preparation procedure is of course essential. While numerous methods such as SEC or native MS are available for the detergent solubilized state, assessing the oligomer after reconstitution is more challenging. Suitable approaches would be for example cw and pulsed EPR (DEER) [76, 77] or AFM [15]. However, these powerful and specialized methods would be too laborious for screening preparation conditions. One alternative is offered by blue native-PAGE analysis, which has turned out to be a convenient way to assess the complex size before and after reconstitution. It is a non-denaturing procedure to analyze protein complexes in a native, folded state [78]. Coomassie Brilliant Blue G250 is used as the charge shift molecule, which retains the oligomeric state of protein complexes and which substitutes lipids in case of membrane proteins. Only a small amount of detergent is added to assist this process but does not solubilize or to disrupt the complex. The observed oligomeric state reflects the complex size within liposomes and agrees well with AFM data of PR in lipid bilayers [71, 72].

The preparation of mixed-labeled proteorhodopsin complexes in this way results in eight differently populated configurations in which the ^{13}C and ^{15}N protomers are arranged within the pentamer. Averaged over the whole ensemble, only 2.5 ^{15}N-^{13}C interfaces occur per pentamer and only 1.25 would have the correct side chain combination to contribute to ^{13}C-^{15}N cross-protomer cross peaks. Fortunately, the expected weak NMR signals from such a reduced number of interacting spins can be compensated by the use of sensitivity enhancement through DNP. Using dipolar through-space ^{15}N-^{13}C correlation experiments (TEDOR), dipolar interactions between coupled ^{13}C and ^{15}N nuclei within a range of 5 Å can be measured. If ^{15}N-containg side chains of residues such as Arg or Lys are within this distance range to ^{13}C-labeled side chains of Asp or Glu (Fig. 5a), characteristic cross peaks in 2D TEDOR spectra occur, which are a spectroscopic representation of a cross-protomer salt bridge. An example is shown in Fig. 5b. The ^{15}N-^{13}C TEDOR spectrum reveals an Arg-Asp cross peak for a mixed-labeled $^{13}C/^{15}N$ PR pentamer. This signal is not observed in uniformly labeled complexes and also disappears upon mutating R51 and D52, providing evidence for a cross-protomer salt bridge formed by these residues, which is important for the pentamer stabilization.

Based on this mixed-labeling approach and by varying the labeling schemes, a number of additional interactions can be identified. One further example is the cross-protomer contact between Trp34 and His75 in green PR (Fig. 5a) [14], which appears of direct functional relevance as outlined below.

5.2 Interface Interactions During the Photocycle

Testing the functional relevance of the homo-oligomers is not a trivial task as this would require performing comparative experiments on the native oligomer and on a monomer under identical conditions. However, the factors which affect the oligomerization (e.g., detergent, pH, certain mutations) might also influence the functional mechanism, and disentangling these effects from each other is difficult. A solution to this problem is to probe interactions at the protomer interfaces during the photocycle utilizing light-induced cryo-trapping as explained above. An example for green PR is shown in Fig. 5a, c: His75 interacts with W34 across the interface but also with the proton acceptor Asp97 within the same protomer [14, 79], which has also been seen for blue PR by X-ray crystallography [80]. Mutations on all three residues have strong effects on the photocycle. Histidine is a highly responsive residue often involved in enzymatic mechanisms because its imidazole side chain can adopt a positively charged and two neutral states (Nε2/τ, Nδ1/π). In solution, τ and π tautomers occur at a 4:1 ratio. His75 is highly conserved within the PR family, which raises the question about its state and role in these proteins. Histidines can be conveniently $^{13}C/^{15}N$ labeled in *E. coli* and a potential background from a His-tag can be prevented by removing it through a TEV cleavage site. The ^{13}C and ^{15}N chemical shifts of the histidine side chain atoms Nδ1, Nε2, Cγ, and Cδ2 are sensitive markers for its charge and tautomeric state and reveal that His75 in the green PR dark state is primarily found as τ with only a small π-subpopulation. The corresponding DNP-enhanced ^{15}N spectrum is shown in Fig. 5c. Surprisingly, upon trapping the M-state, H75 is completely converted into the π-state, and no τ-state signals are detected.

The observed tautomer conversion is also associated with an altered orientation of the imidazole side chain. ^{15}N-^{15}N PDSD through space correlation spectra reveal a clear cross peak between the His75(τ) amid nitrogen and Nδ1 but not with Nε2 in the M-state. However, this peak is not observed in the dark state (Fig. 5c) showing that His75 responds directly to the photocycle by altering its tautomeric state and conformation, which modulates its interactions with Trp34 and Asp97. Such a cross-protomer Asp-His-Trp triad could play a role in regulating photocycle and transport kinetics for example in response to an altered environmental pH. Interestingly, histidines are also found at protomer interfaces of other pentameric rhodopsins such as KR2 [81]. These approaches presented here are therefore of general relevance to understand related microbial rhodopsins.

In summary, solid-state NMR experiments showed for the first time functionally relevant cross-protomer interactions, which affect the photocycle. Especially interesting is the tautomeric and rotameric change of His75 at the interface during the photoreaction. These findings suggest further studies on other rhodopsins to elucidate if similar relationships exist.

5.3 The Use of PREs for Probing Cross-Protomer Interfaces

The approach explained above relies on DNP-enhanced MAS NMR, which is great for cryo-trapping photointermediate states but comes at the cost of lower resolution and requires therefore especially tailored experiments resulting in higher spectral selectivity. An alternative approach is offered by the use of paramagnetic relaxation enhancement (PRE). If the probes are strategically well placed, PREs can be used not only in the context of refining 3D structures but also for analyzing protomer interfaces as demonstrated for the trimeric ASR within lipid bilayers [61]. In this study, a cysteine mutation was introduced in helix A (S26C) and spin-labeled with MTSL (*see* Fig. 5d). As expected, residues in close proximity to the spin label (helices A and B) show the strongest reduction in intensity (Fig. 5d). However, also strong effects are seen in helices D and E, which cannot be explained by intra-protomer PREs since they would be too far away from the MTSL spin label. Thus, they must arise from helices D and E at the interface between adjacent protomers (Fig. 5d). Using these data, the interface can be mapped, and 3D structure calculations can be improved. For unambiguous data interpretation, it is important to carry out dilution experiments with diamagnetic samples, to assess the spin-labeling efficiency and to validate that cys-mutations have no essential structural or functional consequences.

Site-directed spin labeling also offers the possibility to determine electron-electron distances between protomers by cw and pulsed, dipolar EPR spectroscopy (PELDOR/DEER), which has been extensively applied to microbial rhodopsins [76]. In contrast to PREs, only a single distance can be obtained per sample, but these are very valuable because they are long-range (up to 70 Å) and relatively precise so that protomer arrangements can be resolved [82] and NMR structures can be refined [40].

6 Summary and Perspective

Solid-state NMR based on MAS offers versatile experimental possibilities for rhodopsins and two branches of applications have emerged, which differ in terms of experimental design and conditions: Light-induced trapping of photointermediates benefits significantly from the sensitivity boost provided by DNP. The requirement for low temperatures for trapping as well as for DNP makes it a perfect match, but often requires well-tailored experimental schemes including selective isotope labeling to deal with the spectral complexity caused by freezing-induced heterogeneity. In contrast, highly resolved spectra of uniformly labeled samples can be obtained on non-frozen samples of ground state rhodopsins allowing a global analysis of their structure and conformational dynamics. But what is the perspective of solid-state NMR in the context of other spectroscopic and structural methods? An important advantage is certainly the possibility to keep the rhodopsins within proteoliposome preparations and to maintain their native oligomeric state. Furthermore, 3D structures alone are not sufficient for deciphering the subtle molecular details leading to such a variety of functions and which control photocycle kinetics or color adaptation. NMR can provide important additional site-resolve data, e.g., on protonation states of side chains, on sub-Angstrom bond distortions within the chromophore, on the nature of hydrogen bonds or on the electronic environment of sites of interest. NMR parameters such as chemical shifts and dipole-dipole couplings are highly sensitive to these factors and are based on well-established theoretical foundations. NMR offers therefore an excellent link between structure and function. Hence, it would by desirable to connect 3D structures with functional, optical, vibrational, and NMR data. Especially the link with QM/MM approaches appears promising [55, 83–85]. Complex systems become increasingly accessible by methods such as automated fragmentation, which scales linearly with the number of atoms [86] and through which spectroscopic and structural data could be combined in a unified model in the future.

References

1. Grote M, O'Malley MA (2011) Enlightening the life sciences: the history of halobacterial and microbial rhodopsin research. FEMS Microbiol Rev 35(6):1082–1099. https://doi.org/10.1111/j.1574-6976.2011.00281.x
2. Govorunova EG, Sineshchekov OA, Li H, Spudich JL (2017) Microbial rhodopsins: diversity, mechanisms, and optogenetic applications. Annu Rev Biochem 86:845–872. https://doi.org/10.1146/annurev-biochem-101910-144233
3. Beja O, Spudich EN, Spudich JL, Leclerc M, DeLong EF (2001) Proteorhodopsin phototrophy in the ocean. Nature 411(6839):786–789. https://doi.org/10.1038/35081051
4. Beja O, Aravind L, Koonin EV, Suzuki MT, Hadd A, Nguyen LP, Jovanovich SB, Gates CM, Feldman RA, Spudich JL, Spudich EN, DeLong EF (2000) Bacterial rhodopsin: evidence for a new type of phototrophy in the sea. Science 289(5486):1902–1906. https://doi.org/10.1126/science.289.5486.1902
5. Inoue K, Ono H, Abe-Yoshizumi R, Yoshizawa S, Ito H, Kogure K, Kandori H (2013) A light-driven sodium ion pump in marine bacteria. Nat Commun 4:1678. https://doi.org/10.1038/ncomms2689
6. Nagel G, Szellas T, Huhn W, Kateriya S, Adeishvili N, Berthold P, Ollig D, Hegemann P, Bamberg E (2003) Channelrhodopsin-2, a directly light-gated cation-selective membrane channel. Proc Natl Acad Sci U S A 100(24):13940–13945. https://doi.org/10.1073/pnas.1936192100
7. Jung KH, Trivedi VD, Spudich JL (2003) Demonstration of a sensory rhodopsin in eubacteria. Mol Microbiol 47(6):1513–1522. https://doi.org/10.1046/j.1365-2958.2003.03395.x
8. Wang S, Munro RA, Shi L, Kawamura I, Okitsu T, Wada A, Kim SY, Jung KH, Brown LS, Ladizhansky V (2013) Solid-state NMR spectroscopy structure determination of a lipid-embedded heptahelical membrane protein. Nat Methods 10(10):1007–1012. https://doi.org/10.1038/nmeth.2635
9. Carravetta M, Zhao X, Johannessen OG, Lai WC, Verhoeven MA, Bovee-Geurts PH, Verdegem PJ, Kiihne S, Luthman H, de Groot HJ, deGrip WJ, Lugtenburg J, Levitt MH (2004) Protein-induced bonding perturbation of the rhodopsin chromophore detected by double-quantum solid-state NMR. J Am Chem Soc 126(12):3948–3953. https://doi.org/10.1021/ja039390q

10. Lansing JC, Hohwy M, Jaroniec CP, Creemers AF, Lugtenburg J, Herzfeld J, Griffin RG (2002) Chromophore distortions in the bacteriorhodopsin photocycle: evolution of the H-C14-C15-H dihedral angle measured by solid-state NMR. Biochemistry 41(2):431–438. https://doi.org/10.1021/bi011529r
11. Mao J, Aladin V, Jin X, Leeder AJ, Brown LJ, Brown RCD, He X, Corzilius B, Glaubitz C (2019) Exploring protein structures by DNP-enhanced methyl solid-state NMR spectroscopy. J Am Chem Soc 141(50):19888–19901. https://doi.org/10.1021/jacs.9b11195
12. Ding X, Sun C, Cui H, Chen S, Gao Y, Yang Y, Wang J, He X, Iuga D, Tian F, Watts A, Zhao X (2018) Functional roles of tyrosine 185 during the bacteriorhodopsin photocycle as revealed by in situ spectroscopic studies. Biochim Biophys Acta Bioenerg 1859(10):1006–1014. https://doi.org/10.1016/j.bbabio.2018.05.011
13. Ni QZ, Can TV, Daviso E, Belenky M, Griffin RG, Herzfeld J (2018) Primary transfer step in the light-driven ion pump bacteriorhodopsin: an irreversible U-turn revealed by dynamic nuclear polarization-enhanced magic angle spinning NMR. J Am Chem Soc 140(11):4085–4091. https://doi.org/10.1021/jacs.8b00022
14. Maciejko J, Kaur J, Becker-Baldus J, Glaubitz C (2019) Photocycle-dependent conformational changes in the proteorhodopsin cross-protomer Asp-His-Trp triad revealed by DNP-enhanced MAS-NMR. Proc Natl Acad Sci U S A 116(17):8342–8349. https://doi.org/10.1073/pnas.1817665116
15. Shibata M, Inoue K, Ikeda K, Konno M, Singh M, Kataoka C, Abe-Yoshizumi R, Kandori H, Uchihashi T (2018) Oligomeric states of microbial rhodopsins determined by high-speed atomic force microscopy and circular dichroic spectroscopy. Sci Rep 8(1):8262. https://doi.org/10.1038/s41598-018-26606-y
16. Kaur J, Kriebel CN, Eberhardt P, Jakdetchai O, Leeder AJ, Weber I, Brown LJ, Brown RCD, Becker-Baldus J, Bamann C, Wachtveitl J, Glaubitz C (2019) Solid-state NMR analysis of the sodium pump Krokinobacter rhodopsin 2 and its H30A mutant. J Struct Biol 206(1):55–65. https://doi.org/10.1016/j.jsb.2018.06.001
17. Jakdetchai O, Eberhardt P, Asido M, Kaur J, Kriebel CN, Mao J, Leeder AJ, Brown LJ, Brown RCD, Becker-Baldus J, Bamann C, Wachtveitl J, Glaubitz C (2021) Probing the photointermediates of light-driven sodium ion pump KR2 by DNP-enhanced solid-state NMR. Sci Adv 7(11):eabf4213. https://doi.org/10.1126/sciadv.abf4213
18. Shi L, Ahmed MA, Zhang W, Whited G, Brown LS, Ladizhansky V (2009) Three-dimensional solid-state NMR study of a seven-helical integral membrane proton pump--structural insights. J Mol Biol 386(4):1078–1093. https://doi.org/10.1016/j.jmb.2009.01.011
19. Etzkorn M, Martell S, Andronesi OC, Seidel K, Engelhard M, Baldus M (2007) Secondary structure, dynamics, and topology of a seven-helix receptor in native membranes, studied by solid-state NMR spectroscopy. Angew Chem Int Ed Eng 46(3):459–462. https://doi.org/10.1002/anie.200602139
20. Lalli D, Idso MN, Andreas LB, Hussain S, Baxter N, Han S, Chmelka BF, Pintacuda G (2017) Proton-based structural analysis of a heptahelical transmembrane protein in lipid bilayers. J Am Chem Soc 139(37):13006–13012. https://doi.org/10.1021/jacs.7b05269
21. Linser R, Dasari M, Hiller M, Higman V, Fink U, Lopez del Amo JM, Markovic S, Handel L, Kessler B, Schmieder P, Oesterhelt D, Oschkinat H, Reif B (2011) Proton-detected solid-state NMR spectroscopy of fibrillar and membrane proteins. Angew Chem Int Ed Eng 50(19):4508–4512. https://doi.org/10.1002/anie.201008244
22. Ni QZ, Daviso E, Can TV, Markhasin E, Jawla SK, Swager TM, Temkin RJ, Herzfeld J, Griffin RG (2013) High frequency dynamic nuclear polarization. Acc Chem Res 46(9):1933–1941. https://doi.org/10.1021/ar300348n
23. Hall DA, Maus DC, Gerfen GJ, Inati SJ, Becerra LR, Dahlquist FW, Griffin RG (1997) Polarization-enhanced NMR spectroscopy of biomolecules in frozen solution. Science 276(5314):930–932. https://doi.org/10.1126/science.276.5314.930
24. Sauvee C, Rosay M, Casano G, Aussenac F, Weber RT, Ouari O, Tordo P (2013) Highly efficient, water-soluble polarizing agents for dynamic nuclear polarization at high frequency. Angew Chem Int Ed Eng 52(41):10858–10861. https://doi.org/10.1002/anie.201304657
25. Mehler M, Eckert CE, Leeder AJ, Kaur J, Fischer T, Kubatova N, Brown LJ, Brown RCD, Becker-Baldus J, Wachtveitl J, Glaubitz C (2017) Chromophore distortions in photointermediates of proteorhodopsin visualized by dynamic nuclear polarization-enhanced solid-

state NMR. J Am Chem Soc 139(45):16143–16153. https://doi.org/10.1021/jacs.7b05061

26. Leeder AJ, Brown LJ, Becker-Baldus J, Mehler M, Glaubitz C, Brown RCD (2018) Synthesis of isotopically labeled all-trans retinals for DNP-enhanced solid-state NMR studies of retinylidene proteins. J Label Comp Radiopharm 61(13):922–933. https://doi.org/10.1002/jlcr.3576
27. Creemers AF, Kiihne S, Bovee-Geurts PH, DeGrip WJ, Lugtenburg J, de Groot HJ (2002) (1)H and (13)C MAS NMR evidence for pronounced ligand-protein interactions involving the ionone ring of the retinylidene chromophore in rhodopsin. Proc Natl Acad Sci U S A 99(14):9101–9106. https://doi.org/10.1073/pnas.112677599
28. Dawadi PB, Lugtenburg J (2010) Synthesis and use of stable isotope enriched retinals in the field of vitamin A. Molecules 15(3):1825–1872. https://doi.org/10.3390/molecules15031825
29. Makino Y, Kawamura I, Okitsu T, Wada A, Kamo N, Sudo Y, Ueda K, Naito A (2018) Retinal configuration of ppR intermediates revealed by photoirradiation solid-state NMR and DFT. Biophys J 115(1):72–83. https://doi.org/10.1016/j.bpj.2018.05.030
30. Munro RA, de Vlugt J, Ward ME, Kim SY, Lee KA, Jung KH, Ladizhansky V, Brown LS (2019) Biosynthetic production of fully carbon-13 labeled retinal in E. coli for structural and functional studies of rhodopsins. J Biomol NMR 73(1–2):49–58. https://doi.org/10.1007/s10858-019-00225-9
31. Becker-Baldus J, Bamann C, Saxena K, Gustmann H, Brown LJ, Brown RC, Reiter C, Bamberg E, Wachtveitl J, Schwalbe H, Glaubitz C (2015) Enlightening the photoactive site of channelrhodopsin-2 by DNP-enhanced solid-state NMR spectroscopy. Proc Natl Acad Sci U S A 112(32):9896–9901. https://doi.org/10.1073/pnas.1507713112
32. Hatcher ME, Hu JG, Belenky M, Verdegem P, Lugtenburg J, Griffin RG, Herzfeld J (2002) Control of the pump cycle in bacteriorhodopsin: mechanisms elucidated by solid-state NMR of the D85N mutant. Biophys J 82(2):1017–1029. https://doi.org/10.1016/S0006-3495(02)75461-1
33. Lacabanne D, Meier BH, Bockmann A (2018) Selective labeling and unlabeling strategies in protein solid-state NMR spectroscopy. J Biomol NMR 71(3):141–150. https://doi.org/10.1007/s10858-017-0156-z
34. Mehler M, Eckert CE, Busche A, Kulhei J, Michaelis J, Becker-Baldus J, Wachtveitl J, Dotsch V, Glaubitz C (2015) Assembling a correctly folded and functional heptahelical membrane protein by protein trans-splicing. J Biol Chem 290(46):27712–27722. https://doi.org/10.1074/jbc.M115.681205
35. Fan Y, Shi L, Ladizhansky V, Brown LS (2011) Uniform isotope labeling of a eukaryotic seven-transmembrane helical protein in yeast enables high-resolution solid-state NMR studies in the lipid environment. J Biomol NMR 49(2):151–161. https://doi.org/10.1007/s10858-011-9473-9
36. Janke C, Scholz F, Becker-Baldus J, Glaubitz C, Wood PG, Bamberg E, Wachtveitl J, Bamann C (2013) Photocycle and vectorial proton transfer in a rhodopsin from the eukaryote Oxyrrhis marina. Biochemistry 52(16):2750–2763. https://doi.org/10.1021/bi301412n
37. Ward ME, Wang S, Munro R, Ritz E, Hung I, Gor'kov PL, Jiang Y, Liang H, Brown LS, Ladizhansky V (2015) In situ structural studies of Anabaena sensory rhodopsin in the E. coli membrane. Biophys J 108(7):1683–1696. https://doi.org/10.1016/j.bpj.2015.02.018
38. Ullrich SJ, Holper S, Glaubitz C (2014) Paramagnetic doping of a 7TM membrane protein in lipid bilayers by Gd(3)(+)-complexes for solid-state NMR spectroscopy. J Biomol NMR 58(1):27–35. https://doi.org/10.1007/s10858-013-9800-4
39. Becker-Baldus J, Glaubitz C (2018) Cryo-trapped intermediates of retinal proteins studied by DNP-enhanced MAS NMR spectroscopy. eMagRes 7(4):79–91. https://doi.org/10.1002/9780470034590.emrstm1552
40. Milikisiyants S, Wang S, Munro RA, Donohue M, Ward ME, Bolton D, Brown LS, Smirnova TI, Ladizhansky V, Smirnov AI (2017) Oligomeric structure of anabaena sensory rhodopsin in a lipid bilayer environment by combining solid-state NMR and long-range DEER constraints. J Mol Biol 429(12):1903–1920. https://doi.org/10.1016/j.jmb.2017.05.005
41. Voinov MA, Good DB, Ward ME, Milikisiyants S, Marek A, Caporini MA, Rosay M, Munro RA, Ljumovic M, Brown LS, Ladizhansky V, Smirnov AI (2015) Cysteine-specific labeling of proteins with a nitroxide biradical for dynamic nuclear polarization NMR. J Phys Chem B 119(32):10180–10190. https://doi.org/10.1021/acs.jpcb.5b05230
42. Naito A, Makino Y, Shigeta A, Kawamura I (2019) Photoreaction pathways and photointermediates of retinal-binding photoreceptor proteins as revealed by in situ photoirradiation

solid-state NMR spectroscopy. Biophys Rev 11(2):167–181. https://doi.org/10.1007/s12551-019-00501-w

43. Friedrich T, Geibel S, Kalmbach R, Chizhov I, Ataka K, Heberle J, Engelhard M, Bamberg E (2002) Proteorhodopsin is a light-driven proton pump with variable vectoriality. J Mol Biol 321(5):821–838. https://doi.org/10.1016/S0022-2836(02)00696-4
44. Harbison GS, Smith SO, Pardoen JA, Winkel C, Lugtenburg J, Herzfeld J, Mathies R, Griffin RG (1984) Dark-adapted bacteriorhodopsin contains 13-cis, 15-syn and all-trans, 15-anti retinal Schiff bases. Proc Natl Acad Sci U S A 81(6):1706–1709. https://doi.org/10.1073/pnas.81.6.1706
45. Buhl E, Eberhardt P, Bamann C, Bamberg E, Braun M, Wachtveitl J (2018) Ultrafast protein response in channelrhodopsin-2 studied by time-resolved infrared spectroscopy. J Phys Chem Lett 9(24):7180–7184. https://doi.org/10.1021/acs.jpclett.8b03382
46. Eckert CE, Kaur J, Glaubitz C, Wachtveitl J (2017) Ultrafast photoinduced deactivation dynamics of proteorhodopsin. J Phys Chem Lett 8(2):512–517. https://doi.org/10.1021/acs.jpclett.6b02975
47. Jeon J, Thurber KR, Ghirlando R, Yau WM, Tycko R (2019) Application of millisecond time-resolved solid state NMR to the kinetics and mechanism of melittin self-assembly. Proc Natl Acad Sci U S A 116(34):16717–16722. https://doi.org/10.1073/pnas.1908006116
48. Harbison GS, Mulder PPJ, Pardoen H, Lugtenburg J, Herzfeld J, Griffin RG (1985) High-resolution carbon-13 NMR of retinal derivatives in the solid state. J Am Chem Soc 107(17):4809–4816. https://doi.org/10.1021/ja00303a001
49. Englert G (1975) A 13C-NMR. Study of cis-trans isomeric vitamins A, carotenoids and related compounds. Helv Chim Acta 58(8):2367–2390. https://doi.org/10.1002/hlca.19750580817
50. Farrar MR, Lakshmi KV, Smith SO, Brown RS, Raap J, Lugtenburg J, Griffin RG, Herzfeld J (1993) Solid state NMR study of [epsilon-13C]Lys-bacteriorhodopsin: schiff base photoisomerization. Biophys J 65(1):310–315. https://doi.org/10.1016/S0006-3495(93)81065-8
51. Thompson LK, McDermott AE, Raap J, van der Wielen CM, Lugtenburg J, Herzfeld J, Griffin RG (1992) Rotational resonance NMR study of the active site structure in bacteriorhodopsin: conformation of the Schiff base linkage. Biochemistry 31(34):7931–7938. https://doi.org/10.1021/bi00149a026
52. Harbison GS, Herzfeld J, Griffin RG (1983) Solid-state nitrogen-15 nuclear magnetic resonance study of the Schiff base in bacteriorhodopsin. Biochemistry 22(1):1–4. https://doi.org/10.1021/bi00270a600
53. Mao J, Do NN, Scholz F, Reggie L, Mehler M, Lakatos A, Ong YS, Ullrich SJ, Brown LJ, Brown RC, Becker-Baldus J, Wachtveitl J, Glaubitz C (2014) Structural basis of the green-blue color switching in proteorhodopsin as determined by NMR spectroscopy. J Am Chem Soc 136(50):17578–17590. https://doi.org/10.1021/ja5097946
54. Feng X, Verdegem PJE, Lee YK, Sandstrom D, Eden M, BoveeGeurts P, deGrip WJ, Lugtenburg J, deGroot HJM, Levitt MH (1997) Direct determination of a molecular torsional angle in the membrane protein rhodopsin by solid-state NMR. J Am Chem Soc 119(29):6853–6857. https://doi.org/10.1021/ja970710d
55. Bondar AN, Elstner M, Suhai S, Smith JC, Fischer S (2004) Mechanism of primary proton transfer in bacteriorhodopsin. Structure 12(7):1281–1288. https://doi.org/10.1016/j.str.2004.04.016
56. Hong M (1999) Determination of multiple ***φ***-torsion angles in proteins by selective and extensive (13)C labeling and two-dimensional solid-state NMR. J Magn Reson 139(2):389–401. https://doi.org/10.1006/jmre.1999.1805
57. Wishart DS, Sykes BD, Richards FM (1992) The chemical shift index: a fast and simple method for the assignment of protein secondary structure through NMR spectroscopy. Biochemistry 31(6):1647–1651. https://doi.org/10.1021/bi00121a010
58. Shen Y, Delaglio F, Cornilescu G, Bax A (2009) TALOS+: a hybrid method for predicting protein backbone torsion angles from NMR chemical shifts. J Biomol NMR 44(4):213–223. https://doi.org/10.1007/s10858-009-9333-z
59. Otting G (2010) Protein NMR using paramagnetic ions. Annu Rev Biophys 39:387–405. https://doi.org/10.1146/annurev.biophys.093008.131321
60. Nadaud PS, Helmus JJ, Hofer N, Jaroniec CP (2007) Long-range structural restraints in spin-labeled proteins probed by solid-state nuclear magnetic resonance spectroscopy. J Am Chem Soc 129(24):7502–7503. https://doi.org/10.1021/ja072349t

61. Wang S, Munro RA, Kim SY, Jung KH, Brown LS, Ladizhansky V (2012) Paramagnetic relaxation enhancement reveals oligomerization interface of a membrane protein. J Am Chem Soc 134(41):16995–16998. https://doi.org/10.1021/ja308310z
62. Brunger AT, Adams PD, Clore GM, DeLano WL, Gros P, Grosse-Kunstleve RW, Jiang JS, Kuszewski J, Nilges M, Pannu NS, Read RJ, Rice LM, Simonson T, Warren GL (1998) Crystallography & NMR system: a new software suite for macromolecular structure determination. Acta Crystallogr D Biol Crystallogr 54(Pt 5):905–921. https://doi.org/10.1107/s0907444998003254
63. Bardiaux B, Malliavin T, Nilges M (2012) ARIA for solution and solid-state NMR. Methods Mol Biol 831:453–483. https://doi.org/10.1007/978-1-61779-480-3_23
64. Vogeley L, Sineshchekov OA, Trivedi VD, Sasaki J, Spudich JL, Luecke H (2004) Anabaena sensory rhodopsin: a photochromic color sensor at 2.0 A. Science 306(5700):1390–1393. https://doi.org/10.1126/science.1103943
65. Wang S, Shi L, Kawamura I, Brown LS, Ladizhansky V (2011) Site-specific solid-state NMR detection of hydrogen-deuterium exchange reveals conformational changes in a 7-helical transmembrane protein. Biophys J 101(3): L23–L25. https://doi.org/10.1016/j.bpj.2011.06.035
66. Yang J, Aslimovska L, Glaubitz C (2011) Molecular dynamics of proteorhodopsin in lipid bilayers by solid-state NMR. J Am Chem Soc 133(13):4874–4881. https://doi.org/10.1021/ja109766n
67. Etzkorn M, Seidel K, Li L, Martell S, Geyer M, Engelhard M, Baldus M (2010) Complex formation and light activation in membrane-embedded sensory rhodopsin II as seen by solid-state NMR spectroscopy. Structure 18(3):293–300. https://doi.org/10.1016/j.str.2010.01.011
68. Good DB, Wang S, Ward ME, Struppe J, Brown LS, Lewandowski JR, Ladizhansky V (2014) Conformational dynamics of a seven transmembrane helical protein Anabaena Sensory Rhodopsin probed by solid-state NMR. J Am Chem Soc 136(7):2833–2842. https://doi.org/10.1021/ja411633w
69. Henderson R, Unwin PN (1975) Three-dimensional model of purple membrane obtained by electron microscopy. Nature 257(5521):28–32. https://doi.org/10.1038/257028a0
70. Shastri S, Vonck J, Pfleger N, Haase W, Kuehlbrandt W, Glaubitz C (2007) Proteorhodopsin: characterisation of 2D crystals by electron microscopy and solid state NMR. Biochim Biophys Acta 1768(12):3012–3019. https://doi.org/10.1016/j.bbamem.2007.10.001
71. Klyszejko AL, Shastri S, Mari SA, Grubmuller H, Muller DJ, Glaubitz C (2008) Folding and assembly of proteorhodopsin. J Mol Biol 376(1):35–41. https://doi.org/10.1016/j.jmb.2007.11.030
72. Maciejko J, Mehler M, Kaur J, Lieblein T, Morgner N, Ouari O, Tordo P, Becker-Baldus-J, Glaubitz C (2015) Visualizing specific cross-protomer interactions in the homo-oligomeric membrane protein proteorhodopsin by dynamic-nuclear-polarization-enhanced solid-state NMR. J Am Chem Soc 137(28):9032–9043. https://doi.org/10.1021/jacs.5b03606
73. Hoffmann J, Aslimovska L, Bamann C, Glaubitz C, Bamberg E, Brutschy B (2010) Studying the stoichiometries of membrane proteins by mass spectrometry: microbial rhodopsins and a potassium ion channel. Phys Chem Chem Phys 12(14):3480–3485. https://doi.org/10.1039/b924630d
74. Iizuka A, Kajimoto K, Fujisawa T, Tsukamoto T, Aizawa T, Kamo N, Jung KH, Unno M, Demura M, Kikukawa T (2019) Functional importance of the oligomer formation of the cyanobacterial H(+) pump Gloeobacter rhodopsin. Sci Rep 9(1):10711. https://doi.org/10.1038/s41598-019-47178-5
75. Hussain S, Kinnebrew M, Schonenbach NS, Aye E, Han S (2015) Functional consequences of the oligomeric assembly of proteorhodopsin. J Mol Biol 427(6 Pt B):1278–1290. https://doi.org/10.1016/j.jmb.2015.01.004
76. Klare JP, Steinhoff HJ (2009) Spin labeling EPR. Photosynth Res 102(2–3):377–390. https://doi.org/10.1007/s11120-009-9490-7
77. Edwards DT, Huber T, Hussain S, Stone KM, Kinnebrew M, Kaminker I, Matalon E, Sherwin MS, Goldfarb D, Han S (2014) Determining the oligomeric structure of proteorhodopsin by Gd3+-based pulsed dipolar spectroscopy of multiple distances. Structure 22(11):1677–1686. https://doi.org/10.1016/j.str.2014.09.008
78. Wittig I, Braun HP, Schagger H (2006) Blue native PAGE. Nat Protoc 1(1):418–428. https://doi.org/10.1038/nprot.2006.62
79. Hempelmann F, Holper S, Verhoefen MK, Woerner AC, Kohler T, Fiedler SA, Pfleger N, Wachtveitl J, Glaubitz C (2011) His75-Asp97 cluster in green proteorhodopsin. J Am Chem

Soc 133(12):4645–4654. https://doi.org/10.1021/ja111116a

80. Ran T, Ozorowski G, Gao Y, Sineshchekov OA, Wang W, Spudich JL, Luecke H (2013) Cross-protomer interaction with the photoactive site in oligomeric proteorhodopsin complexes. Acta Crystallogr D Biol Crystallogr 69 (Pt 10):1965–1980. https://doi.org/10.1107/S0907444913017575
81. Kovalev K, Astashkin R, Gushchin I, Orekhov P, Volkov D, Zinovev E, Marin E, Rulev M, Alekseev A, Royant A, Carpentier P, Vaganova S, Zabelskii D, Baeken C, Sergeev I, Balandin T, Bourenkov G, Carpena X, Boer R, Maliar N, Borshchevskiy V, Buldt G, Bamberg E, Gordeliy V (2020) Molecular mechanism of light-driven sodium pumping. Nat Commun 11(1):2137. https://doi.org/10.1038/s41467-020-16032-y
82. Stone KM, Voska J, Kinnebrew M, Pavlova A, Junk MJ, Han S (2013) Structural insight into proteorhodopsin oligomers. Biophys J 104(2):472–481. https://doi.org/10.1016/j.bpj.2012.11.3831
83. Gansmuller A, Concistre M, McLean N, Johannessen OG, Marin-Montesinos I, Bovee-Geurts PH, Verdegem P, Lugtenburg J, Brown RC, Degrip WJ, Levitt MH (2009) Towards an interpretation of 13C chemical shifts in bathorhodopsin, a functional intermediate of a G-protein coupled receptor. Biochim Biophys Acta 1788(6):1350–1357. https://doi.org/10.1016/j.bbamem.2009.02.018
84. Toker Y, Langeland J, Gruber E, Kjaer C, Nielsen SB, Andersen LH, Bonin VA, Schapiro I (2018) Counterion-controlled spectral tuning of the protonated Schiff-base retinal. Phys Rev A 98(4):043428. https://doi.org/10.1103/PhysRevA.98.043428
85. Hoffmann M, Wanko M, Strodel P, Konig PH, Frauenheim T, Schulten K, Thiel W, Tajkhorshid E, Elstner M (2006) Color tuning in rhodopsins: the mechanism for the spectral shift between bacteriorhodopsin and sensory rhodopsin II. J Am Chem Soc 128(33):10808–10818. https://doi.org/10.1021/ja062082i
86. He X, Wang B, Merz KM Jr (2009) Protein NMR chemical shift calculations based on the automated fragmentation QM/MM approach. J Phys Chem B 113(30):10380–10388. https://doi.org/10.1021/jp901992p

Chapter 10

FTIR and Raman Spectroscopy of Rhodopsins

Hideki Kandori and Yasuhisa Mizutani

Abstract

Vibrational spectroscopy such as FTIR and Raman spectroscopy is a powerful, sensitive, and informative method for studying protein structural changes in rhodopsins during their functions. The usefulness has been historically proven for the study of bacteriorhodopsin and bovine rhodopsin before their structural determination of rhodopsins. We now have atomic structures of many animal and microbial rhodopsins, and it is now important to know the structural dynamics of rhodopsins for function. FTIR and Raman spectroscopy provides useful information for this aim. In this chapter, we introduce the methods of FTIR and resonance Raman spectroscopy applied to rhodopsins. These vibrational methods offer deeper understanding on the mechanism how rhodopsins change their structures for function.

Key words Molecular vibration, Structural dynamics, Photointermediate, Hydrogen bond

1 Introduction

Biological systems utilize light as the source of energy and cellular signaling. These light-induced functions are governed by photoreceptive proteins, which contain an embedded chromophore molecule that senses light. Rhodopsins are photoreceptive proteins that contain a retinal chromophore in animals and microbes [1]. Animal and microbial rhodopsins possess 11-*cis* and all-*trans* retinal, respectively. As cofactors bound with their animal and microbial rhodopsin (seven transmembrane α-helices) environments, 11-*cis* and all-*trans* retinal undergo photoisomerization into all-*trans* and 13-*cis* retinal forms as part of their functional cycle. While animal rhodopsins are G-protein-coupled receptors, the function of microbial rhodopsins is highly divergent [1, 2]. Many of the microbial rhodopsins are able to transport ions in a passive or an active manner. These light-gated channels or light-driven pumps represent the main tools for respectively effecting neural excitation and silencing in the emerging field of optogenetics [3].

Valentin Gordeliy (ed.), *Rhodopsin: Methods and Protocols*, Methods in Molecular Biology, vol. 2501,
https://doi.org/10.1007/978-1-0716-2329-9_10,

To understand molecular mechanism of rhodopsin function, structure and structural changes during function are needed. Methods of structural biology require positional acquisition of atoms (or components) of biomolecules. Crystal structures of animal and microbial rhodopsins serve as the molecular basis of our understanding on the mechanism of the functions of rhodopsins. Other structural techniques such as NMR and electron microscopy are also being widely utilized. Vibrational spectroscopy such as IR and Raman spectroscopy has not been regarded as an experimental tool of structural biology, as it cannot determine atomic positions of biomolecules. In addition, in case of IR spectroscopy, strong absorption of water results in difficulties during the measurements. However, if water content of the sample is significantly reduced without losing its function, IR spectroscopy is able to capture even protein-bound functional water molecules in rhodopsins [4]. Resonance Raman spectroscopy in the visible and UV region is also extremely sensitive to the structure of the retinal chromophore and aromatic residues such as Trp and Tyr, respectively [5].

Usefulness of vibrational spectroscopy can be seen from the history of the study of bacteriorhodopsin (BR), one of the best studied rhodopsin [6, 7]. BR is a light-driven proton pump in *Halobacterium salinarum*, whose high-resolution structure in crystal was determined at around 2000 [1]. Absorption of light triggers a cyclic reaction that comprises a series of intermediates, designated as the J, K, L, M, N, and O states. These intermediate states were first identified by transient absorption spectroscopy in the visible region. Next, resonance Raman spectroscopy revealed the structure of the retinal chromophore, where the all-*trans* to 13-*cis* photoisomerization is the primary event in the proton pump of BR, and the reverse 13-*cis* to all-*trans* isomerization occurs in the N-to-O transition thermally. In addition, resonance Raman spectroscopy showed that the retinal Schiff base is deprotonated in the M intermediate and reprotonated in the N intermediate, indicating that the Schiff base proton is conducted into the extracellular side upon M formation, and the cytoplasmic side supplies a proton to the Schiff base upon M decay. Then, Fourier transform infrared (FTIR) spectroscopy revealed how protons are captured by specific amino acids during the pumping process, for which the C=O stretching vibrations of protonated carboxylic acids in the 1800–1700 cm^{-1} region provided fruitful information. Protonation of Asp96 in BR, its transient deprotonation in N, and transient protonation of Asp85 in M, N, and O were observed, indicating that a proton is translocated via Asp96, Schiff base, and Asp85. This knowledge was gained in 1970–1980, much earlier than the atomic structure [1].

FTIR and Raman spectroscopy both monitors vibrations of sample, while they differ in some key fundamental ways. FTIR spectroscopy depends on a change in the dipole moment, whereas Raman spectroscopy depends on a change in polarizability of a molecule in the non-resonance condition. In resonance condition, intensity enhancements of Raman bands depend on structural differences between electronic ground and excited states in the Franck-Condon mechanism. Raman spectroscopy detects relative frequencies at which a sample scatters radiation while FTIR spectroscopy that detects absolute frequencies at which a sample absorbs radiation. FTIR and Raman spectroscopy provides complementary information of vibrations in general. Therefore, combination of FTIR and Raman spectroscopy deepens our understanding of rhodopsins. In this chapter, we introduce the methods of vibrational spectroscopy such as FTIR and resonance Raman spectroscopy, which was applied to rhodopsins. These vibrational methods offer deeper understanding on the mechanism how rhodopsins change their structures for function.

2 Materials

Rhodopsins are membrane proteins and are normally purified after solubilizing in detergent. Bacteriorhodopsin (BR), a light-driven proton pump in *Halobaterium salinarum*, is an exceptional example, as BR forms native two-dimensional crystal, which is called purple membrane (PM). PM is generally purified without solubilization, by sucrose gradient method. Preparation of other rhodopsins starts from the cells of native specimen or heterologous expression. Bovine rhodopsin (bRh) is the typical example of the former, whose preparation starts from cattle eyeballs, and rod outer segments in the retina are isolated by sucrose gradient, followed by solubilization and purification using concanavalin A (ConA) column. BR and bRh have been the best studied microbial and animal rhodopsins, respectively. In case of many other rhodopsins, it is difficult to obtain samples from native cells. Therefore, heterologous expression is normally used for the preparation of these proteins. Even in case of BR and bRh, preparation of site-directed mutants requires heterologous expression. In heterologous expression such as *E. coli* and mammalian cells, tags such as His-tag are introduced normally at the C-terminal side for the purification of proteins.

The purified rhodopsin solution in detergent is highly transparent, which is ideal for Raman spectroscopy. On the other hand, water has strong absorption in the IR region, which results in difficulties during measurements. Therefore, reduction of water absorption has to be taken into account in FTIR spectroscopy. Hydrated films are conventionally used in the FTIR measurements, where the sample is first dried and rehydrated by adding water drop next to the film [4, 8]. It is known that many proteins, either soluble or transmembrane, are functional in hydrated films, and the hydrated film methods have been established for FTIR measurements of microbial and animal rhodopsins [4, 8]. When the rhodopsin solution in detergent is dried, the drying process significantly increases detergent concentration. To avoid such harsh conditions, purified rhodopsin samples in detergent are reconstituted into lipids, which are dried into a film with an aspirator. The reconstituted rhodopsin sample into lipids is closer to native environment than the rhodopsin solution in detergent.

The reconstituted rhodopsin sample into lipids is generally turbid. Transparent samples are suited for Raman spectroscopic measurements, still turbid samples are applicable for the measurements with a disadvantage of reduced signal-to-noise ratio. In order to obtain spectra with high signal-to-noise ratio, nanodiscs are often used, where rhodopsin is solubilized transiently with a detergent in the presence of phospholipids and an encircling amphipathic helical protein belt that is referred to as a membrane scaffold protein (MSP). Upon removal of the detergent, rhodopsin simultaneously assembles with phospholipids into a discoidal bilayer. Nanodiscs are transparent, being suitable for Raman spectroscopy.

Figure 1a shows the sample preparation of hydrated films. The sample is first dried, and for hydration, a small aliquot (~1 μL) of either H_2O or its stable isotope (D_2O, $H_2{}^{18}O$ or $D_2{}^{18}O$) is added next to the film on the IR window, which is then sealed with the second IR window and a rubber ring and cooled. Thus, in this case, hydration stops when water becomes frozen. In contrast, if the measurements are performed at temperatures >260 K, hydration is continuous and difficult to control. Therefore, a dry film is typically hydrated by placing ~1 μL of 20% glycerol/water (v/v) next to the film on the IR window to avoid continuous hydration. Most protein molecules are randomly oriented on the IR window. In contrast, BR molecules display an orientational preference on the IR window due to their propensity to form PM, the two-dimensional crystal. Polarized FTIR spectroscopy of the oriented BR film can determine particular angles of the dipole moment of the vibration in question, in addition to its frequency. If a higher water content is needed to facilitate protein–protein interaction or enzyme turnover, the sample is re-dissolved (Fig. 1b). In this case, water drop is placed not next to the film, but onto the film. Spectral comparison between Fig. 1a, b clearly shows that re-dissolved sample contains much more water molecules than the hydrated films.

Continuous irradiation by laser light for Raman measurements can damage the proteins in the sample; therefore, it is necessary to avoid continuously exposing the same volume of the sample to the excitation light. When the sample volume is sufficient, the flow cell (Fig. 2a) should be used to measure resonance Raman spectra. Samples can be recycled if the rhodopsin exhibits a photocyclic reactions. When the sample volume is insufficient (a few mL or

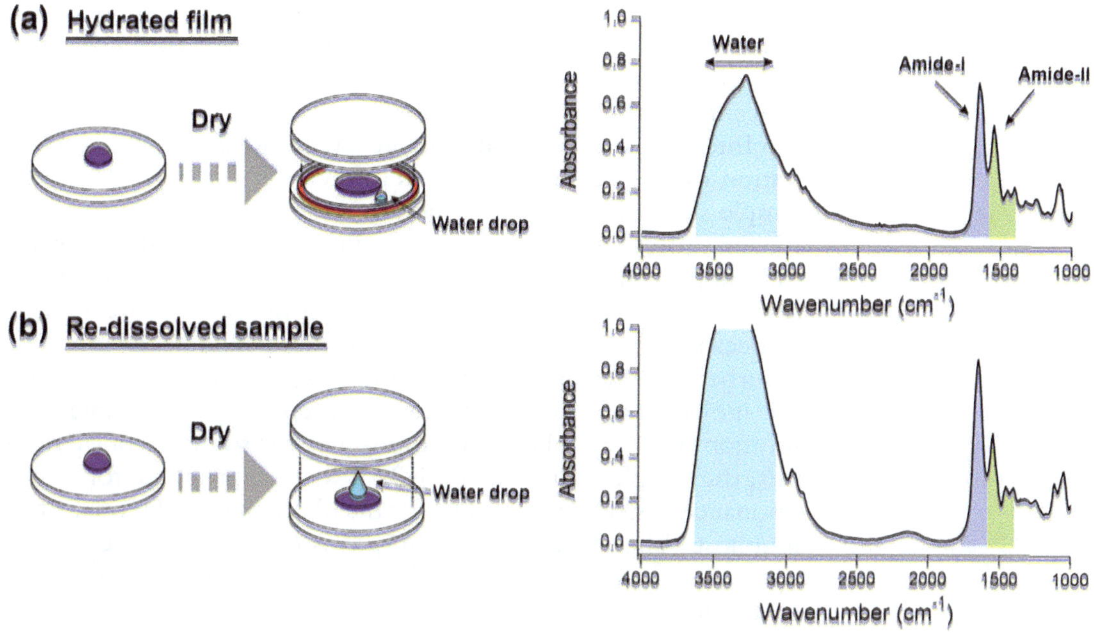

Fig. 1 Summary of sample preparations of rhodopsins for light-induced difference FTIR spectroscopy, together with absorption spectra in IR region. To prepare hydrated film (**a**) or re-dissolved sample (**b**), the sample solution is first dried on an IR window (BaF_2), followed by addition of buffer next to or onto the film, respectively. Water content in the sample is much higher for re-dissolved sample than for hydrated film, as is clearly shown in the 3700–3000 cm^{-1} region (light blue) in each spectrum. Amide-I and –II correspond to C=O stretch and C-N stretch/N-H bending of peptide backbone, while the former also contains O-H bending of water. This figure is modified from Ref. [8]

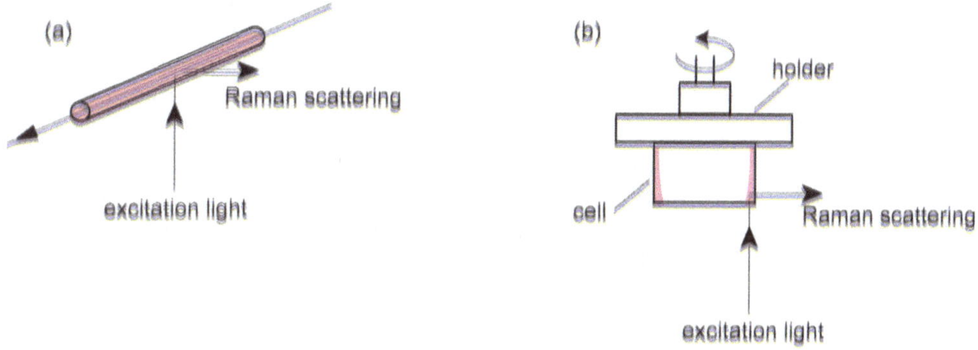

Fig. 2 Spectroscopic cells often used for Raman measurements of protein samples. (**a**) a flow cell, (**b**) rotating cell

less), the use of a rotating cell is effective. In this case, more attention must be paid to the contribution of the photoreaction products. As shown in Fig. 2b, a cylindrical cell with a diameter of 10–50 mm and a height of 10–20 mm is mounted on a motor through a metal holder and rotated at a speed of 500–2000 rpm. Another advantage of the rotating cell is that it is possible to gain height even with a small amount of sample since the liquid sample inside the cell sticks to the wall of the cell due to centrifugal force. For the use of a rotating cell, it is important to minimize the off-center vibration of the rotation.

Intensity of Raman scattering generally increases as the concentration of the sample increases. However, in the case of colored sample such as solutions of rhodopsins, the laser beam to excite Raman scattering does not penetrate into the sample sufficiently for the sample where the concentration of the protein is too high, resulting in the decrease of Raman scattering of the proteins. This is because the excitation light and the Raman scattering are absorbed by the sample. The optimal concentration depends on the molecule of interest and can be found by measurements of solutions with several concentrations. For the solutions of rhodopsins, the sample solutions of 10–30 μM are appropriate for visible resonance Raman spectroscopy for the observations of retinal chromophore. The solution of 30–50 μM is often used for UV resonance Raman spectroscopy for the observations of aromatic amino acid residues.

3 Methods

3.1 FTIR Spectroscopy

3.1.1 Low-Temperature FTIR Spectroscopy

The spectra in Fig. 1a, b originate from IR absorption of the whole rhodopsin sample, but the structural information is very limited. In contrast, difference FTIR spectra upon light absorption provides highly accurate and reproducible spectra, which allows us to calculate the difference spectrum with minute spectral changes ($\sim 10^{-6}$) caused by photo reaction in rhodopsins [4, 6, 8–11]. They correspond to single vibrations of the retinal chromophore, protein, lipid, and protein-bound water molecules, and consequently, fruitful structural information is gained. Difference FTIR spectroscopy of rhodopsins is based on a commercial FTIR machine, and photointermediates are trapped at low temperatures or time-resolved by rapid-scan or step-scan techniques. Difference spectra derived by subtracting the photointermediate spectrum to its resting state spectrum (light minus dark spectrum) is the standard data type in light-induced difference FTIR spectroscopy.

Low-temperature light-induced difference FTIR spectra of rhodopsins are measured using commercial FTIR spectrophotometers, to which a cryostat is equipped with liquid nitrogen as a coolant [4, 8, 10]. Figure 3 illustrates the optical alignment of an FTIR spectrophotometer. He-Ne laser is used to measure exactly the position of the movable mirror, resulting in precise measurements of wavelength. A liquid N_2-cooled Mercury Cadmium Telluride (MCT) detector is normally used. Illumination light of rhodopsins is introduced either in parallel or perpendicular to the IR beam. The FTIR spectrometer has a movable mirror to take out

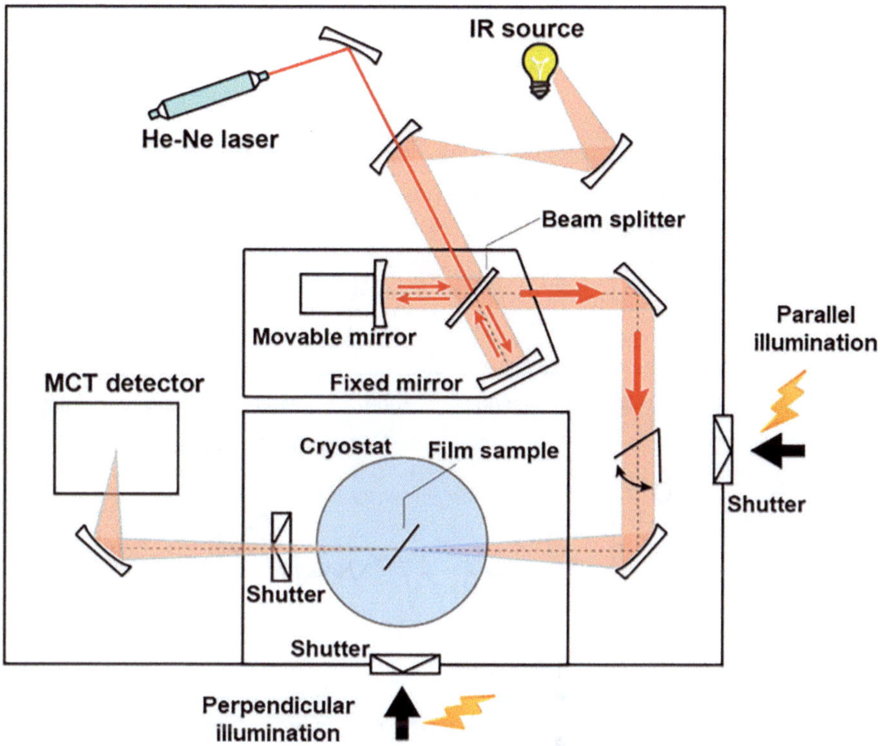

Fig. 3 Typical experimental setup for light-induced difference FTIR spectra at low temperatures. A cryostat is equipped in a commercial FTIR machine, and liquid nitrogen is used as a coolant. A tungsten-halogen lamp or a Xenon lamp is used as the illumination light source, and several optical filters are used. Open/close of shutters is controlled by the switch box. Illumination light is introduced either parallel or perpendicular to the IR beam. This figure is modified from Ref. [8]

the IR beam after the interferometer, whose position changes in order to accommodate illumination. In the case of parallel illumination, a shutter behind the sample is closed during illumination, and the difference of the after-minus-before illumination is measured. For perpendicular illumination, the sample, which is tilted by 45°, is illuminated from the side. In this case, the difference of the during-minus-before illumination can be measured, which is advantageous, as it can capture short-lived intermediates. Polarized FTIR spectroscopy is applied to the oriented rhodopsins films by placing an IR polarizer in front of the detector, from which the dipolar orientation can be obtained in addition to the frequency.

During the photoactivation processes of animal and microbial rhodopsins, many intermediate states are formed such as batho, lumi, meta-I, and meta-II for animal rhodopsins and K, L, M, N, and O for microbial rhodopsins [1]. Many of these transient photointermediates can be stabilized at low temperatures. Figure 4a shows absorption spectra of hydrated films of monkey green pigment (MG) in D_2O hydration at 77 K [8, 12]. The spectra before

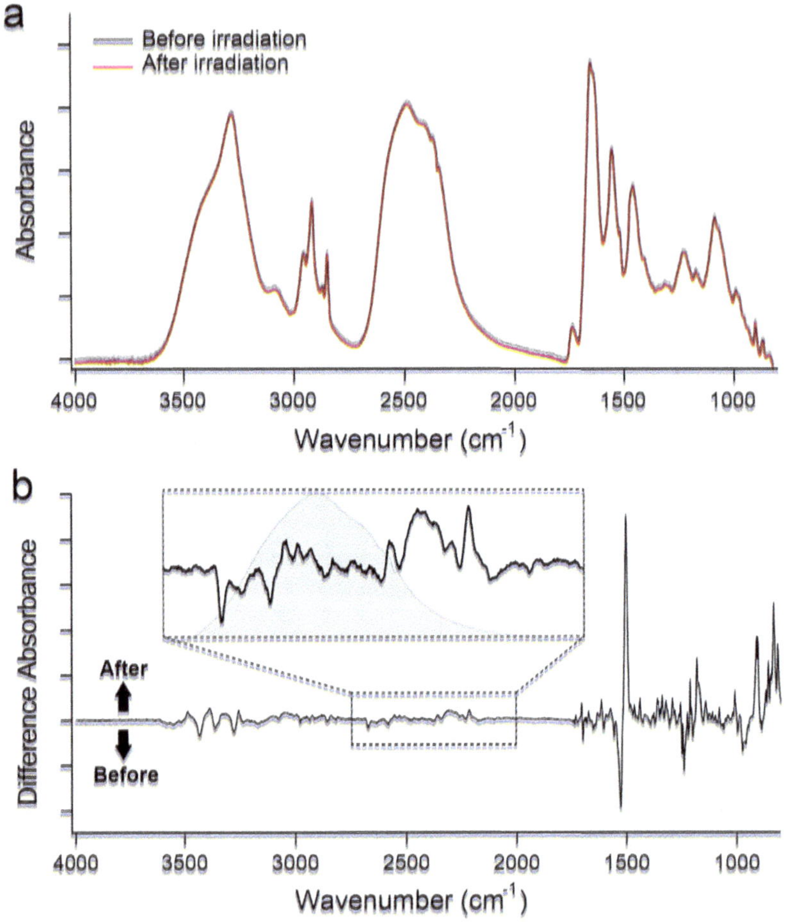

Fig. 4 (**a**) Absorption spectra of hydrated films of monkey green pigment (MG), being hydrated with D_2O. Black and red curves represent the spectra before and after light irradiation at 77 K. One division of the *y*-axis corresponds to 0.1 absorbance unit. (**b**) Light-induced difference FTIR spectra (after-minus-before illumination) of MG. The hydrated film was illuminated by a 501 light, which was supplied with a combination of a halogen-tungsten lamp (1 kW) and an interference filter. One division of the *y*-axis corresponds to 0.001 absorbance unit. Inset shows the expanded spectrum in the X-D stretch region for protein-bound water molecules. This figure is modified from Ref. [8]

(black) and after (red) light irradiation coincide very well with each other, but the difference of the after-minus-before illumination provides reproducible spectral features (Fig. 4b). This is the light-induced (light-minus-dark) difference FTIR spectrum, whose one division corresponds to 0.001 absorbance units. Batho-intermediate is the state stabilized at 77 K, so that the positive and negative bands originate from vibrational signals of batho-intermediate and MG, respectively. It should be noted that MG is an animal rhodopsin that bleaches upon light absorption. However, at 77 K, illumination of batho-intermediate at >610 nm causes reversion into the original MG, and such photochromic property is highly advantageous to improve the signal/noise ratio by accumulation. In case of MG, 128 interferograms were accumulated for each measurement, and 64 recordings were averaged to obtain the difference spectrum in Fig. 4b. The obtained spectrum is accurate and reproducible not only in the conventional 1800–800 cm^{-1} region, but also at >1800 cm^{-1}.

The strongest peak pair in Fig. 4b appears at around 1550–1500 cm^{-1}, being assignable to the C=C stretch of the retinal chromophore. As the C=C stretch of the retinal chromophore has the correlation with visible absorption maxima, downshifted frequency in the C=C stretch corresponds to the spectral red-shift upon formation of the batho-intermediate. Strong peaks at 1250–1100 cm^{-1} and 900–800 cm^{-1} originate from C-C stretch and hydrogen out-of-plane (HOOP) vibrations of the retinal chromophore, respectively. Thus, strong peaks in the difference FTIR spectra come from the retinal chromophore, which are unclear in the absolute spectra in Fig. 4a. This is reasonable, as difference FTIR spectra extract the vibrations whose frequencies are altered, and retinal chromophore is responsible for the conversion of light into protein structural changes. In addition, vibrational signals due to protein are observable in Fig. 4b, though batho-intermediate is the primary intermediate after retinal photoisomerization. The C=O stretching vibrations of protonated carboxylic acids appear in the 1800–1700 cm^{-1} region, which provided fruitful information about proton transfer.

In Fig. 4a, broad bands at 3600–3000 cm^{-1} or at 2700–2200 cm^{-1} correspond to O-H/N-H or O-D/N-D stretching vibrations, respectively, while sharp peaks at 3000–2800 cm^{-1} are C-H stretching vibrations. O-H and N-H groups are generally deuterated in D_2O, so that the bands at 3600–3000 cm^{-1} or at 2700–2200 cm^{-1} in Fig. 4b originate from H/D unexchangeable X-H stretches or X-D stretches, respectively. Functionally important protein-bound water molecules are deuterated in D_2O, and their O-D stretches are lowered when hydrogen bond becomes strong. Gray line in the inset of Fig. 4b exhibits absorption spectra of deuterated water (D_2O). Water vibration is identified by comparing the spectra between D_2O and $D_2{}^{18}O$, and water O-D stretches exhibit isotope shift. This method was successfully applied to BR, where six water O-D stretches were observed at 2690–2170 cm^{-1}. Of particular note is that an O-D stretch of water was observed at 2171 cm^{-1} in BR, which is very unique [13]. Absorption spectrum of D_2O is clearly seen in Fig. 5 (light gray line), where weak, moderate, and strong hydrogen-bonded water possess the O-D stretches at >2600 cm^{-1}, 2600–2400 cm^{-1}, and < 2400 cm^{-1}, respectively. As the frequency at 2171 cm^{-1} (BR_{AT} in Fig. 5) is located at the low-frequency tail of light gray spectrum, the water forms extremely strong hydrogen bond. The

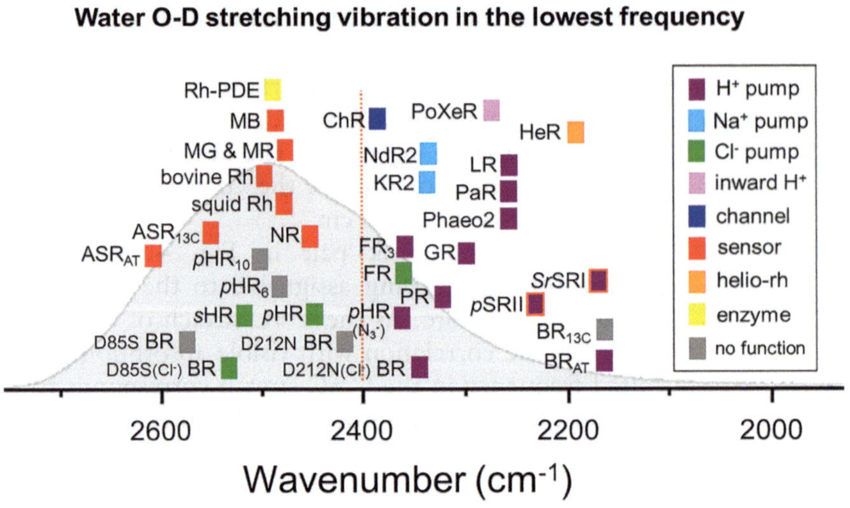

Fig. 5 Detection of strongly hydrogen-bonded water molecules. In the light-induced difference FTIR spectra of various rhodopsins at 77 K, the lowest frequency among the observed water O-D stretching vibrations is plotted for the unphotolyzed state. Purple squares represent the results of the proton pump, while light-blue, green, pink, dark-blue, red, orange and yellow squares are those of the sodium pump, chloride pump, inward proton pump, cation channel, sensor, heliorhodopsin, and enzyme, respectively. The proteins represented by dark gray squares have no function. Light gray spectrum shows O-D stretching vibration of D_2O, and proton-pumping rhodopsins possess strongly hydrogen-bonded water (O-D stretch at <2400 cm^{-1}). This figure is modified from Ref. [4]

origin was investigated by use of mutants, and the 2171-cm^{-1} band was determined to the bridged water between the Schiff base and Asp85. Interestingly, mutations of Asp85 or Asp212 diminish both proton-pump function and strongly hydrogen-bonded water (OD stretch at <2400 cm^{-1}), while D212N mutant restores proton-pump function together with strong hydrogen-bonded water in the presence of chloride ion (Fig. 5).

Through several studies conducted on BR, it has been well established that strong hydrogen-bonded water molecules play a crucial role in its function. The analysis was extended to various microbial rhodopsins to elucidate whether they contain strong hydrogen-bonded water or not in their resting state (O-D stretch at <2400 cm^{-1}). As a result, strong hydrogen-bonded water molecules were found in all light-driven proton pumps, but not for chloride pumps and sensors. While strong hydrogen-bonded water molecules were observed only for proton-pumping rhodopsins until 2012, the emergence of new rhodopsin functions complicated this correlation. As shown in Fig. 5, strong hydrogen-bonded water molecules were also observed for other rhodopsins such as light-driven sodium pumps, a eubacterial chloride pump, an inward proton pump, channelrhodopsin, and heliorhodopsin. These results clearly show the wide presence of strong hydrogen-bonded water molecules in microbial rhodopsins. As the Schiff base is protonated and its positive charge is stabilized by counterion(s), it is likely that such water molecules hydrate a negatively charged counterion.

3.1.2 Time-Resolved FTIR Spectroscopy

It is generally accepted that intermediates of rhodopsins trapped at low temperatures display absorptions essentially equivalent to their room temperature counterparts. Then, what about protein structure and protein-bound water molecules? To answer these questions, room temperature measurements using time-resolved FTIR spectroscopy is needed. Rapid-scan measurements are based on the standard FTIR spectroscopy, where fixed and movable mirrors are used to obtain an interferogram for FTIR spectra. Therefore, time-resolution of rapid-scan FTIR spectroscopy is limited to sub-second time resolution. In contrast, step-scan FTIR spectroscopy significantly improved the time resolution into 10 ns, where the movable mirror is stopped at each data collection point. When laser irradiation is synchronized to the data acquisition, time-resolved FTIR measurements at high temporal resolution are possible. Step-scan time-resolved FTIR spectroscopy is also based on the commercial FTIR spectrophotometers, where synchronization of data acquisition and laser triggering is performed with a delay pulse generator. Single-beam spectra recorded before laser excitation are averaged and used to transform single-beam spectra to absorbance spectra.

Figure 6 shows step-scan time-resolved FTIR difference spectra of BR at room temperature, where the L and M intermediates dominate the difference spectra at early (2.5–20 μs) and late (700 μs to 5.3 ms) timescale, judged from the low-frequency region (right panel) [14]. The time-resolved spectra in the 3800–2800 cm^{-1} region contain broad bilobed spectra with a positive absorbance between 3700 and 3500 cm^{-1} and a negative absorbance between 3500 and 2400 cm^{-1}. This broad band originates from transiently heated bulk water, as it coincides well with the temperature-induced spectral change (blue line in Fig. 6). This suggests that the local heat from the sample volume is released on the time scale similar to that of the L-M transition. Time-resolved room temperature FTIR difference spectra for the BR photocycle showed that the room temperature M-minus-BR spectrum was almost identical to that at 230 K, even in the frequency region of water stretching, suggesting the same hydrogen-bonding network at room and low temperatures.

Time-resolved FTIR spectroscopy has proven to be extremely useful in studying the proton release group (PRG) in the extracellular domain of BR. Although two glutamic acids at the extracellular surface, Glu204 and Glu194, were good candidates for the PRG, a deprotonation signal of carboxylic acid was not clearly detectable. Previous time-resolved FTIR studies reported a broad

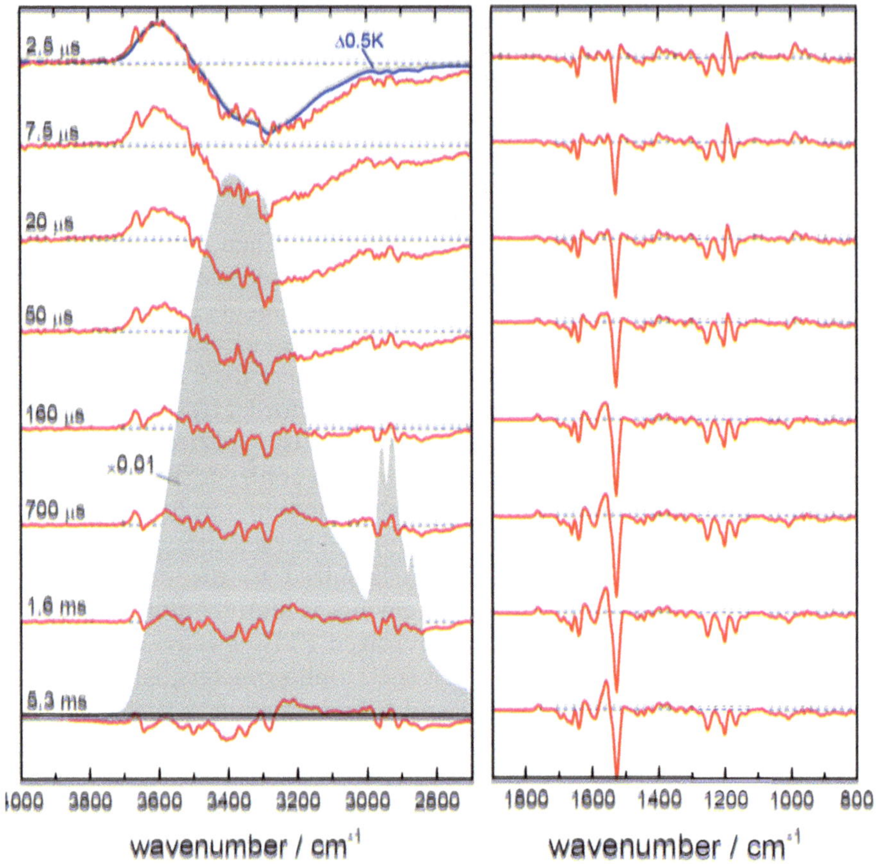

Fig. 6 Time-resolved FTIR difference spectra of BR. Left and right panels show the 3800–2800 cm^{-1} and 1850–800 cm^{-1} regions, respectively. This figure is modified from Ref. [14]

negative IR continuum band at 2000–1800 cm^{-1} in the M-minus-BR difference spectrum and proposed that the continuum band originates from a protonated water cluster [15]. It is now established that the PRG in BR is a protonated water cluster, although the structure of the water cluster and its mechanism of high pK_a are still under investigation.

3.1.3 Attenuated Total Reflection FTIR Spectroscopy

In difference FTIR spectroscopy, light is an ideal stimulus, and thus the obtained spectra even enable the analysis of protein-bound water molecules. Although it has not been easily applied for other stimuli such as ion- and ligand-binding, attenuated total reflection (ATR)-FTIR spectroscopy allows the accurate detection of difference FTIR spectra. In the case of ATR-FTIR spectroscopy of rhodopsins, samples are reconstituted into a membrane and physically adsorbed onto an ATR prism (Fig. 7a), and two buffers, A and B, are introduced alternately, from which difference FTIR spectra can be obtained (Fig. 7b). Thus, it is possible to obtain ion- and ligand-binding-induced difference spectra. It should be

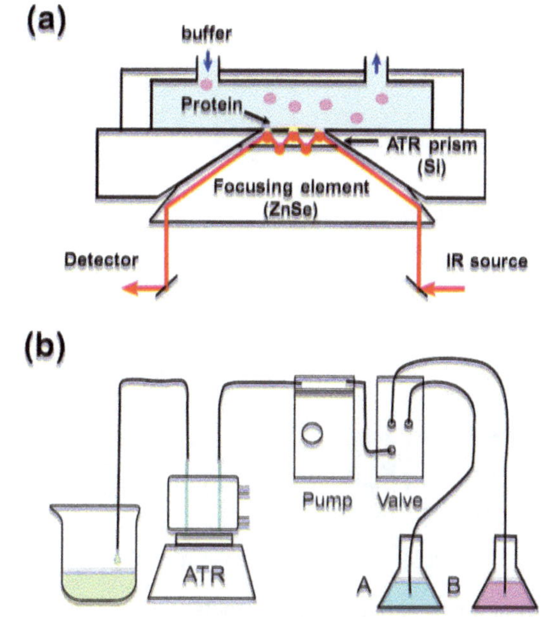

Fig. 7 Experimental setup for stimulus-induced difference ATR-FTIR spectroscopy. (**a**) Sample compartment for membrane proteins reconstituted into lipids. Samples are attached to the ATR cells, which are covered by a glass jacket to maintain the temperature and to introduce buffer solutions to the sample. (**b**) Buffer exchange system. The exchange of buffer solutions is regulated by a chromatographic system with a computer-controlled solenoid valve. Using this system, it is possible to obtain the difference FTIR spectra between buffers A and B. This figure is modified from Ref. [4]

noted that "physical adsorption" does not require any special treatment for samples in the case of membrane proteins. In the case of soluble proteins and membrane proteins solubilized by detergents, samples are easily detached from the ATR cells upon buffer exchange. In contrast, a weak interaction between membrane proteins in lipids and the ATR cell surface is sufficient for adsorption during measurements.

ATR-FTIR spectroscopy was applied to a light-driven sodium pump KR2 upon its discovery [16]. As the retinal molecule is a good internal probe, a color change following changes in the concentration of ions is used to judge specific binding of ions to rhodopsins. KR2 does not show any change in color between the presence and absence of sodium ions, suggesting no binding. However, if the sodium binding site of KR2 is distant from retinal, the binding of sodium to KR2 cannot be studied by conventional UV-visible absorption spectroscopy. This was indeed the case for KR2, and ATR-FTIR spectroscopy clearly observed conformational changes of KR2 upon sodium binding [16]. This was confirmed by the X-ray structure of KR2 [17].

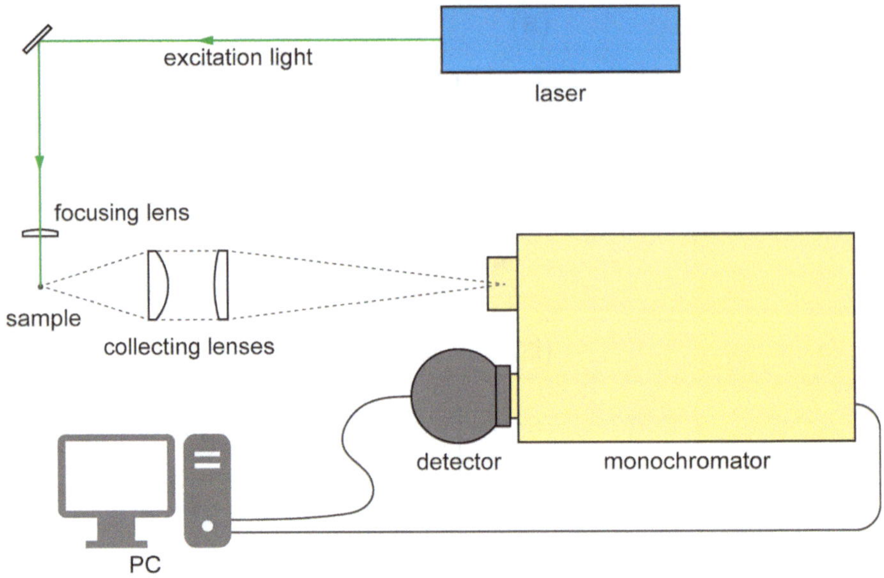

Fig. 8 Overview of a typical Raman spectrometer

3.2 Raman Spectroscopy

3.2.1 Resonance Raman Spectroscopy

Figure 8 shows an overview of a typical Raman spectrometer. Main components of the spectrometer are a laser for Raman excitation, collecting optics, a monochromator, and an optical detector. The excitation light is focused to protein sample. Raman scattering is collected and focused onto the entrance slit of the monochromator. Then it is dispersed in the monochromator and finally detected by the optical detector.

The excitation light is focused into the sample solution by ~100-μm width in the flow cell or the rotating cell. Since the scattering volume is small, fine adjustment of the position of the scattering is necessary to deliver the Raman scattering into the monochromator. For this purpose, the cell is usually placed on a translational stage that its position can be moved finely in *x*-*y*-*z* directions. It should be noted that a broad Raman bands due to glass may be observed in the 300–500 cm^{-1} region. Fine adjustment of the cell position reduces the contribution of the glass to the observed Raman spectra. The contribution of the spectral component of the glass cell can also be reduced by using cell with thin glass.

In the measurements of the resonance Raman spectra, photochemical and/or photophysical products are generated by the irradiation of the excitation light to the sample because the wavelength of the excitation light is close to the absorption maximum wavelength of the sample. Therefore, the care is needed not to observe spectral components due to the photoproducts in the Raman spectra. This can be achieved by reducing the intensity of the excitation light. Only one photon is required for the generation of Raman scattering in the electronic ground state, while two photons are needed for the electronic excitation and the excitation of Raman scattering of the photoproducts. Accordingly, the spectral contributions of the photoproducts become negligible as the excitation intensity is reduced.

Dependence of the observed spectra on the power of the excitation light is examined to check whether the observed resonance Raman spectra contain no spectral contamination of the photoproducts. The power of the excitation light is chosen in the range where the Raman band intensity increases in proportional to the power and where the relative intensities of the Raman bands due to the proteins in the spectrum have not changed. Figure 9 shows an example of the check for the absence of the spectral components due to the photoproducts for heliorhodopsins 48C12 [18]. The UV-visible absorption spectra of the samples are measured before and after the resonance Raman measurements to check the integrity of the sample.

In order to obtain accurate Raman spectra from the measured data, it is necessary to calibrate the horizontal axis (i.e., Raman shift) of the Raman spectra. For example. Using 500-nm excitation light, a difference in wavenumber of 1 cm^{-1} corresponds to a difference in wavelength of 0.025 nm. Therefore, an accuracy of 0.025 nm in wavelength determination is required to determine Raman shift of the spectrum with an accuracy of 1 cm^{-1}. Such a high degree of accuracy in wavelength determination is beyond reproducibility of the wavelength scan of the monochromator. Therefore, it is required to calibrate Raman shift based on standard spectra.

Indene is one of the compounds for which accurate Raman shifts have been reported as a standard [19]. It gives strong and sharp Raman bands in the 500–1700 cm^{-1} region, making it a good reference for the calibration. However, since indene is readily oxidized in air and the oxidized indene emits strong fluorescence, it is necessary to use distilled and purified indene in a sealed tube. The purified and sealed samples can be used for about 1 year. Common organic solvents can be practically used for the calibration of Raman shift although the accuracy of the wavenumber determination by organic solvents is less than that by indene. For example, carbon tetrachloride and cyclohexane are used for the calibration in 200–1000 cm^{-1} region. For the 1000–1800 cm^{-1} region, cyclohexane and acetone are appropriate for the calibration.

An example of the Raman shift calibration is shown in Fig. 10. Light is spatially dispersed in the focal plane on the exit side in proportion to the wavelength in a Czerny-Turner-type monochromator. A pixel number-wavelength correlation is linear and thus more appropriate for least-squares fitting than a pixel number-Raman shift correlation. Therefore, we first calculate the wavelength of Raman scattering of the observed bands based on the absolute wavenumber of the excitation light in vacuum and the

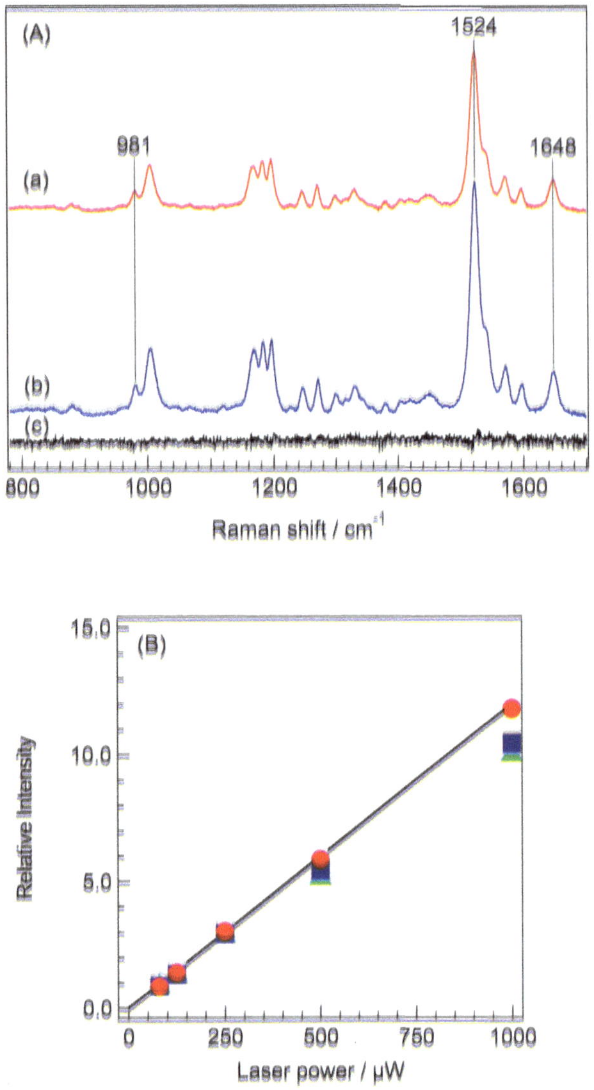

Fig. 9 Dependence of the probe power on the resonance Raman intensities of heliorhodopsins 48C12. Panels A displays resonance Raman spectra obtained with different laser power. Traces a and b show the spectra measured with the probe power of 80 and 125 μW, respectively. Trace c shows difference spectrum obtained by (trace a)—0.68 × (trace b). The intensities of trace c were magnified by three times. Panel B displays relative intensities of the resonance Raman bands of the protein with respect to the probe power; green triangles, blue squares, and red circles indicate the relative intensities of the ν(C=C) (1524 cm^{-1}) and the ν(C=N) (1648 cm^{-1}) bands of the retinal chromophore and that due to the sulfate ion (981 cm^{-1}) added as an internal intensity standard, respectively. The intensity for each Raman band in panel B was normalized with the intensity measured with the probe power of 80 μW. The black line represents linear fits for the intensities of the sulfate band. The linearity of the resonance Raman intensities against the probe power was observed in the probe power below 125 μW. Accordingly, the probe power of 80 μW was weak enough to avoid the contamination of spectral contributions of

reported Raman shift of the reference material. In the monochromator, the light is dispersed in proportion to the wavelength in air, but the error of the Raman shift is less than 0.01 cm^{-1} when calibrated using the wavelength in vacuum. Since a linearity holds between the wavelength and the pixel numbers of the band peaks, the least-squares fitting is used to obtain a calibration curve between the wavelength and the pixel numbers (Fig. 10a). From the calibration curve, the values of the Raman shift corresponding to each pixel are calculated (Fig. 10b). After the calculation, Raman shifts of the bands of the reference material should be read and confirmed if deviations are less than 1.0 cm^{-1} (Fig. 10c). A correlation coefficient in the least-squares fitting is an appropriate quantity to check if the calibrated Raman shift is reliable.

The Raman scattering from the sample goes through the collecting optics and the mirrors and gratings in the monochromator before it reaches the detector. Because of ununiform characteristics in transmission and reflectance of the optics as a function of wavelength, relative intensities in Raman scattering at the detector may be different from those at the sample. In addition, photodetection elements of the optical detector have ununiformity in the sensitivity. Therefore, correction of relative intensity in each pixel of the detector is necessary.

A reasonable approximation for the emission spectrum of quinine as a function of absolute wavenumber was obtained in the 17,000–25,000 cm^{-1} absolute wavenumber range [20]. A comparison of a fluorescence spectrum of quinine measured in the identical optical arrangement of Raman measurements and the standard fluorescence spectrum calculated from the approximation formula gives factors to correct the relative intensity in measured Raman spectra. Since the excitation light of Raman scattering can also be used as the excitation light of fluorescence, the fluorescence spectra can be measured in the identical optical arrangement as the Raman measurements simply by replacing the sample with a quinine solution, which facilitates accurate calibration. Figure 11 depicts an example of the intensity correction carried out for resonance Raman spectra of M intermediate of a sodium-ion pumping rhodopsin, KR2 [21].

Figure 12 shows resonance Raman spectra of the retinal chromophore in the unphotolyzed state of KR2 probed at 532 nm [21]. The bands at 1640, 1532–1600, 1382–1445, 1272–1356, 1165–1243, 1006, and 822–971 cm^{-1} are ascribed to the C=N stretching modes at the Schiff base, the ethylenic C=C stretching modes, deformations of the methyl groups, the C–CH/C–NH rocking modes, the skeletal C–C stretching modes, the methyl

Fig. 9 (continued) the photolyzed species. (Reproduced with permission from J. Phys. Chem. Lett. 2018 9, 6431–6436, DOI: 10.1021/acs.jpclett.8b02741. Copyright 2018 American Chemical Society)

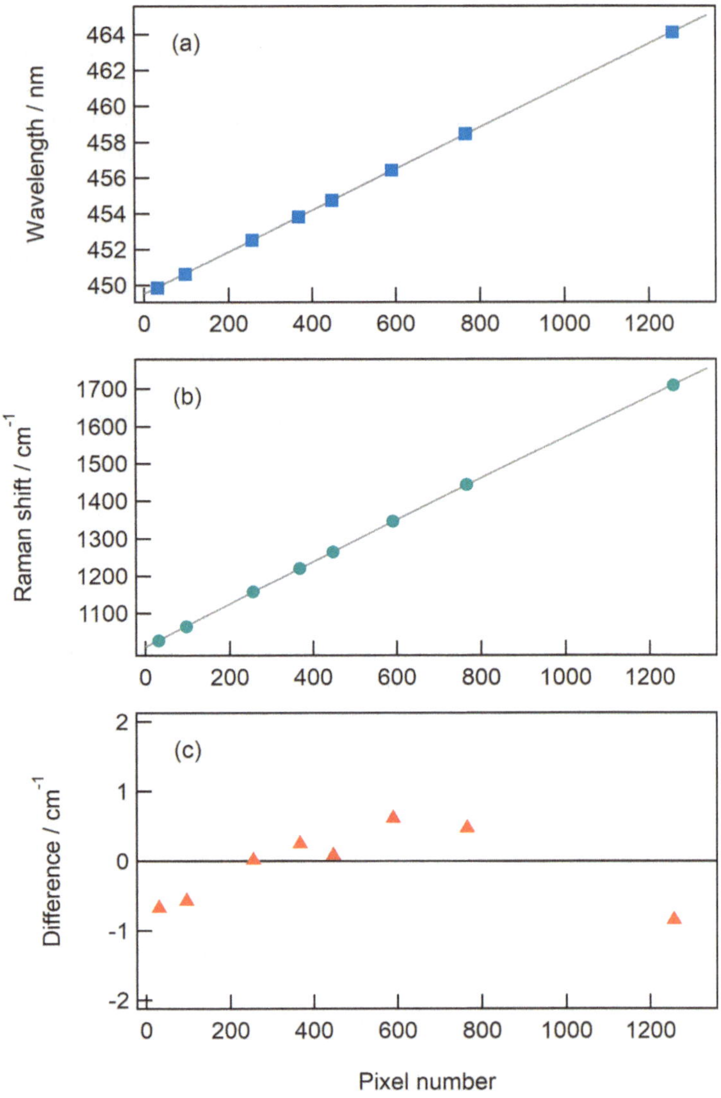

Fig. 10 An example of the Raman shift calibration. Raman bands at 1443.9, 1347.1, 1265.5, 1157.6, and 1027.9 cm^{-1} of cyclohexane and those at 1709.6, 1221.4, and 1065.9 cm^{-1} of acetone were applied to the calibration

rocking modes, and the hydrogen-out-of-plane (HOOP) wagging modes, respectively [22]. The C=N stretching band is sensitive to the protonation state of the Schiff base and the hydrogen bond strength of the protonated Schiff base. Frequencies of the ethylenic C=C stretching modes reflect delocalization of π electrons in the chromophore. The skeletal C–C stretching bands are sensitive to the configuration of the chromophore. The HOOP bands are resonantly enhanced when the polyene chain is twisted; thus, their intensities are good structural markers for distortion of the chromophore.

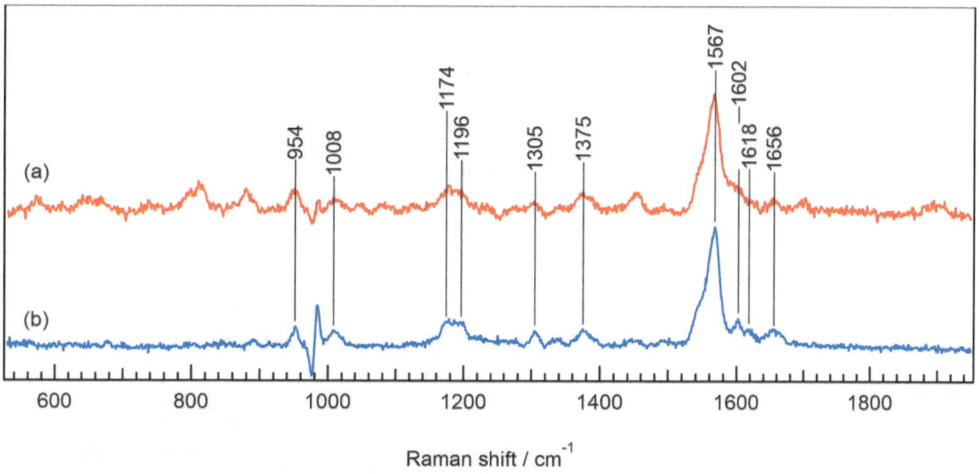

Fig. 11 An example of the intensity correction carried out for resonance Raman spectra of L intermediate of a sodium-ion pumping rhodopsin, KR2. Traces a and b indicate raw and corrected spectra, respectively. It should be noted that the ununiformity in the sensitivity of the photodetection elements results in false bands in raw spectra as shown in trace a when the fluorescence background is high

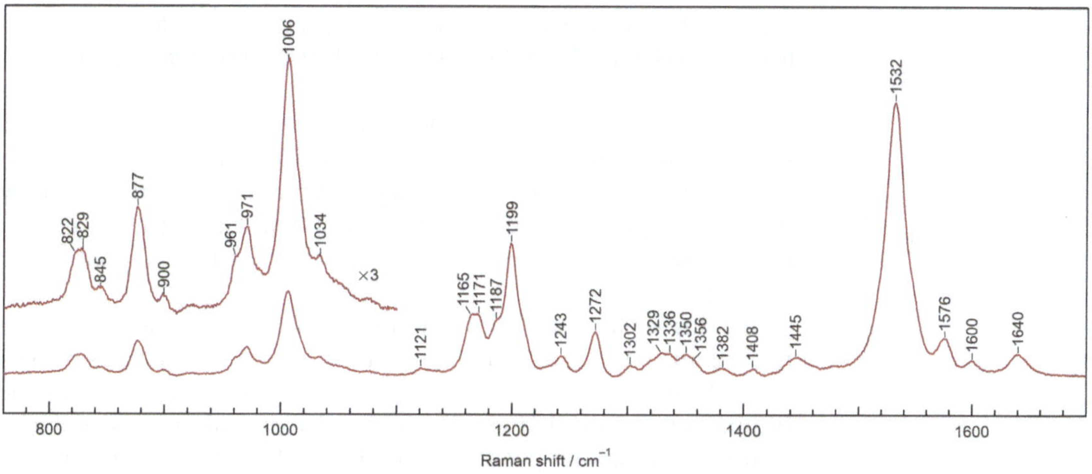

Fig. 12 Resonance Raman spectrum for the unphotolyzed state of KR2 probed at 532 nm

3.2.2 Time-Resolved Resonance Raman Spectroscopy

For time-resolved measurements, two types of methods are adopted. One is pump-probe method, in which a pump pulse initiates the photoreaction of the retinal chromophore and a probe pulse excites Raman scattering of the protein. The setup at the sample position is shown in Fig. 13a. The pump pulses were spatially overlapped on the probe pulse at the focal point on the sample cell. To measure time-resolved spectra on the picosecond to early nanosecond timescales with a few picosecond resolution, pump-probe measurements are carried out by using 2 ps pulses generated from output pulses of a picosecond laser [4, 23]. Timing between the probe and pump pulses was controlled with a

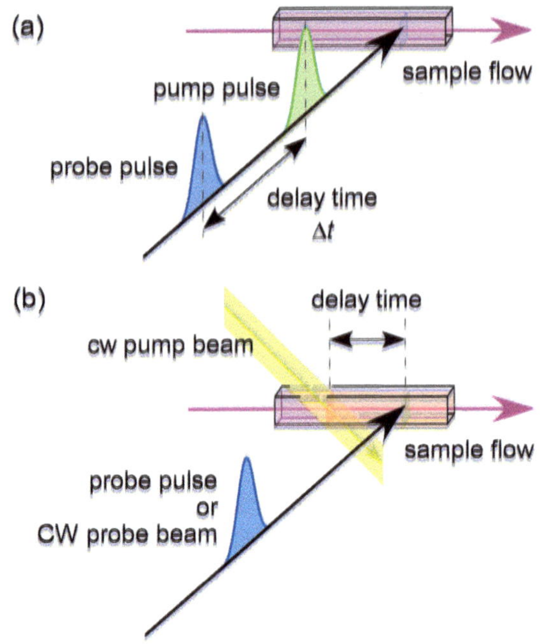

Fig. 13 Setups at the sample position for the pump-probe methods. (**a**) Pump-probe method using 2 ns pulses, and (**b**) dual-beam rapid-flow method

computer-controlled optical delay line. To measure time-resolved spectra on the nanosecond to early microsecond timescales with 50-ns resolution, pump-probe measurements are carried out by using 2 ns pulses generated by two lasers. Timing between the probe and pump pulses was controlled with a digital pulse generator.

A dual-beam rapid-flow method is used to obtain spectra on the late microsecond to millisecond timescales with 100-μs resolution. Time separation between the two-color pump and probe beams was controlled by changing the spatial separation. The setup at the sample position is shown in Fig. 13b. A continuous wave laser as a pump beam initiates the photoreaction in the sample. The pump beam was introduced to the sample solution and focused into a line on the sample using a cylindrical lens. Timing between the probe and pump beams was controlled by changing the irradiating position of the pump beam using a computer-controlled optical stage.

References

1. Ernst OP, Lodowski DT, Elstner M, Hegemann P, Brown LS, Kandori H (2014) Microbial and animal rhodopsins: structures, functions, and molecular mechanisms. Chem Rev 114:126–163. https://doi.org/10.1021/cr4003769
2. Govorunova EG, Sineshchekov OA, Li H, Spudich JL (2017) Microbial Rhodopsins: diversity, mechanisms, and Optogenetic applications. Annu Rev Biochem 86:845–872. https://doi.org/10.1146/annurev-biochem-101910-144233
3. Deisseroth K, Hegemann P (2017) The form and function of channelrhodopsin. Science 357:eaan5544. https://doi.org/10.1126/science.aan5544
4. Kandori H (2020) Structure/function study of photoreceptive proteins by FTIR spectroscopy. Bull Chem Soc Jpn 93:904–926. https://doi.org/10.1246/bcsj.20200109
5. Mizutani Y (2017) Time-resolved resonance Raman spectroscopy and application to studies on ultrafast protein dynamics. Bull Chem Soc Jpn 90:1344–1371. https://doi.org/10.1246/bcsj.20170218
6. Siebert F (1990) Resonance Raman and infrared difference spectroscopy of retinal proteins. Methods Enzymol 189:123–136. https://doi.org/10.1016/0076-6879(90)89283-n
7. Mathies RA (1995) Biomolecular vibrational spectroscopy. Methods Enzymol 246:377–389. https://doi.org/10.1016/0076-6879(95)46018-7
8. Kandori H, Katayama K (2020) Light-induced difference Fourier-transform infrared spectroscopy of photoreceptive proteins. In: Ozaki Y, Baranska M, Lednev IK, Wood BR (eds) Vibrational spectroscopy in protein research. From purified proteins to aggregates and assemblies. Academic Press, San Diego, CA, pp 23–57
9. Gerwert K (1999) Molecular reaction mechanisms of proteins monitored by time-resolved FTIR-spectroscopy. Biol Chem 380:931–935. https://doi.org/10.1515/BC.1999.115
10. Kandori H (2000) Role of internal water molecules in bacteriorhodopsin. Biochim Biophys Acta 1460:177–191. https://doi.org/10.1016/s0005-2728(00)00138-9
11. Kottke T, Lorenz-Fonfria VA, Heberle J (2017) The grateful infrared: sequential protein structural changes resolved by infrared difference spectroscopy. J Phys Chem B 121:335–350. https://doi.org/10.1021/acs.jpcb.6b09222
12. Katayama K, Furutani Y, Imai H, Kandori H (2010) An FTIR study of monkey green- and red-sensitive visual pigments. Angew Chem Int Ed 49:891–894. https://doi.org/10.1002/anie.200903837
13. Shibata M, Tanimoto T, Kandori H (2003) Water molecules in the schiff base region of bacteriorhodopsin. J Am Chem Soc 125:13312–13313. https://doi.org/10.1021/ja037343s
14. Lórenz-Fonfría VA, Furutani Y, Kandori H (2008) Active internal waters in the bacteriorhodopsin photocycle. A comparative study of the L and M intermediates at room and cryogenic temperatures by infrared spectroscopy. Biochemistry 47:4071–4081. https://doi.org/10.1021/bi7024063
15. Rammelsberg R, Huhn G, Lübben M, Gerwert K (1998) Bacteriorhodopsin's intramolecular proton-release pathway consists of a hydrogen-bonded network. Biochemistry 37:5001–5009. https://doi.org/10.1021/bi971701k
16. Inoue K, Ono H, Abe-Yoshizumi R, Yoshizawa S, Ito S, Kogure K, Kandori H (2013) A light-driven sodium ion pump in marine bacteria. Nat Commun 4:1678. https://doi.org/10.1038/ncomms2689
17. Gushchin I, Shevchenko V, Polovinkin V, Kovalev K, Alekseev A, Round E, Borshchevskiy V, Balandin T, Popov A, Gensch T, Fahlke C, Bamann C, Willbold D, Büldt G, Bamberg E, Gordeliy V (2015) Crystal structure of a light-driven sodium pump. Nat Struct Mol Biol 22:390–395. https://doi.org/10.1038/nsmb.3002
18. Otomo A, Mizuno M, Singh M, Shihoya W, Inoue K, Nureki O, Béjà O, Kandori H, Mizutani Y (2018) Resonance Raman investigation of the chromophore structure of Heliorhodopsins. J Phys Chem Lett 9:6431–6436. https://doi.org/10.1021/acs.jpclett.8b02741
19. Hamaguchi H (1988) Calibrating multichannel Raman spectrometers. Appl Spectrosc Rev 24:137–174. https://doi.org/10.1080/05704928808060455
20. Iwata K, Hamaguchi H, Tasumi M (1988) Sensitivity calibration of multichannel Raman spectrometers using the least-squares-fitted fluorescence spectrum of quinine. Appl Spectrosc 42:12–14
21. Nishimura N, Mizuno M, Kandori H, Mizutani Y (2019) Distortion and a strong hydrogen bond in the retinal chromophore enable sodium-ion transport by the sodium-ion pump

KR2. J Phys Chem B 123:3430–3440. https://doi.org/10.1021/acs.jpcb.9b00928

22. Smith SO, Braiman MS, Myers AB, Pardoen JA, Courtin JML, Winkel C, Lugtenburg J, Mathies RA (1987) Vibrational analysis of the all-trans-retinal chromophore in light-adapted bacteriorhodopsin. J Am Chem Soc 109: 3108–3125. https://doi.org/10.1021/ja00244a038

23. Mizuno M, Mizutani Y (2015) Protein response to chromophore isomerization in microbial Rhodopsins revealed by picosecond time-resolved ultraviolet resonance Raman spectroscopy: a review. In: Recent progress in colloid and surface chemistry with biological applications, vol 5., Chap 16. American Chemical Society, Washington, DC, pp 329–353. https://doi.org/10.1021/bk-2015-1215.ch016

Chapter 11

Single-Cell Resolution Optogenetics Via Expression of Soma-Targeted Rhodopsins

Changyang Linghu, I-Wen Chen, Dimitrii Tanese, Valeria Zampini, and Or A. Shemesh

Abstract

Optogenetics allows control of neural activity in genetically targeted neuron populations by light. Optogenetic control of individual neurons in neural circuits would enable powerful, causal investigations of neural connectivity and function at single-cell level and provide insights into how neural circuits operate. Such single-cell resolution optogenetics in neuron populations requires precise sculpting of light and subcellular targeting of optogenetic molecules. Here we describe a group of methods for single-cell resolution optogenetics in neuron cultures, in mouse brain slices, and in mouse cortex in-vivo, via patterned light and soma-targeted optogenetic molecules.

Key words Single-cell optogenetics, Single-cell control, Functional connectivity, Synaptic transmission, Cell-body-targeted opsins, Somatic optogenetics, Computer-generated holography, Two-photon stimulation

1 Introduction: Targeted Optogenetic Molecules

1.1 General Principles: How Do Opsins Work?

In biological sciences, genetically encoded light switches termed opsins are widely used to control the voltage across the membrane of cells. There are two types of opsins. Type I opsins are light-sensitive membrane proteins that constitute either an ion channel or an ion pump and originate from archaea, bacteria, algae, and fungi. Upon light stimulation, a type I opsin either opens its channel to allow ions to pass through or activates its pump to transport ions through the cell membrane. All the commonly used membrane voltage actuators are type I opsins. For example, light activated ion channels open passive conductance to either cations such as Na^+, Ca^{2+}, K^+, and H^+ [1] or anions such as Cl^- [2–4]. Light-activated ion pumps transport cations such as H^+ or Na^+ out of the cell [5–7], cations such as H^+ into the cell [8], or anions such as Cl- into the cell [9, 10]. Together, these type I opsins

Valentin Gordeliy (ed.), *Rhodopsin: Methods and Protocols*, Methods in Molecular Biology, vol. 2501,
https://doi.org/10.1007/978-1-0716-2329-9_11,

are versatile tools for increasing and decreasing membrane potentials, which are useful for a wide variety of applications including activation and inhibition of the electrical activity in neurons. Type II opsins are G-protein-coupled receptors originating from animalia. Upon light stimulation, a type II opsin initiates a G-protein signal transduction, which leads to altered membrane potential among other biochemical changes.

Light sensitivity of opsins requires a retinal molecule. A single retinal molecule naturally and covalently binds to the transmembrane domain 7 of an opsin molecule in cells. When a photon within a certain range of wavelengths interacts with the retinal molecule, the retinal isomerizes and triggers a conformational change of the opsin, which in turn leads to ion channel opening or ion pump activation.

Currently there exists an abundance of optogenetic molecules for neuron activation or inhibition. Below we describe how type I opsins can be used for perturbing neuron-spiking in dense neural networks at single cell resolution.

1.2 Opsins for Single-Cell Resolution Optogenetics

To optogenetically stimulate entire neural populations, one can express opsins in a large volume of interest in the brain. Illumination of these opsins will cause a synchronous activation (for excitatory opsins) or deactivation (for inhibitory opsins) of action potentials in the illuminated tissue [11]. However, the brain operates at much finer scales than entire populations. To give some examples, neural codes vary from neuron to neuron even within the same neural populations, with neighboring neurons of the same type potentially exhibiting highly differing neural codes [12], and sets of neurons firing sparsely in complex patterns to represent information. It was also found that the firing of even single neurons is capable of altering behavior [13, 14]. To stimulate neurons at a single-cell resolution using optogenetics, one can, in principle, express opsins in neurons and photostimulate them by shining light on individual cell bodies. This optical stimulation of single cells can be achieved by patterning the light. Several light patterning techniques have been developed to precisely tailor light over specific structure of interest [15]. In scattering tissues such as the brain, photostimulating cell bodies using 2-photon (2P) microscopy is the tool of choice. It is possible to precisely pattern 2P illumination over individual cell bodies, opening the way for an optical manipulation of neural circuits at the single-cell level [16, 17]. However, illuminating single-cell bodies expressing wild-type opsins in the brain will not result in single-cell photostimulation. This is because opsins express through the entirety of cellular membranes, from the cell body through axons and dendrites. Since neurons in the brain are densely packed together, cell neurites come in contact with cell bodies of other neurons. Focusing light on a cell body (soma) of interest will lead to artifactual activation of its neighbors via photostimulation of their neurites, a situation termed "crosstalk." We quantified this artifactual photostimulation in cultures and brain slices, by patching an opsin expressing cell and activating its opsin-expressing neighbors. We found that photostimulation of neighboring cell bodies was likely to depolarize the patched cell of interest [18]. The solution for the crosstalk problem was "molecular focusing." By restricting the opsin to the soma, one can avoid activating neurites, resulting in the activation of individual cells only.

Several cell body-targeted opsins, also called somatic opsins, exist today. One such opsin is ChR2 targeted to the cell body. The advantage of using a somatic-ChR2 is that non-targeted ChR2 is heavily used and its expression has been optimized and tested on multiple preparations and animal models. The soma targeted ChR2 was engineered by fusing ChR2 with $K_V2.1$-motif. This is a 65 amino acid sequence [19] from the intracellular loop between transmembrane domains IV and V of the potassium channel $K_V2.1$ [20]. The resultant ChR2- $K_V2.1$-motif enabled single-cell resolution optogenetics in mouse brain slices [21]. Nevertheless, ChR2 is not the most efficient excitatory opsin available and even its soma targeted version may provide relatively low photocurrent, which may result in long latency of dozens of millisecond between the start of the stimulation and the firing of an action potential [21]. Low photocurrent is a limiting factor for a reliable 2P photostimulation. This might lead to a small number of cells that can be activated at the same time. Therefore, high-photocurrent opsins might be more suitable for 2P applications.

Another opsin that can be used for single-cell resolution purposes is CoChR. For similar light powers, non-targeted CoChR has a photocurrent larger than that of ChR2 by a factor of approximately 10 [22]. For this reason, it is suitable for 2P multicell photostimulation as it will require lower light powers to bring the membrane to threshold voltage. By fusing CoChR with the 150 amino acid long N-terminal domain out of the kainate receptor subunit KA2 [23, 24], a somatic CoChR (soCoChR) was produced. SoCoChR effectively minimized crosstalk in cultures and mouse cortical slices, therefore enabling single-cell resolution activation of multiple neurons [18].

For applications that will demand high spiking rates, soma-targeted ChroME (ST-ChroME) can be used. ChroME is an opsin that was generated to increase the photocurrent of Chronos by mutating its pore region. By fusing ChroME to the KV2.1-motif, ST-ChroME was targeted to the cell body [25]. This soma-targeted opsin has a photocurrent larger than that of ChR2 by a factor of 10, while maintaining temporal characteristics and spike fidelity similar to that of soma-targeted Chronos [25].

For inhibition of neural spiking, one can choose from two different soma targeted opsins based on the anion channel rhodopsin (ACR) from the algae *Guillardia theta,* GtACR1 and GtACR2, which have peak-photocurrents at 520 nm and 470 nm, respectively [26]. The soma-targeted opsin based on GtACR1 (ST-eGtACR1) was created by fusing the light activated anion channel GtACR1 [26] to an endoplasmic reticulum (ER) escape motif, thereby generating enhanced eGtACR1 (eGtACR1). This was followed by fusion of eGtACR1 to the $K_V2.1$-motif, which resulted in a soma-targeted eGtACR1. ST-eGtACR1 had its highest photocurrent (termed spectral peak) at 1040 nm, and enabled in vivo suppression of more than 90% of the neural spikes on the target cells. The soma-targeted opsin based on GtACR2 (ST-GtACR2) was created by fusing the light-activated anion channel GtACR2 [26] to an ER escape motif described above followed by fusion to the $K_V2.1$-motif, which caused GtACR2 to be soma targeted. The investigators used single-photon (1P) excitation and found that illumination of cells (at 4.5 mW/mm^2, 470 nm) expressing non-targeted GtACR2 led to spurious action potentials that came from the neurites into the soma (antidromic action potentials). In contrast, in cells expressing ST-GtACR2, the probability for antidromic action potentials decreased (from approximately 45% in GtACR2 expressing hippocampal neurons to approximately 5% in ST-GtACR2 expressing hippocampal neurons [27]).

To achieve photostimulation in deep regions of the brain, opsins can be excited by inserting optical devices capable of delivering light into the volume of interest. This was achieved by using micro-objectives, tapered fibers, multi-light emitting diode (multi-LED) arrays and thin GRIN lenses [28, 29]. Recently, 2P-patterned illumination through gradient index optics (GRIN) lenses was demonstrated, allowing cell-resolved photoactivation [30, 31]. Although these endoscopic approaches are popular [32, 33], they often require an invasive surgery that can cause edema, inflammation, and sepsis [34–36], which might damage or alter the interrogated brain circuits. In addition, some optogenetic experiments aim at stimulating a large population of neurons, which occupy extended volumes, inaccessible to the inserted optical probes. To achieve large-volume and deep-brain optogenetic stimulation, light should reach into opsin-expressing cells within the brain tissue with minimal attenuation. Light penetration is limited by both absorption (mainly by water molecules, oxygenated hemoglobin, and pigments [37]) and scattering, with both absorption and scattering strongly depending on wavelength. For visible light, the optimal range of light penetration is at approximately 600–700 nm [37, 38]. However, most available opsins for optogenetics have a 1P absorption spectral peak at blue to green wavelengths (450 nm–540 nm). Therefore, there was a need for optogenetic molecules with a spectral peak at the red part of the spectrum (termed "red-shifted opsins"). One such red-shifted opsin, which also enables single-cell control is the soma targeted

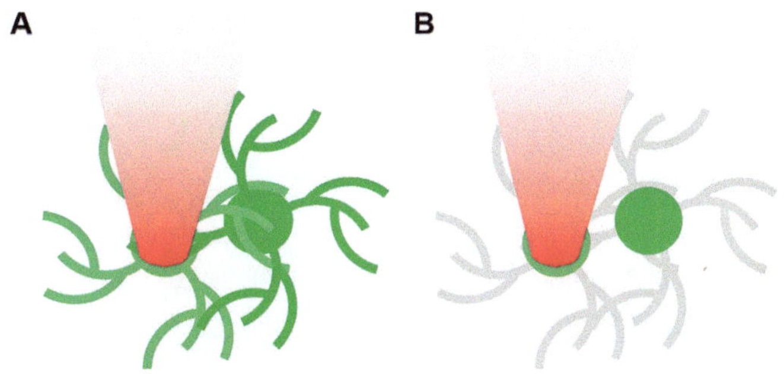

Fig. 1 A schematic of soma-targeted optogenetics. (**a**) Neurons (light and dark green) expressing untargeted opsins throughout the membranes of entire neurons. When photostimulating a single neuron at the soma (red beam), the opsin-bearing neurites from neighboring neurons will also be stimulated, causing unwanted stimulation of these neurons. (**b**) Neurons (gray) expressing soma-targeted opsins (light and dark green) enable single-cell resolution photostimulation

ChRmine (ST-ChRmine) [39] from the ciliate *Tiarina fusus*. ST-ChRmine was able to drive sustained spiking in cortical neurons and has a spectral peak at 585 nm. Under 2P excitation, red-shifted opsins are optimally excited at long infrared wavelength (1000–1200 nm), compared to blue absorbing opsin (excited at around 900 nm), and these wavelengths undergo less scattering and can be better focused on cell bodies in deeper brain regions, up to hundreds of microns.

One particular benefit of single-cell optogenetics is the ability to study at a cellular resolution neurons connected by synapses. Synapses are of particular interest in neuroscience research as they connect neural networks, and since their structural integrity, densities and components are compromised in various diseases of the nervous system (see reviews [40, 41]). For this reason, we dedicate this chapter for demonstrating functional connectivity experiments at single-cell resolution using somatic opsins.

Below we describe how to combine patterned illumination together with expression of somatic opsins, to analyze synaptic connectivity in densely labeled neural networks, at single-cell resolution (Fig. 1). We will describe three protocols for detecting synaptic connections in the rodent brain. One for neuron cultures, one for cortical brain slices, and one for the cortex in vivo.

2 Protocol: Detecting Synaptic Partners Using Single-Cell Resolution Optogenetics in Culture (Fig. 2)

2.1 Materials

2.1.1 Microscopy and Photostimulation

An inverted widefield microscope, where the sample stage can be approached by a microelectrode for electrophysiology, with objective lenses of 5×,10×, 20× (40× optional).

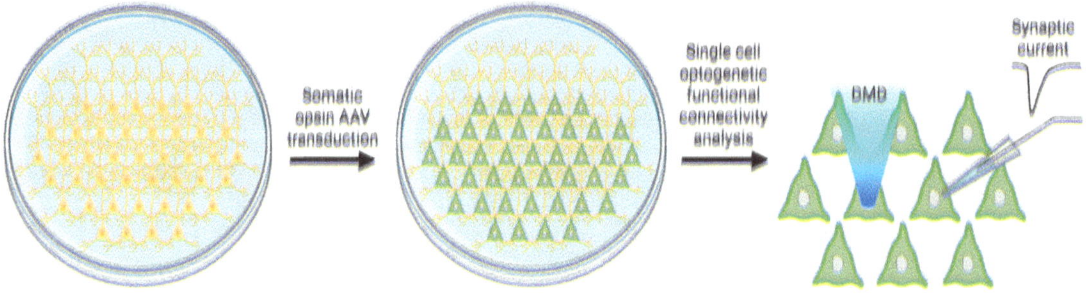

Fig. 2 A schematic of the protocol of single-cell optogenetic functional connectivity analysis in cultured neurons. Cultured neurons (left) are transduced with AAV of soma-targeted opsin and express the opsin (middle) within 10–14 days. Individual cell bodies of opsin expressing neurons (green triangles; right) are located under fluorescent microscope, and a neuron of interest is then electrically probed by patch clamp (glass pipette with wire; right). Selected neurons are then photostimulated at the cell bodies by a digital micromirror device (blue beams from DMD; right) and the electrical responses of the patched neuron are recorded

A digital micromirror device (DMD), which receives light from a light source (light emitting diode, laser, arc lamp) and triggers the electrophysiology system at the onset of photostimulation, with the ability to simultaneously (or sequentially within milliseconds) deliver spatially patterned illumination on multiple regions of interest (ROIs) within the sample.

Optical filters. For blue light excitation, 470 nm ± 20 single-band bandpass filter. For yellow to red excitation, 590 nm ± 20 nm single-band bandpass filter. For imaging the opsin in the red channel (e.g., mCherry) and activation at the green channel (e.g., ChR2), an mCherry filter set with a multiple-band bandpass excitation filter for both 470 nm and 590 nm light can be used.

A digital camera, such as a Complementary Metal Oxide Semiconductor, scientific Complementary Metal Oxide Semiconductor, or Coupled Charged Device (sCMOS, CMOS, or CCD) camera, for fluorescent imaging of the fluorescent proteins fused to the opsins.

LED with nominal wavelength at 470 nm and minimal power of 1–10 mW/mm^2 [22] for blue light opsins such as CoChR, ChR2, Chronos. LED with nominal wavelength at 590 nm and minimal power of 5 mW/mm^2 for yellow to red light opsins such as Chrimson, Halo [6, 22].

2.1.2 Expressing Single-Cell Opsins

Adeno-associated viral vectors, packing genes that encode somatic opsins including, but not limited to SoCoChR, ChR2-Kv2.1, ST-ChroME, and stGtACR2 [18, 21, 25].

Promoter: for all neurons Syn or CAG, for excitatory neurons, CaMKII or EF1alpha, for inhibitory neurons Dlx or h56D promoters [42].

Titers of AAVs can be between 10^{11} and 10^{12} genome copies/mL. Titers at 10^{13} genome copies/mL or above genome copies/mL can lead to toxicity in culture.

2.1.3 Neural Culture

3 cm glass slide, carrying a 1 mm round coverslip.

Matrigel coating for neurons.

10,000–50,000 hippocampal or cortical neurons plated.

Neurons transduced with AAV at 4–9 days in vitro (DIV) which have been expressing the somatic opsin for at least 10–14 additional days (see detailed protocol in Subheading 2.2.1).

2.1.4 Electrophysiology

Head-stage, amplifier, digitizer, and a PC running a recording software.

Borosilicate Glass pipette that was pulled to reach a resistance of 5–10 MΩ and filled with internal solution.

Patch pipette puller.

Reference electrode, made of silver (Ag), coated with AgCl

Internal solution containing 125 mM K-gluconate, 8 mM NaCl, 0.1 mM $CaCl_2$, 0.6 mM $MgCl_2$, 1 mM EGTA, 10 mM HEPES, 4 mM Mg-ATP, and 0.4 mM Na-GTP. pH at 7.3 (adjust with NaOH) and osmolarity at 298 mOsm (adjust with sucrose). Keep frozen at −20 °C for 2 weeks. Keep frozen at −80 °C for up to a year.

Tyrode solution: 125 mM NaCl, 2 mM KCl, 3 mM $CaCl_2$, 1 mM MgC_{l2}, 10 mM HEPES, 30 mM glucose, pH 7.3 (adjusted using NaOH), osmolarity 300 mOsm (adjust using sucrose). Warm up to 37 °C, 1 h prior to experiment, 10 mL of Tyrode per plate.

2.2 Methods

2.2.1 Expressing Somatic Opsins in Neural Culture

Choose opsin of interest, as described in Subheading 1. Subclone it under promoter of interest. CAG promoter will give expression ~3 days post transduction and enable experimenting ~10 days post transduction, but can lead to neural toxicity due to overexpression. Synapsin (Syn) promoter will give expression after 7 days and enable experimenting 14 days post transduction, with minimal toxicity.

Choose a fluorescent protein and fuse it downstream to the somatic opsin. For experiments that involve electrophysiology for the detection of synaptic activity, we suggest using a green fluorescent protein (GFP) due to its brightness and to compatibility with microscope systems.

Aliquot 0.5 μL of opsin AAV at 10^{11}–10^{12} genome copies/mL. Add the AAV to a neural culture dish inside a biosafety hood and return the plate to the 37 °C incubator.

Monitor the expression of the somatic opsin by imaging the fluorescent protein (FP) fused to it. This can be done using an epifluorescent low-magnification culture microscope.

Monitor the health of the neurons using their morphology. This can be done on the culture microscope, using transmitted light. The cell bodies of healthy excitatory neurons usually have pyramidal-shaped boundaries (Fig. 2) as opposed to round-shaped boundaries.

With a Syn promoter, neurons should be ready for experimentation within 14 days. We suggest preparing culture neurons solely devoted to test the maturation of the opsin using electrophysiology.

2.2.2 Preparing the Experimental Setup

At least 45 min prior to the experiment, thaw an aliquot of internal solution on ice.

At least 30 min prior to the experiment, warm up Tyrode solution in a 37 °C bath (water bath or a bead bath).

Prepare patch pipettes using a pipette puller for a pipette resistance of 5–10 MΩ.

Turn on the microscope and set the 5× objective. Test the excitatory 470 nm light (in case of blue light opsins) or 590 nm light (in case of yellow- and red-shifted opsins) to confirm it is operational by turning the light on and off.

Turn on the electrophysiological system, including the digitizer and the amplifier.

Turn on the micromanipulator (if powered by electricity).

Turn on the DMD and the software that controls it.

Perform a "sham" experiment: Take a glass bottom dish, fill it with Tyrode solution. Fill a micropipette with an internal solution. Place the micropipette inside the solution, so that its tip barely touches the solution. Insert the reference electrode into the solution. Check that you can administer a brief (10 ms) current pulse through the electrode that can be recorded through the electrophysiology system. Move the micromanipulator to make sure you full range motion in the x, y, z dimensions. Look at the tip of the micropipette through the microscope eyepiece to verify you can track its tip. Switch to a 20× magnification, at which you will perform the experiment, and photostimulate several regions of interest (ROIs) in sequence.

2.2.3 Finding the Cells of Interest

Take a glass bottom plate. Fill it with 2.5 mL of Tyrode solution at 37 °C. Insert it into the biosafety hood.

Place the 24-well plate with neural cultures in the biosafety hood. Using tweezers, pick up a single 1 cm round coverslip and insert it into the glass bottom plate. For ideal electrophysiological recordings, the experiment should be held within the next 30 min.

Place the plate under the microscope, and focus on the neurons at 5× using the transmitted light at low intensity to avoid unwanted photostimulation on the opsin-expressing neurons.

Fill an electrode with internal solution and mount it on the headstage. Lower both the micropipette and the reference electrode into the solution. The reference electrode should be place away from the neural culture to allow maneuverability of the micropipette. See that the micropipette is placed far above the neural culture. Do not lower the micropipette to the plane of the cells at this point.

Switch magnification to 20× and find the cells of interest using low brightness transmitted light. As a general rule, the mid area of the coverslip should be the most neuron dense, while the rim of the coverslip should be neuron sparse.

Take an image of the fluorescence of neurons. If the opsin is labeled with GFP, you are photostimulating during exposure minimize exposure time. A 200 ms exposure should be enough for getting a high signal to noise ratio image of the cells. The image should give you a good estimate which cells not to stimulate (low opsin expression, or very high opsin expression, or altered morphology). Locate the cell body that you will patch and choose the cells around it for photostimulation.

Draw ROIs inside the cell bodies. Make sure ROIs are smaller than the diameter of cells, to avoid photostimulating cells adjacent to the cell of interest, i.e., for neurons with an average diameter of 20 μm, draw ROIs of 15 μm.

Choose a stimulation light intensity and duration. We recommend an intensity high enough to produce an action potential with a 100% probability, and a photostimulation of up to 2 ms.

Run the sequence of stimulation once.

Return to a 5× magnification.

2.2.4 Detecting Presynaptic Cells at Single-Cell Resolution Optogenetically

Bring the electrode ~10 μm above the cell of interest, switch the microscope to a 20× magnification.

Take an image of the neurons of interest again, in order to ensure that there was no significant *x*–*y* translation of the neurons of interest. Correct the position of ROIs if needed.

Switch the amplifier to voltage clamp mode, to deliver a square voltage step that will allow you to see an increase in resistance once the pipette hits the neural membrane. We recommend using 10 mV and 5–10 ms steps, at 4–10 Hz. Patch the neuron of interest.

Make sure that the electrophysiological trace is stable for 1 min before starting your experiment.

Photostimulate the patched cell. If done properly, an inward photocurrent with a fast rise time and a slow decay time should be detected. The rise and decay times may vary, based on the opsin used. Make sure that the photocurrent has a good signal to noise ratio, ideally larger than 50.

Run the sequence of ROI stimulation using the DMD. Have the DMD trigger the electrophysiological system. Have the interstimulation interval at 5 s.

Record the synaptic currents (Fig. 2), save the image of neurons, and the coordinates of the ROIs.

3 Protocol: Detecting Synaptic Partners Using Single-Cell Resolution Optogenetics in Brain Slices (Figs. 3 and 4)

3.1 Materials

3.1.1 Microscopy and Photostimulation

The system includes three imaging pathways (1P and 2P), and one 2P photostimulation pathway for single-cell targeting. The components are described below.

The imaging pathways are: (1) Transmitted light imaging (Fig. 3a, orange): *This imaging provides contrast images of the cell membrane and the patch pipette.* An infrared light source, either a filtered white lamp or an infrared light emitting diode (IR-LED), is sent through the slice in a tilted illumination or DIC configuration. The transmitted light is then collected through a CMOS or a dedicated IR CCD camera. **(2) Widefield epifluorescence imaging** (Fig. 3a, green). *This imaging system is used to quickly evaluate the opsin expression over the slice, although it is limited to the superficial layers of the tissue or to preparation with very sparse opsin staining.* A filtered white lamp or multiple LED source can provide different wavelength illumination of different fluorophores. The 1P excited fluorescence is collected through a tube lens and separated from the excitation light using a dichroic mirror and detected by a CMOS camera after passing through a visible bandwidth filter. **(3) 2P scanning imaging** (Fig. 3a, purple): *This imaging system is critical to determine the exact location of opsin expressing cells deep in the tissue and target them for photostimulation.* The light source is a pulsed femtosecond laser, typically a Titanium-Sapphire laser with a tunable emission wavelength. The laser beam is focused on the sample via a pair of XY galvo or galvo resonant mirrors, and the generated fluorescence is collected by photomultiplier tubes (PMTs). The objectives should have a high numerical aperture ($\geq$0.8 NA) and a long working distance (>2 mm), to guarantee a good resolution and signal detection while still allowing the insertion of the patch pipette.

The 2P photostimulation path (Fig. 3a, red) for single- or multiple-cell stimulation is based on a patterned illumination technique named Computer Generated Holography (CGH) and on the use of a liquid crystal Spatial Light Modulator (SLM) [43]. Briefly, the propagation of an optical beam is determined by its *phase.* Holography is based on the modulation of the phase of a laser beam by the SLM, in order to redirect light on specific region of interest in the sample. CGH allows the generation of microscale spots to precisely tailor the shape of the illumination to a cellular structure of interest [17]. Here, CGH is used to pattern the beam of an infrared pulsed laser to generate single or multiple illumination spots whose lateral and axial dimension match neural cell bodies located over the field of view (Fig. 3b–d) in order to stimulate single or multiple neurons simultaneously.

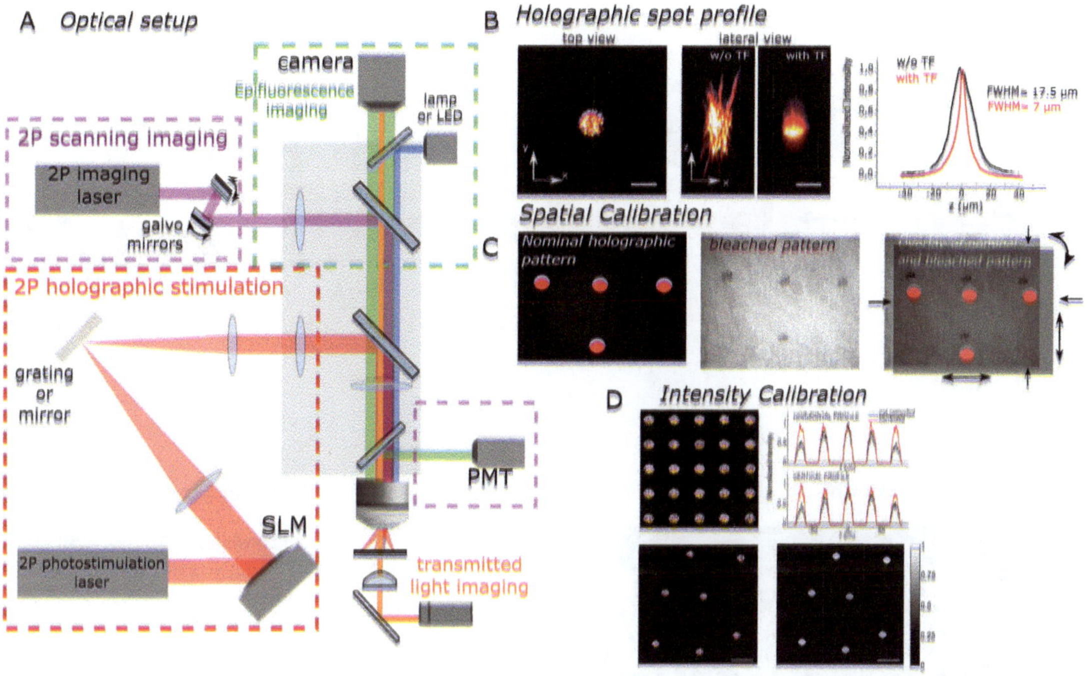

Fig. 3 Microscopy setup for single-cell resolved 2P photostimulation. (**a**) A schematic of a microscope for 2P holographic photostimulation of single cells: multiple imaging methods are combined with an SLM-based patterned holographic illumination. (**b**) Reconstruction of the excitation volume of a 10 μm diameter circular holographic spot. We show the top and lateral view of the fluorescence induced by the spot on a thin rhodamine layer with and without the use of temporal focusing (TF). Scale bar: 10 μm. On the right, the corresponding integrated axial profile of the fluorescence showing the improved confinement of the stimulation on the *z*-axis thanks to the use of temporal focusing. (**c**) A schematic of the spatial calibration procedure between the imaging and the stimulation systems. The nominal holographic pattern (left) is projected at high intensity on a rhodamine layer, generating a corresponding bleaching pattern, imaged then under 2P scanning (middle). A transformation (made of stretches, rotations, translations) is then applied to overlay the two images and spatially register the holography and imaging coordinates (right). (**d**) Intensity calibration of holographic spots. Top: Max projection of the fluorescence generated by a sequence of single holographic spots displaced on a square grid (left). The generation of an additional spot of variable size (as described in [18]) is applied to correct for the position dependent diffraction efficiency (right). Bottom: Multispot illumination pattern with uniform light intensities between spots (left). A proper input pattern to the algorithm of phase calculation (right) allowed to adjust the relative intensities between spots and to compensate for diffraction efficiency by assigning higher intensities to peripheral spots. Scale bar: 30 μm. Panels **b** and **d** are adapted from [18]

Practically, the user will: (1) define a desired illumination pattern to photostimulate cell bodies of interest; (2) feed the pattern into an iterative algorithm (such as a weighted Gerchberg-Saxton algorithm [44]) to calculate the required phase modulation to impose on the beam; (3) send the phase modulation to the SLM to obtain the desired illumination pattern on the sample and photostimulate the target neurons. Additionally, CGH is combined with a technique named *temporal focusing* (TF), which, by manipulating the duration of the laser pulses, improves the axial confinement of

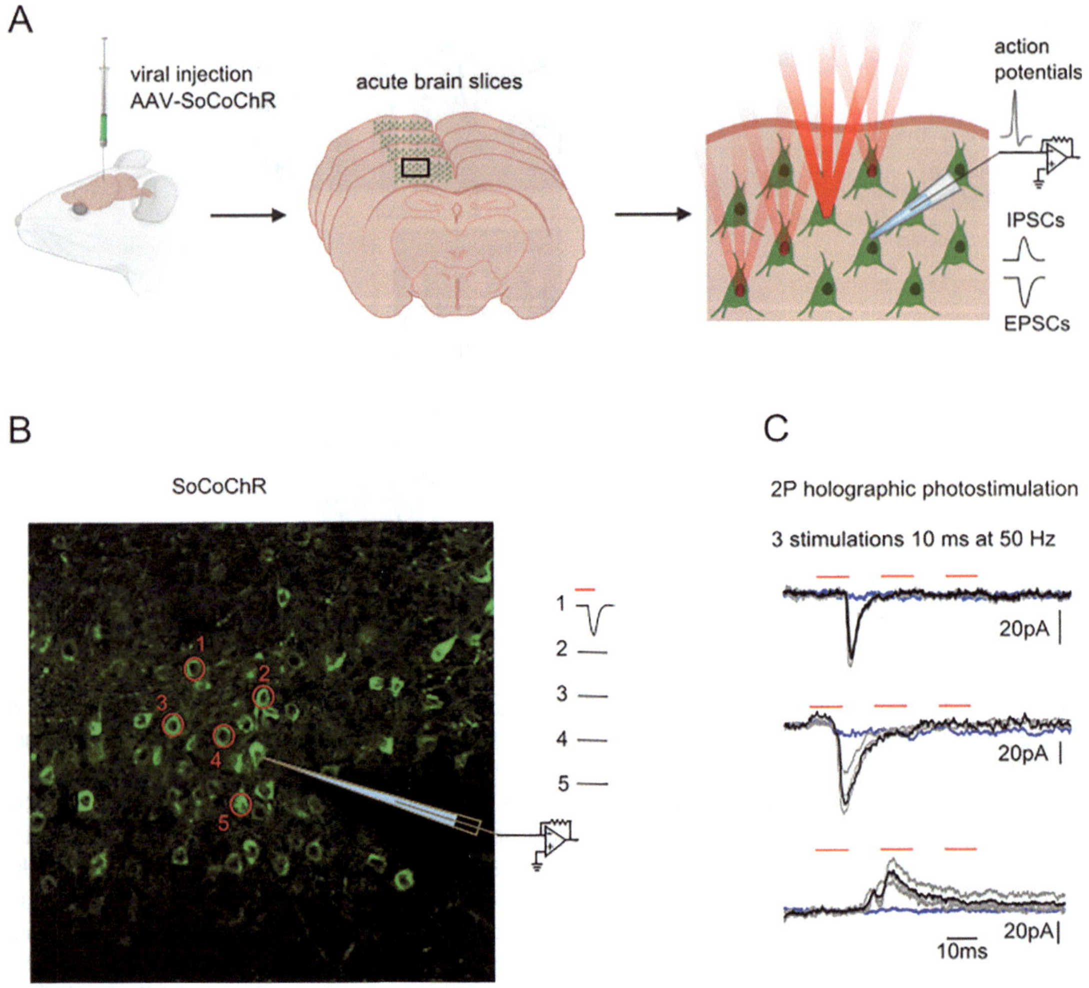

Fig. 4 Detection of synaptic partners using single-cell resolution optogenetics in brain slices. (**a**) A Schematic of experimental protocol: stereotaxic viral injection of a soma targeted opsin in mouse V1; acute brain slice preparation (inset underlies the area of photostimulation and patch-clamp); patch-clamp recording of a post-synaptic cell. The potential pre-synaptic cells are photostimulated with holographic illumination in 3D. Action potentials, inhibitory or excitatory post synaptic currents (IPSCs, EPSCs, respectively) are recorded in the patched cell. (**b**) An image showing a brain slice with cells expressing the opsin SoCoChR-GFP. The illustration to the right shows a schematic of a typical experiment where a cell is patch-clamped and neighboring cells are sequentially photostimulated by holographic spots. On the right is a schematic of electrophysiological recordings that can be obtained, wherein only cell #1 is connected to the patched cell. (**c**) Sample EPSCs and IPSCs recorded in whole-cell configuration from three post-synaptic cells while sequentially positioning the holographic spot on neighboring cells. Neighboring cells were photostimulated at a power of 150 $\mu W/\mu m^2$, $\lambda = 1030$ nm. Red bars represent the photostimulation periods (three light pulses at 50 Hz of 10 ms duration, repeated 3–5 times). Gray traces: currents recorded before the perfusion of receptor blockers APV, NBQX and gabazine. Black trace: average of three trials. Blue traces: currents recorded after the perfusion of receptors blockers (average of three trials on each spot position)

the illumination patterns (Fig. 3b), reduces scattering and therefore aid in achieving single-cell resolved photoactivation [45, 46].

Spatial and Intensity Calibration of the Holographic Illumination

The mapping of synaptically connected cells requires a precise control of sizes, locations, and intensities of the illumination spots. Below we describe the spatial and intensity calibrations needed.

Characterization of the Holographic Spots

Start by visualizing the holographic spots. This will allow to reconstruct the illumination patterns on the $x-y$ plane (Fig. 3b, left) and verify that its shape, size, and location correspond to the defined ones.

- Place a thin fluorescent layer at the focus of the microscope objective, i.e., a layer of rhodamine-6G spin coated and dried on a glass coverslip.
- Define one or multiple circular spots of the size of a soma (10–15 μm), calculate the corresponding phase profile and send it to the SLM.
- Activate the modulated laser beam on the fluorescent layer and observe the generated fluorescent pattern using the camera of the epifluorescence imaging path.

To fully reconstruct the 3D shape of the spots, an additional imaging path is required. A compact inverted microscope (made of an objective, a tube lens, and a camera) can be mounted below the main objective to image the fluorescent layer from the bottom [44]. By scanning the z-position of the upper objective with a piezo, different horizontal sections of the spot can be visualized and its full 3D profile can be reconstructed (Fig. 3b, center).

This procedure is critical to characterize the axial confinement of the excitation spots (Fig. 3b, right) and verifies their correct location along the z axis to assure a precise cell targeting. The inverted microscope can be removed after the characterization.

Spatial Calibration

After this initial spot characterization, the system requires a precise spatial registration between the photostimulation and the 2P scanning imaging. This calibration will allow selecting opsin expressing cells based on the 2P fluorescent images and targeting them with holographic spots with micrometric precision. This procedure needs to be repeated every day prior to the experiments.

- Define a multispot pattern, project it on the fluorescent layer and increase laser power to locally bleach the fluorophores.
- Image the bleaching patterns with the 2P scanning imaging system. The bleached spots will appear as dark regions on an otherwise homogenous fluorescence plane.

- Compare the nominal pattern initially defined with the observed bleaching pattern (Fig. 3c). Extract the transformation (x–y stretch, translation and rotation) that allows overlaying the nominal and observed locations of the spots.
- This transformation represents the correction to apply between the coordinates of the holographic and imaging systems. Practically, during each photostimulation experiment apply this transformation to the coordinates of the 2P scanning images before defining holographic spots to target the cells of interest. Use an automatized script to implement this correction.

Intensity Calibration

Ideally, when photostimulating multiple neurons, each of them is stimulated with similar light intensity. However, holographic patterns suffer from position dependent *diffraction efficiency*, meaning that the intensity of a spot decreases the more is moved away from the center of the field of view (FOV) [47]. This means that neurons in the periphery of the FOV will be exposed to lower light intensities compared to neurons at the center. To avoid this effect, specific correction needs to be implemented, as follows.

At first, diffraction efficiency needs to be characterized. To achieve this: generate sequentially single holographic spots placed on a squared grid all over the FOV; project them on the fluorescent layer and extract their intensities from the fluorescence camera image; reconstruct the intensity decrease along the x and y axis (Fig. 3d, top).

The strategy to correct for the diffraction efficiency will depend on the temporal characteristics of photostimulation protocols applied during the experiment.

- *For sequential targeting of single neurons with a single spot*, such as during mapping of synaptic connections, the diffraction efficiency can be compensated by constantly adjusting the intensity of the beam at the laser output so that peripheral neurons are illuminated at higher laser powers. To avoid this continuous laser power adjustment, one can generate, together with the spot that targets the cell of interest, an additional spot of variable size [18]. This will allow to redistribute the total laser power between the two spots in order to maintain the same photostimulation intensities on the target cells, independently on their locations (Fig. 3d, top).
- *For simultaneous stimulation of multiple neurons*, the correction is based on the modification of the input pattern to the algorithm for the phase calculation [18]. By adjusting the relative intensity between the spots at the input (Fig. 3d, bottom), it is possible to direct more power to the peripheral spots and compensate for their lower diffraction efficiency. This will allow to simultaneously excite multiple neurons with the same intensity.

Suppression of the Zero Order of Diffraction

Typically, available SLMs modulate only 80–90% of the incoming light. This means a fraction of the laser light, named zero order of diffraction, is not redirected on the user-defined holographic spots. Zero order of diffraction is simply reflected by the SLM and is focused on a point-like spot at the center of the FOV. This spot could induce unwanted neural activation or damage the sample. To suppress the effect of such a spot, a convenient method exists based on the use of a cylindrical lens placed in front of the SLM, as described in [48]. The lens distributes the power of the zero-order spot over a much larger surface, significantly decreasing its light intensity and suppressing any induced 2P excitation.

3.1.2 Opsin Expression

Adeno-associated viral vectors for soma-targeted expression of opsins (SoCoChR, ST-ChroME, soma targeted ChRmine) coupled with a fluorescent reporter. Serotypes: AAV9, AAV8, AAV1. Promoters: hSyn, EF1a, CaMKII, CAG.

Choose the proper viral serotype and promoter to express the soma-targeted opsin in cortical neurons of the mouse brain at the desired time window and in the desired population of cells. Optimal opsin expression window can vary depending on serotype, promoter, and viral titer used. Typically, cortical neuron expression window can vary from 3 to 5 weeks without showing signs of toxicity, when infusing viral preparations with a titer of 10^{12} genome copies/mL. For longer time windows, monitor the appearance of neuronal toxicity.

3.1.3 Animal Preparation

Viral vector injections can be performed in wild-type or transgenic male or female mice (Janvier Labs).

Equipment for surgery and viral vectors delivery in mouse brain: stereoscope (Olympus, SZ61); stereotaxic frame (Kopf); heating pad (Ugo Basile); sterile surgical instruments (Fine Science Tools); driller (Microtorque II); tungsten driller tip (0.5 Carbide US No. 1/4; Phymep); microsyringe (Hamilton, 10 μL); silica injection cannula (Phymep); syringe pump (kds LEGATO 100); heated recovery chamber; anesthetic (ketamine/xylazine); buprenorphine (Buprecare); atipamezole (Antisedan); lidocaine solution 2%; sterile NaCl 0.9% solution; eye ointment; saline; betadine solution; absorption surgical triangles (Kettenbach); surgical glue; surgical suture.

Equipment for acute brain slices preparation from mice at 4–15 weeks after viral injection: vibratome (VT1200S Leica Biosystems, Germany); ice; dissection tools (Fine Science Tools); recovery chamber at 35 °C; cyanoacrylate glue.

3.1.4 Electrophysiology

Equipment for patch-clamp recordings: amplifier, digitizer, headstage, and recording software (Molecular Devices); micromanipulator for pipettes approach (Luigs and Neumann); pipette puller (Sutter Instrument); borosilicate glass pipettes (i.d.0.86 mm, o.

d.1.5 mm; Harvard Apparatus; pipettes final resistance in the bath: 4.5–5 mΩ); AgCl pellet as reference electrode; heater for extracellular solution; peristaltic pump for perfusion of extracellular solution; microsyringe for localized perfusion of synaptic receptors antagonists; system for aeration of solutions with 95% O_2 and 5% CO_2; a recovery chamber for acute slices; a heated water bath; anesthetic (isoflurane 5%); metallic harp with nylon grid for holding slices during the recordings.

Intracellular solution (in mM): 130 potassium gluconate, 7 KCl, 4 Mg-ATP, 0.3 mM Na-GTP, 10 sodium phosphocreatine, 10 mM HEPES (pH adjusted to 7.28 with KOH; osmolarity 280 mOsm).

Extracellular ACSF solution (in mM): 125 NaCl, 2.5 KCl, 26 $NaHCO_3$, 1.25 NaH_2PO_4, 1 $MgCl_2$, 1.5 $CaCl_2$, 25 glucose, 0.5 ascorbic acid (310 mOsm).

Slicing solution (in mM): 85 NaCl, 2.5 KCl, 0.5 $CaCl_2$, 4 $MgCl_2$, 65 sucrose, 25 glucose, 0.5 ascorbic acid (320 mOsm).

Synaptic receptor antagonists when identification of receptor subtypes is required (gabazine, NBQX, APV from Tocris or Abcam).

3.2 Methods

3.2.1 Expressing Somatic Opsins in Mouse Visual Cortex: Stereotactic Viral Injections

Here we describe the procedure for viral infusion in V1 of 4 weeks old C57BL/6J mice (Janvier Labs) (Fig. 4a).

Clean the benchtop with a disinfectant and prepare an operatory camp with sterilized instruments and materials.

Centrifuge the viral aliquot and load the infusion syringe and cannula.

Anesthetize the animal with intraperitoneal injection of mixture ketamine (100 mg/kg body weight)-xylazine (10 mg/kg body weight) prepared in sterile NaCl 0.9% solution.

Inject buprenorphine subcutaneously (0.05–0.1 mg/kg).

Shave the mouse head and clean with EtOH 70%, betadine, and EtOH 70% the incision site.

Inject lidocaine subcutaneously on the head.

Place the mouse in the stereotactic frame, warm, and cover the animal.

Protect eyes with an eye ointment.

Cut the skin with scalpel, pull skin with bent needles. Remove connective tissues on top of the cranium gently.

Clean with some drops of sterile saline solution.

Measure coordinates of the region of interest and mark the injection site.

Drill carefully by avoiding vessels. Create a small craniotomy (about 0.7 mm in diameter).

Remove all residual bones on top of the skull and at the edge of the craniotomy.

Slowly insert the cannula at the desired depth.

Infuse 0.5–1 μL of viral vector solution at the speed of 70–100 nL/min.

At the end of viral infusion, retract the needle of about 50–100 μm and wait for 5 min.

Retract the cannula and suture skin. Cover the wound with a thin layer of surgical glue.

For mouse recovery: perform an intraperitoneal injection of antipamezole hydrochloride (1 mg/kg), a subcutaneous injection of 0.1 mL warm saline solution for rehydration, place the animal in a warm recovery chamber.

Monitor the animal during the next 2–3 days and inject analgesics during the following days if needed.

3.2.2 Acute Brain Slice Preparation

Photostimulation experiments are performed in acute slices prepared from mice, 3–15 weeks after viral infusion in the brain through stereotactic injections (Fig. 4a).

Prepare solutions and aerate them with 95% O_2–5% CO_2 for at least 40 min.

Cool down slicing solution and the cutting plate.

Fill the recovery chamber with 20% sucrose solution and 80% ACSF, adjust aeration and warm at 35 °C. Ensure that the net is free of air bubbles.

Set up the slicer, align the blade.

Prepare dissection tools.

Deeply anesthetize the mouse with isoflurane (5% in air). Once fully anesthetized, decapitate the animal.

Keep the head in ice-cold solution for 5–8 s.

Remove quickly the brain region of interest, taking care of avoiding any pressure on the brain, and glue it on the specimen holder.

Cut slices at the desired angle and thickness (about 250–300 μm) in aerated and ice-cold slicing solution.

Move slices in the warm recovery chamber for 45 min. Then keep them at room temperature.

3.2.3 Preparing the Experimental Setup

Switch on devices and lasers.

Measure the laser power range for photostimulation at the exit of the objective.

Assure alignment of the 2P imaging field and the photostimulation field by performing the spatial registration procedure described in Subheading 3.1.1.1.

Thaw intracellular solution and filter it. Keep it on ice. Prepare a syringe with tip to fill the recording pipettes.

Pull pipettes.

Load warm ACSF solution (35 °C), and perfuse it into the recording chamber.

Place a slice in the recording chamber, and hold it with a metallic harp.

3.2.4 Detection of Synaptic Connections in Slices with Holographic Photostimulation at Single-Cell Resolution

Electrophysiology is used as readout of synaptic connections. The postsynaptic cell is monitored using the patch-clamp technique while presynaptic cells are photoactivated with holographic spots of 15 μm (Fig. 4b).

Find a region where cells are in optimal health using IR transmitted light and monitor expression of the fluorescent reporter by using the 1P illumination with a 10× objective.

Visualize opsin expressing cells in the FOV with the 2P scanning imaging system.

Load the patch pipette with intracellular solution. Under IR transmitted light illumination, approach the cell of interest with the pipette by applying a pressure of 25–40 mbar. Monitor the pipette resistance with a voltage step test of −5 or −10 mV.

By releasing pressure and applying gentle suction, achieve the Gigaohm seal and switch to whole-cell configuration. Acquire cell's electrophysiological parameters (R_m, C_m, V_m, I_{leak}).

Acquire a *z*-stack of the fluorescence emission in the volume around the patched cell in order to identify neighboring positive cells.

In slices expressing somatic opsin cells are clearly distinguishable, since their somata are predominantly fluorescent with minimal neuropil fluorescence. In the volume around the patched cells, identify and select an ensemble of cells to target with 10–15 μm holographic spots.

Generate the corresponding holographic patterns and start the stimulation protocol.

For illumination, use a laser power that guarantees the generation of an action potential in the presynaptic cell.

Potential presynaptic cells can be stimulated by delivering trains of light stimuli (3–5 stimuli of 5–10 ms at 10–50 Hz). Repeat the stimulation sequence 5–10 times.

Synaptic currents can be recorded in voltage-clamp configuration at the desired holding potential. Depolarizations or hyperpolarizations of the membrane potential of the patched cell generated by the synaptic currents can be recorded in current-clamp configuration (Fig. 4c).

Monitor the reproducibility of synaptic events in the stimulation trials to identify a synaptic connection. Good indications of a monosynaptic connection are a millisecond timescale latency between the beginning of the photostimulation and the onset of the synaptic event; reproducible event kinetics.

Perfuse ACSF with synaptic receptor antagonists at the end of the experiment to block synaptic currents and identify the receptors involved. Repeat the photostimulation sequence until the synaptic currents disappear.

Perfuse artificial cerebrospinal fluid (ACSF) while repeating the photostimulation sequence, until recovery of the synaptic currents.

4 Protocol: Single-Cell Resolution Optogenetics in Mouse Cortex In Vivo (Fig. 5)

4.1 Materials

4.1.1 Microscopy and Photostimulation

The optical system and its operation are similar to the one described in Subheading 3.1.1 for brain slices, except for the following differences. For in vivo mouse experimentation, the transmitted/DIC imaging is not available and the pipette positioning is mainly guided by epifluorescent or 2P scanning imaging. Additionally, due to the larger tissue depth in an experiment, some further corrections to the laser intensities need to be introduced to compensate the attenuation of the illumination of spots due to scattering [49]. Finally, movement of the animal during hours-long experimental protocols could affect the cell-targeting precision. Frequent acquisition of 2P reference images and on-the-fly correction of the position of the holographic spot might be required.

4.1.2 Expressing Single-Cell Opsins

Viral vectors of somatic opsins based on CoChR, ChroME, ChRmine as templates [18, 25, 50]: adeno-associated viral vectors of different serotypes (i.e., AAV1, AAV9) and promoters (hSyn, EF1a, CaMK2, CAG) can be used for expressing somatic opsins coupled with a fluorescent reporter in the mouse visual cortex.

Serotypes would affect the expression efficiency, and promoters would influence the cell type of expression. Enhanced specificity of opsin expression could be obtained by implementing a FLEX sequence in the viral construct for Cre recombinase-dependent expression. Subcloning a P2A sequence into the viral constructs for separating the membrane expression of opsins from the cytosol expression of fluorescent reporter enable the visualization of opsin-positive cells in the brain sample.

4.1.3 Animal Preparation

Experimental animals: wild-type C57BL/6J mice (Janvier Labs) and transgenic mouse line such as GP4.3 for GCaMP6 expression (Jackson Laboratory).

Reagents for mouse brain surgery: anesthetics of ketamine, xylazine, isoflurane; analgesics of lidocaine, buprenorphrine; betadine; eye ointment; antibiotics ointment; saline; 70% ethanol.

Materials for mouse brain surgery: agarose; cover glass; cyanoacrylate glue; absorption triangles (Kettenbach); dental cement and solvent (GC America); 27-gauge needle (Terumo); bottled oxygen (Air Liquide).

Equipment for mouse brain surgery: stereoscope (Olympus); stereotaxic frame with mouse adaptor (Kopf); animal holder (custom-built); surgical tools (Fine Science Tools); Peristaltic pump (Gilson); Anesthesia unit (Kent Scientific); Homeothermic system for small animal (Harvard Apparatus); Hand-held driller (Microtorque II).

Equipment for viral infusion: microliter syringe (Hamilton); syringe pump unit (kd Scientific).

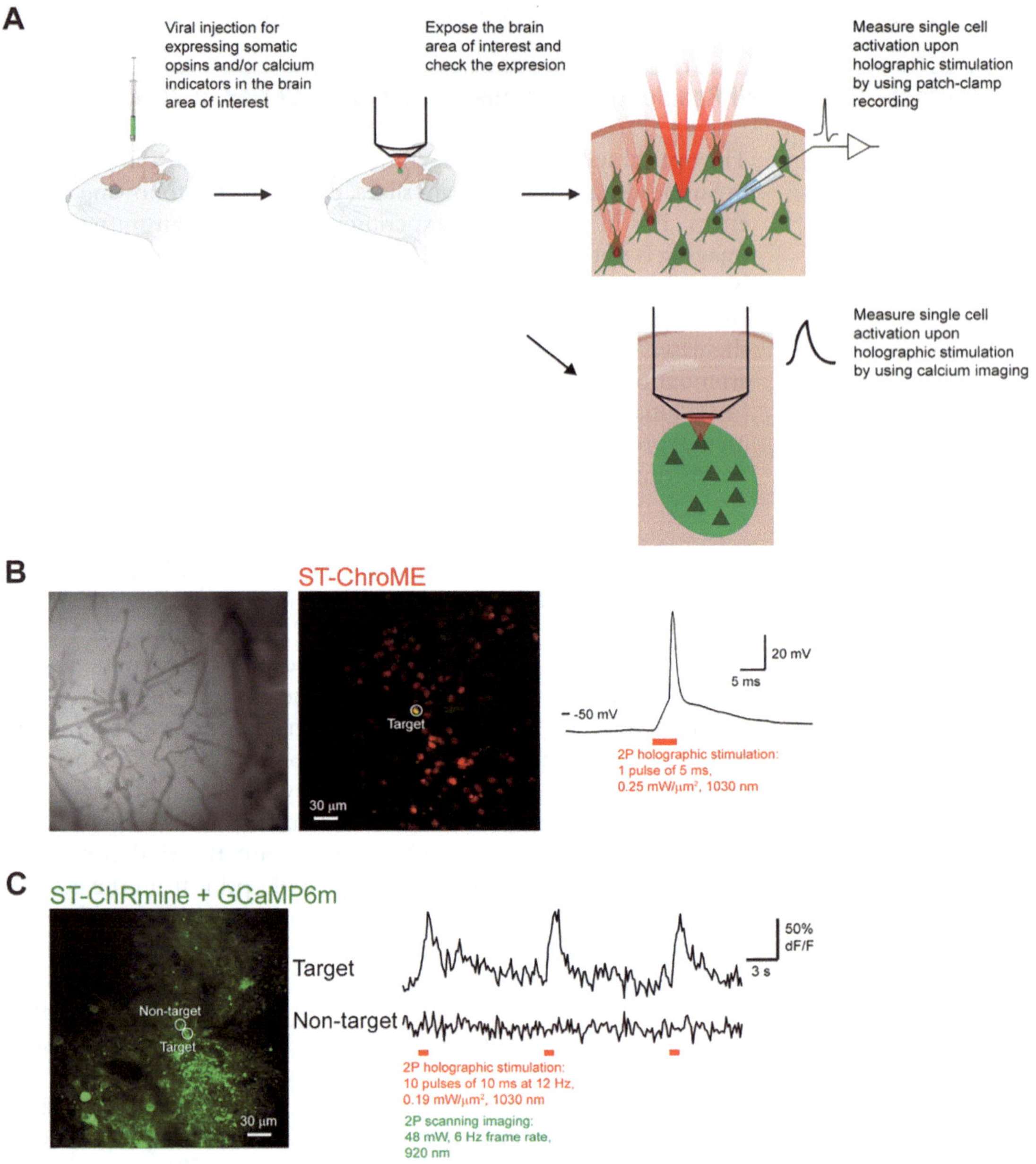

Fig. 5 In vivo single-cell photostimulation in mouse visual cortex using a somatic opsin together with 2P holographic stimulation. (**a**) A schematic of the photostimulation experiment in the mouse brain. For expressing somatic opsins and/or calcium indicators, viral vectors of somatic opsin and/or calcium indicators are injected into the brain region of interest prior to the experiment. During the experiment, the brain region of interest is exposed and the expression area can be identified using 2P scanning imaging or 1P widefield imaging. Upon photostimulation, activation of opsin expressing cells (green cells) can be read out using patch-clamp recording (right, top) or calcium imaging (right, bottom). (**b**) In the mouse visual cortex (widefield image on the left), a target cell expressing ST-ChroME at layer 2/3 (red cells in the 2P image in the middle) is stimulated by a brief (5 ms) holographic light pulse to yield an action potential (trace at the right). (**c**) Neurons co-expressing ST-ChRmine and the calcium indicator GCaMP6m (shown in green is the GCaMP6m expression in neurons co-expressing ST-ChRmine). Single-cell activation can be read out optically. Upon 2P imaging and illumination of a train of holographic light pulses, the "target" cell, but not the neighboring "non-target" cells, shows calcium transients (traces to the right)

4.1.4 Electrophysiology

Materials and reagents: Borosilicate capillary with filament (Harvard Apparatus); 4-mm syringe filter of 0.2 μm pore size (Thermo Fisher Scientific).

Extracellular solution (Ringer's solution): 145 mM NaCl, 5.4 mM KCl, 10 mM HEPES, 1 mM Mg_2Cl, 1.8 mM $CaCl_2$; adjust to pH 7.3; 290–310 mOsm.

Pipette internal solution: 135 mM K-gluconate, 10 mM HEPES, 10 mM Na-phosphocreatine, 4 mM KCl, 4 mM Mg-ATP, 0.3 mM Na_3GTP, 25–50 μM AlexaFlour-488 or AlexaFlour-594; adjust to pH 7.3; 280–290 mOsm.

Equipment: Patch-clamp amplifier (Molecular Devices, MultiClamp700B), Digitizer (Molecular Devices, Digidata 1550A); Motorized micromanipulators (Luigs and Neumann); Pipette puller (Sutter Instrument); Oscilloscope (Tektronix).

Software: Electrophysiology acquisition software pCLAMP (Molecular Devices), which coordinates electrophysiological recording, imaging acquisition, and triggering photostimulation.

4.2 Methods

4.2.1 Expressing Somatic Opsins in Mouse Visual Cortex

Infusing viral vectors for expressing somatic opsins in the mouse brain is performed prior to photostimulation experiment (Fig. 5a). The time period required for opsin expression is typically 3–8 weeks. Here we describe a procedure of stereotaxic viral injection for 2P holographic photostimulation and 2P imaging similar to [51] and [52]. Three phases of virus infusion, wound closure, and recovery are involved. We also describe a procedure to implant a cranial window after virus infusion for chronic experiments.

1. Viral infusion.
   (a) Sterilize the surgical tools and disinfect the surgical area with 70% ethanol.
   (b) Intraperitoneal injection of ketamine/xylazine mixture (ketamine/xylazine mixture of 0.1 mg ketamine and 0.01 mg xylazine/g body weight) in mice of ~4 weeks old.
   (c) Place the mouse on the heating pad and fix the head in the stereotaxic frame.
   (d) Apply the eye ointment to prevent dry eyes.
   (e) Apply 70% ethanol and betadine at the head for disinfection.
   (f) Subcutaneously inject the analgesic lidocaine at the head.
   (g) Use a scalpel blade to make a midline incision above the skull and scrap away the surrounding tissue on top of the skull, but not apply too much pressure onto the skull to prevent damage.

(h) Bend 2–3 needles for making hooks to fix the skin around the incision and expose the skull. Use saline to clean the skull.

(i) Mark the brain area of interest according to the respective stereotaxic coordinates on the skull (e.g., mouse visual cortex: 3.5 mm posterior to the bregma and then 2.5 mm lateral from the midline).

(j) Thin the skull over the area of interest gently by using a hand-held drill under the stereoscope until the skull can be perforated by the needle.

- For an acute experiment, drill a small hole and carefully penetrate the dura with a needle.
- For a chronic experiment requiring implanting a window afterwards, drill a circular groove of ~2 mm diameter; use forceps to remove the bone within the thinned groove, and penetrate the dura with a needle.

(k) Load the Hamilton syringe with viral vectors.

(l) Position the syringe needle first on top of Bregma and then move the syringe using the stereotaxic manipulators till on top of the brain area of interest.

(m) Insert the syringe needle slowly using the stereotaxic manipulator to the target depth (~250 μm for L2/3 mouse cortex).

(n) Wait for ~2 min for the brain tissue to relax from the pressure of needle insertion.

(o) Infuse the virus slowly (e.g., 50–100 nL/min, 0.5–1 μL total volume).

(p) Wait for ~2 min and slowly withdraw the injection needle.

(q) Repeat the above procedures if multiple injection sites are required.

2. Wound closure.

(a) For acute preparation:

- Suture the wound after cleaning the skull with saline.
- Apply cyanoacrylate glue to facilitate the closure of the wound.

(b) For chronic preparation:

- Clean the craniotomy with saline and then dry the exposed brain and the surrounding skull.
- Apply the round cover glass directly on top of the craniotomy (optionally, first apply a thin layer of agarose).

- Apply cyanoacrylate glue at the margin of cover glass (avoid excess glue infusing into the craniotomy) and wait for a few minutes until the glue is dried.
- Apply a thin layer of cyanoacrylate glue to the rest of the skull and the wounded margins of skin and muscles and wait until the glue is dried.
- Seal the cranial window with dental cement on top of the cyanoacrylate glue.

3. Recovery.
   (a) Keep the animal warm during recovery (normally it requires ~30 min for the animal to wake up from the anesthesia).
   (b) Inject saline (~0.15 mL/g body weight) under the skin to prevent dehydration.
   (c) Apply antibiotic ointment.
   (d) Inject analgesic (buprenorphrine 0.05 mg/kg body weight).
   (e) Monitor the recovery of animal for at least 1 week.

4.2.2 Preparing the Experimental Setup

1. Switch on both the photostimulation laser and the imaging laser at least 30 min before experiment so that their output powers are stabilized.
2. Measure laser power out of the objective for both the imaging and photostimulation path.
3. Assure alignment of the 2P imaging field and the photostimulation field by performing the spatial registration procedure described in Subheading 3.1.1.

4.2.3 Holographic Photoactivation of Single or Multiple Cells

Two approaches are used for monitoring the spiking activity upon holographic photostimulation. One approach is the electrophysiological readout of membrane potential for monitoring single target neuron (see A below) [53] (Fig. 5b), and the other is the imaging readout of the signal from a calcium indicator such as GCaMP [54] for monitoring multiple target and non-target neurons (see B below) (Fig. 5c). Relevant results are summarized in [55]. We use a stimulation laser at 1030 nm to actuate the somatic opsins and an imaging laser at 920 nm to measure the GCaMP activity.

Notably, due to the overlapping 2P excitation spectra of opsins and calcium indicators such as GCaMP, 2P scanning for calcium imaging could induce spurious activation of opsins. Excessive scanning-induced activation of opsins may result in unspecific network activation. To minimize the degree of scanning-induced activation of opsins, several strategies can be implemented including (1) selecting opsin/indicator that minimally overlap in their 2P excitation spectra; (2) using somatic opsins to avoid activating neurites during 2P scanning; (3) reducing the scanning laser power to decrease the probability of action potential generation in opsin expressing cells.

A. Electrophysiological readout using 2P-guided patch-clamp in an acute preparation:

1. Sterilize the surgical tools and disinfect the surgical area with 70% ethanol.
2. Anesthetize the animal with isoflurane (2% for induction and 0.5–1% for experiment) and place it on a heating pad.
3. Use scissors to remove the skin on top of the brain area of interest, i.e., where the viral injection was performed.
4. Remove the tissue and muscles surrounding the injection site, which should be visible on the skull, and mark the area of interest.
5. Use cyanoacrylate glue and dental cement to attach the headplate with its opening centered around the area of interest.
6. Wait until the glue has dried and the dental cement has solidified (~20 min).
7. Fix the headplate to a stable animal holder.
8. Use the hand-held driller to thin out a groove of approximately ~2 mm diameter circle centering the brain area of interest until the groove is thin enough to be perforated by a needle.
9. Use forceps to remove the skull within the thinned groove.
10. Gently remove the dura mater with fine tip forceps.
11. Identify the brain area with opsin expression by visualizing the expression of fluorescence reporter with 2P scanning imaging or with 1P wide field illumination from a LED source.
12. Melt 0.5–2% of agarose in Ringer's solution. Wait until the agarose cools down and apply it on top of the exposed brain.
13. Apply a rectangular coverslip on top of the agarose covering the expression area.
14. Apply Ringer's solution on top of the brain to create a bath of external solution.
15. Fill a patch pipette (5–10 MΩ) with 5–10 μL of internal solution.
16. Apply a pipette pressure > 150 mbar, lower the pipette tip into the solution on top of the brain, position the pipette tip at the margin of the coverslip and at the height between the brain surface and the cover glass (till almost at the brain

surface), while visualizing the pipette position with the CCD camera and monitoring the pipette resistance from an oscilloscope.

17. Switch to 2P scanning and monitor the pipette opening based on the fluorescence at the tip of the pipette.
18. Penetrate the pia mater by rapidly stepping forward the pipette (10–20 μm/step, 2–3 steps/s; advance for ~100 μm for targeting upper cortical layers).
19. Lower the pipette pressure to ~50 mbar when the pipette is in the mouse brain and move the pipette tip to the target depth (~200 μm from the brain surface for targeting L2/3).
20. Select a potential positive cell as the target cell (the expression pattern varies between opsin constructs) in the vicinity of pipette tip.
21. Approach the target cell with the pipette and lower the pipette pressure to ~20 mbar when the distance is within ~50 μm.
22. When the pipette tip is in immediate contact of the target cell, move the pipette toward the cell center at different sites until the movement corresponds to a steady decrease in pipette resistance of 30–50%.
23. Release the positive pipette pressure, and swiftly apply a suction of <10 mbar to facilitate the sealing of cell membrane to the pipette tip.
    (a) If a Giga-ohm seal is obtained, a ramp suction is applied to rupture the cell membrane for the whole-cell configuration; switch to the current-clamp mode to record the membrane potential.
    (b) If a loose seal is obtained, switch directly to cell-attached configuration in the current-clamp mode.
24. Once the patch recording (whole-cell or cell-attached) is established, acquire a high-resolution image (400×400 pixels) of the field of view (FOV; 300×300 μm^2) where the target cell is located at the FOV center.
25. Generate a 12-μm circular holographic spot over the target soma according to this reference image.
26. Trigger the stimulation laser to illuminate holographic spot onto the target cell while monitoring its membrane potential response.
27. Determine the threshold intensity of holographic light for inducing single spike of millisecond latency (Fig. 5b).
28. Induce a train of spikes by using a train of regular or irregular holographic illuminations (for the latter, for

example, design a train of illuminations whose onsets are the same as the peak latencies of a train of spontaneous spikes).

29. The spike latency can be determined as the average peak latency of photo-induced spikes and the jitter as their standard deviation.

B. Imaging readout of GCaMP signal through a cranial window in a chronic preparation:

1. Sterilize the surgical tools and disinfect the surgical area with 70% ethanol.
2. Anesthetize the animal with an implanted window by using isoflurane (2% for induction and 0.5–1% for experiment) and place it on a heating pad.
3. Use the hand-held driller to smoothen the dental cement surrounding the implanted round cover glass.
4. Attach the headpost to the margin of cranial window with cyanoacrylate glue and dental cement.
5. Fix the headpost in the animal holder.
6. Polish the dental cement around the round cover glass such that it holds a bath of Ringer's solution for immersion objective.
7. Identify the area with fluorescence expression under LED wide field illumination.
8. Switch to 2P scanning, acquire a high-resolution reference image at L2/3 (we acquire an image of 400 × 400 pixels in a 300 × 300 μm^2 FOV).
9. Select one or multiple target neurons, each covered with a circular holographic spot of 12-μm diameter according to this reference image.
10. Trigger the stimulation laser whose power is adjusted for simultaneously activating all the target neurons at the threshold intensity, while reading out the GCaMP calcium signal from all cells in the FOV (we acquire functional images of 128 × 128 pixels at 6 Hz frame rate) by scanning the imaging laser beam. We stimulate the target cells with a train of 10 light-pulses of 5 or 10 ms at 12 Hz (Fig. 5c).
11. To perform holographic stimulation in the cortex of awake and head-fixed mice (habituated to the experimental setup), gradually reduce the anesthesia level to zero, and repeat the above **steps 8–10**.

Acknowledgments

We thank the IHU FOReSIGHT grant (Grant P-ALLOP3-IHU-000), the National Institute of Health (Grant NIH 1UF1NS107574 - 01), the "Agence National de la Recherche" ANR (project PRCI), and the Axa research funding.

References

1. Boyden ES, Zhang F, Bamberg E et al (2005) Millisecond-timescale, genetically targeted optical control of neural activity. Nat Neurosci 8:1263–1268
2. Kohler M, Hirschberg B, Bond CT et al (1996) Small-conductance, calcium-activated potassium channels from mammalian brain. Science (80-) 273:1709–1714
3. Nagel G, Ollig D, Fuhrmann M et al (2002) Channelrhodopsin-1: a light-gated proton channel in green algae. Science 296: 2395–2398
4. Wietek J, Beltramo R, Scanziani M et al (2015) An improved chloride-conducting channelrhodopsin for light-induced inhibition of neuronal activity in vivo. Sci Rep 5:14807
5. Inoue K, Ono H, Abe-Yoshizumi R et al (2013) A light-driven sodium ion pump in marine bacteria. Nat Commun 4:1678
6. Chow BY, Han X, Dobry AS et al (2010) High-performance genetically targetable optical neural silencing by light-driven proton pumps. Nature 463:98–102
7. Michel H, Oesterhelt D (1976) Light-induced changes of the pH gradient and the membrane potential in *H. halobium*. FEBS Lett 65: 175–178
8. Inoue K, Ito S, Kato Y et al (2016) A natural light-driven inward proton pump. Nat Commun 7:13415
9. Han X, Boyden ES, Huhn W et al (2007) Multiple-color optical activation, silencing, and desynchronization of neural activity, with single-spike temporal resolution. PLoS One 2: e299
10. Zhang F, Wang L-P, Brauner M et al (2007) Multimodal fast optical interrogation of neural circuitry. Nature 446:633–639
11. Häusser M (2014) Optogenetics: the age of light. Nat Methods 11:1012–1014
12. Dombeck DA, Harvey CD, Tian L et al (2010) Functional imaging of hippocampal place cells at cellular resolution during virtual navigation. Nat Neurosci 13:1433–1440
13. Li C-YT, Poo M-M, Dan Y (2009) Burst spiking of a single cortical neuron modifies global brain state. Science 324:643–646
14. Houweling AR, Brecht M (2008) Behavioural report of single neuron stimulation in somatosensory cortex. Nature 451:65–68
15. Ronzitti E, Ventalon C, Canepari M et al (2017) Recent advances in patterned photostimulation for optogenetics. J Opt 19:113001
16. Emiliani V, Cohen AE, Deisseroth K et al (2015) All-optical interrogation of neural circuits. J Neurosci 35:13917–13926
17. Bovetti S, Fellin T (2015) Optical dissection of brain circuits with patterned illumination through the phase modulation of light. J Neurosci Methods 241:66–77
18. Shemesh OA, Tanese D, Zampini V et al (2017) Temporally precise single-cell-resolution optogenetics. Nat Neurosci 20: 1796–1806
19. Wu C, Ivanova E, Zhang Y et al (2013) rAAV-mediated subcellular targeting of optogenetic tools in retinal ganglion cells in vivo. PLoS One 8:e66332
20. Lim ST, Antonucci DE, Scannevin RH et al (2000) A novel targeting signal for proximal clustering of the Kv2.1 K+ channel in hippocampal neurons. Neuron 25:385–397
21. Baker CA, Elyada YM, Parra A et al (2016) Cellular resolution circuit mapping with temporal-focused excitation of soma-targeted channelrhodopsin. Elife 5:e14193
22. Klapoetke NC, Murata Y, Kim SS et al (2014) Independent optical excitation of distinct neural populations. Nat Methods 11:338–346
23. Ren Z, Riley NJ, Garcia EP et al (2003) Multiple trafficking signals regulate Kainate receptor KA2 subunit surface expression. J Neurosci 23: 6608–6616
24. Valluru L, Xu J, Zhu Y et al (2005) Ligand binding is a critical requirement for plasma membrane expression of heteromeric kainate receptors. J Biol Chem 280:6085–6093
25. Mardinly AR, Oldenburg IA, Pégard NC et al (2018) Precise multimodal optical control of

neural ensemble activity. Nat Neurosci 21: 881–893

26. Govorunova EG, Sineshchekov OA, Janz R et al (2015) Neuroscience. Natural light-gated anion channels: A family of microbial rhodopsins for advanced optogenetics. Science 349:647–650
27. Mahn M, Gibor L, Malina KC-K et al (2017) High-efficiency optogenetic silencing with soma-targeted anion-conducting channelrhodopsins. Nat Commun 9:4125
28. Pisano F, Pisanello M, De Vittorio M et al (2019) Single-cell micro- and nano-photonic technologies. J Neurosci Methods 325: 108355
29. Miyamoto D, Murayama M (2016) The fiber-optic imaging and manipulation of neural activity during animal behavior. Neurosci Res 103: 1–9
30. Accanto N, Chen I-W, Ronzitti E et al (2019) Multiplexed temporally focused light shaping through a gradient index lens for precise in-depth optogenetic photostimulation. Sci Rep 9:7603
31. Jennings JH, Kim CK, Marshel JH et al (2019) Interacting neural ensembles in orbitofrontal cortex for social and feeding behaviour. Nature 565:645–649
32. Aharoni D, Khakh BS, Silva AJ, et al (2019), All the light that we can see: a new era in miniaturized microscopy
33. Aravanis AM, Wang L-P, Zhang F et al (2007) An optical neural interface: *in vivo* control of rodent motor cortex with integrated fiberoptic and optogenetic technology. J Neural Eng 4: S143–S156
34. Nimmerjahn A, Kirchhoff F, Helmchen F (2005) Resting microglial cells are highly dynamic Surveillants of brain parenchyma in vivo. Science (80-) 308:1314
35. Schmidt S, Horch K, Normann R (1993) Biocompatibility of silicon-based electrode arrays implanted in feline cortical tissue. J Biomed Mater Res 27:1393–1399
36. Polikov VS, Tresco PA, Reichert WM (2005) Response of brain tissue to chronically implanted neural electrodes. J Neurosci Methods 148:1–18
37. König K (2000) Multiphoton microscopy in life sciences. J Microsc 200:83–104
38. Tromberg BJ, Shah N, Lanning R et al (2000) Non-invasive in vivo characterization of breast tumors using photon migration spectroscopy. Neoplasia 2:26–40
39. Marshel JH, Kim YS, Machado TA et al (2019) Cortical layer-specific critical dynamics triggering perception. Science (80-) 365:eaaw5202
40. Sheng M, Sabatini BL, Südhof TC (2012) Synapses and Alzheimer's disease. Cold Spring Harb Perspect Biol 4:10
41. Guang S, Pang N, Deng X, Yang L, He F, Wu L, Chen C, Yin F, Peng J. Synaptopathology Involved in Autism Spectrum Disorder. Front Cell Neurosci. 2018 Dec 21;12:470
42. Mehta P, Kreeger L, Wylie D. C, Pattadkal J. J, Lusignan T, Davis M. J, Turi G. F., Li, W. K., Whitmire, M. P., Chen, Y., Kajs, B. L., Seidemann, E., Priebe, N. J., Losonczy, A., & Zemelman, B. V. (2019). Functional Access to Neuron Subclasses in Rodent and Primate Forebrain. Cell reports, 26(10), 2818–2832. e8. https://doi.org/10.1016/j.celrep.2019.02.011
43. Lutz C, Otis T, DeSars V et al (2008) Holographic photolysis of caged neurotransmitters. Nat Methods 5:821–827
44. Hernandez O, Papagiakoumou E, Tanese D et al (2016) Three-dimensional spatiotemporal focusing of holographic patterns. Nat Commun 7:11928
45. Papagiakoumou E, Ronzitti E, Emiliani V (2020) Scanless two-photon excitation with temporal focusing. Nat Methods 17:571–581
46. Papagiakoumou E, Bègue A, Leshem B et al (2013) Functional patterned multiphoton excitation deep inside scattering tissue. Nat Photonics 7:274–278
47. Oron D, Papagiakoumou E, Anselmi F et al (2012) Two-photon optogenetics. Prog Brain Res 196:119–143
48. Hernandez O, Guillon M, Papagiakoumou E et al (2014) Zero-order suppression for two-photon holographic excitation. Opt Lett 39:5953–5956
49. Oheim M, Beaurepaire E, Chaigneau E et al (2001) Two-photon microscopy in brain tissue: parameters influencing the imaging depth. J Neurosci Methods 111:29–37
50. Marshel JH, Kim YS, Machado TA et al (2019) Cortical layer–specific critical dynamics triggering perception. Science (80-) 365:eaaw5202
51. Cetin A, Komai S, Eliava M et al (2007) Stereotaxic gene delivery in the rodent brain. Nat Protoc 1:3166–3173
52. Holtmaat A, Bonhoeffer T, Chow DK et al (2009) Long-term , high-resolution imaging

in the mouse neocortex through a chronic cranial window. Nat Protoc 4:19–22

53. Kitamura K, Judkewitz B, Kano M et al (2008) Targeted patch-clamp recordings and single-cell electroporation of unlabeled neurons in vivo. Nat Methods 5:61–67

54. Chen TW, Wardill TJ, Sun Y et al (2013) Ultrasensitive fluorescent proteins for imaging neuronal activity. Nature 499:295–300

55. Chen I-W, Ronzitti E, Lee BR et al (2019) In vivo sub-millisecond two-photon optogenetics with temporally focused patterned light. J Neurosci 39:1785–1718

Chapter 12

Electrophysiological Characterization of Microbial Rhodopsin Transport Properties: Electrometric and ΔpH Measurements Using Planar Lipid Bilayer, Collodion Film, and Fluorescent Probe Approaches

Tatyana I. Rokitskaya, Nina L. Maliar, Sergey A. Siletsky, Valentin Gordeliy, and Yuri N. Antonenko

Abstract

Electrophysiological approaches to the study of the activity of retinal-containing protein bacteriorhodopsin (bR) or other proteins of this family are based usually on measurements of electrical current through a planar bilayer lipid membrane (BLM) with proteoliposomes attached to the BLM surface at one side of the membrane. Here, we describe the measurements of the pumping activity of bR and channelrhodopsin 2 (ChR2) with special attention to the study of voltage dependence of the light-induced currents. Strong voltage dependence of ChR2 suggests light-triggered ion channel activity of ChR2. We also describe electrophysiological measurements with the help of collodion film instead of BLM for the measurements of fast stages of a rhodopsin photocycle as well as the estimation of the activity of proteoliposomes without a macro membrane using fluorescent probes such as oxonol VI or 9-aminoacridine.

Key words Electrophysiology, Rhodopsins, Ion channel, Proton pump, Proteoliposomes, Planar lipid membrane

1 Introduction

Retinal-containing protein bacteriorhodopsin (bR) from halophilic archaea *Halobacterium salinarum* is the most studied member of a wide family of bacterial proteins mainly functioning as ion transporters and photosensors [1–3]. bR contains seven transmembrane α-helices and a co-factor retinal bound to the lysine side chain covalently. bR functions as a proton pump generating electrochemical difference of hydrogen ions across the bacterial membrane which fuels the bacterial energetics. The activity of isolated and reconstituted proteins of this family in lipid vesicles (liposomes) is measured usually by light-induced shift in pH in the aqueous

Valentin Gordeliy (ed.), *Rhodopsin: Methods and Protocols*, Methods in Molecular Biology, vol. 2501, https://doi.org/10.1007/978-1-0716-2329-9_12,

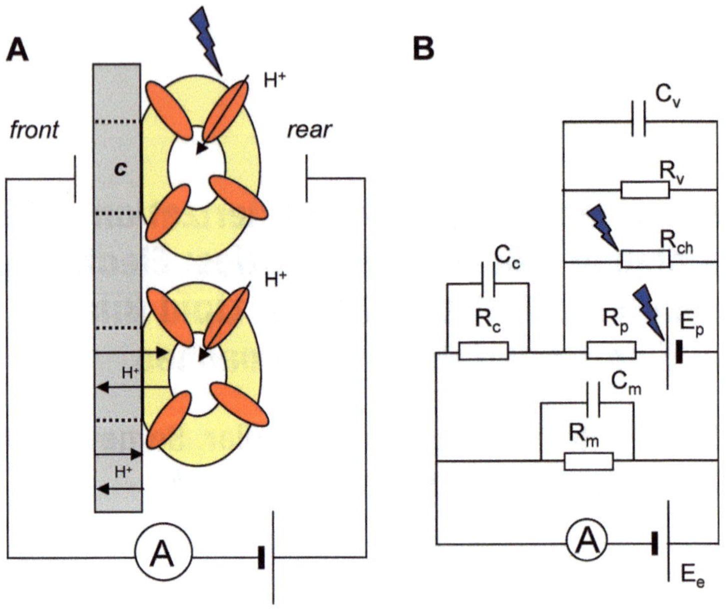

Fig. 1 Scheme of the planar bilayer with attached liposomes having protein pumps (**a**) and an equivalent electrical scheme corresponding to this system (**b**). R_m and C_m are the resistance and capacitance of that part of the planar membrane which is not occupied by vesicles, R_v and C_v are vesicle resistance and capacitance, R_c and C_c are the resistance and capacitance of the vesicle-membrane contact, R_p and R_{ch} are the proton pump or channel resistance, E_p is the voltage of the pump equivalent voltage generator

solution or by electrophysiological methods. One of the most used methods is the measurements of light-induced electrical current through planar bilayer lipid membrane (BLM) with proteoliposomes attached to one side of this macro membrane [4, 5]. Figure 1a shows a scheme of the planar bilayer with attached liposomes having protein pumps and an equivalent electrical scheme corresponding to this system (Fig. 1b). Electrical current is accessed by two electrodes located at two sides of the BLM forming a layer of electrical insulator of very high resistance. It is important that liposomes (a single (unilamellar) lipid vesicles) are not fused with the BLM, and the proteins can pump protons across the membranes of liposomes only. Therefore, the pumping activity can induce transient electrical responses on BLM upon the excitation of the system by visible light with negligible steady-state currents [6–8]. Regarding the equivalent electrical scheme, the transient current can be attributed to the existence of a capacitance of the BLM (C_m) and the membrane of liposomes (C_v, Fig. 1b). The system cannot generate stationary current without ionic conductance through the BLM and the measurement of the activity of bR or similar proteins consists in the analysis of the current transients with durations of less than a second. The addition of a protonophore to the system can lead to preferential permeabilization of the areas of the BLM covered with liposomes (c-regions) without shunting other parts of liposomes. Under these conditions, the excitation induces the bR-mediated H^+ transfer across the membrane of liposomes with subsequent H^+ transfer across the BLM mediated by the protonophore, thus leading to the induction of a steady-state electrical current. In other words, the addition of a protonophore allows the electrical access of the inner liposome aqueous phase to that on the opposite side of the planar membrane lacking liposomes.

In this chapter, we describe the protocols of electrophysiology measurements for two proteins, bacteriorhodopsin (bR) and channelrhodopsin 2 (ChR2). ChR2 exhibits poor proton pumping activity while can act as a light-triggered ionic channel [9, 10]. We are going to show that the ion channel activity of ChR2 leads to a strong voltage dependence of the light-induced currents and to describe how the measurement should be performed for identification of channels. In opposite, in case of bR, an ion pump, the voltage-dependence is very weak, and one should apply over a hundred millivolts to change the sign of the electrical current of bR in the presence of a protonophore [6, 8]. In addition, in this chapter, we describe a protocol for electrical measurements of the activity of proteoliposomes using lipid-impregnated collodion film as a macro membrane (Subheading 3.3). This modification of the system is useful for the measurements of fast stages of the photocycle of bR and other rhodopsins (up to 100 ns) [11, 12]. Finally, in Subheading 3.4, we describe the protocol of light-induced activity of proteoliposomes without a macro membrane, instead using fluorescent probes such as oxonol VI [13–15] or 9-aminoacridine [16, 17]. The last approach is substantially less sensitive compared to electrophysiological methods; however, it can be useful for initial characterizations of the preparations of the proteins.

2 Materials

2.1 Proteoliposomes Preparation

1. Phospholipids: asolectin from soybean.
2. Asolectin solvent: chloroform ultrapure.
3. Solvent evaporation: pear-shaped glass flask 10 mL, rotary evaporator (e.g., Heidolph, Germany), laboratory water bath (usually supplied with rotary evaporator).
4. Vacuum pump (e.g., Welch Chemstar Dry, USA).
5. Ultrasonic bath or probe (e.g., Elmasonic S30H, Germany).
6. Detergent-absorbing beads: we used Amberlite XAD-2 but other detergent-absorbing beads may be also accepted (e.g.,

Bio-Beads SM Hydrophobic and Polar Interaction Adsorbents). Amberlite XAD-2 should be stored in water. If using other detergent-absorbing beads, see manufacturer instruction manual (*see* **Note 1**).

7. Degassing beads slurry: parafilm (*see* **Note 1**).
8. Microcentrifuge tubes (e.g., Eppendorf) 2 mL, scissors, lint-free cloths.
9. Laboratory tube rotator.
10. Resuspension buffer: NaCl (0.1 M), sodium cholate (2% (w/v) solution in water). Use purified deionized water with a conductivity over 18 MΩ-cm (at 25 °C).
11. Channelrhodopsin-2 (ChR2): express ChR2 in *Leishmania tarentolae*, isolate and purify as reported [18]. Protein concentration should be not less than 50 mg/mL.
12. Bacteriorhodopsin (bR): solubilize bR from purple membranes as reported [19]. Protein concentration should be not less than 50 mg/mL.
13. Foil for wrapping the tubes (*see* **Note 2**).
14. Cold room (*see* **Note 3**).
15. Clean the glass flask manually with any labware detergent, thoroughly rinse with water and then with methanol to remove residual lipids. Dry before the next use.

2.2 Functional Characterization of Retinal Bearing Proteins on Planar Lipid Bilayers

1. Phospholipids: 1,2-di-phytanoyl-sn-glycero-3-phosphocholine, 1,2-dimyristoyl-sn-glycero-3-ethylphosphocholine.
2. Special cell for BLM measurements (e.g., Warner Instruments).
3. Patch-clamp amplifier (e.g., Warner Instruments, model BC-525C).
4. Output signal digitizer: for example, NI-DAQmx (National Instruments).
5. Personal computer with the WinWCP Strathclyde Electrophysiology Software designed by J. Dempster (University of Strathclyde) or another appropriate program.
6. Light source: halogen lamp (e.g., Novaflex, World Precision Instruments, Sarasota, FL).
7. Protonophore: tetrachloro-2-(trifluoromethyl)benzimidazole (TTFB).
8. Magnetic stirrers.

2.3 Light-Induced Electrogenic Steps of Retinal-Containing Proteins

1. Collodion solution in isoamyl acetate.
2. Asolectin solution in *n*-decane.

3. Special cuvette with two symmetrical compartments having a hole of 4 mm in diameter at facing sides holding the collodion filter.
4. Magnetic stirrers.
5. Cuvette solution: 25 mM HEPES-TRIS, pH 7.5.
6. Laser (frequency-doubled YAG, 532 nm; pulse half-width, 15 ns; pulse energy, 20 mJ; e.g., Quantel, Les Ulis, France).

2.4 Fluorescent Tools for Initial Characterization of Proteoliposome Functionality

1. Oxonol VI solution in DMSO.
2. Valinomycin solution in Ethanol.
3. Fluorescent cuvette solution containing 10 mM Tris, 10 mM MES, 100 mM NaCl, pH 6.0.
4. Fluorimeter (e.g., Panorama, Lumex, St. Petersburg, Russia or Fluoromax, Horiba, Japan).
5. Illuminating laser module (532-nm) or a halogen lamp (e.g., Novaflex, World Precision Instruments, Sarasota, FL) with a green filter.
6. 9-Aminoacridine solution in Ethanol.

3 Methods

3.1 Asolectin Proteoliposomes Preparation

1. To prepare 1 mL of proteoliposomes, weigh 10 mg of asolectin in a pear-shaped glass flask and dissolve in 1 mL of chloroform.
2. Fill the bath of the rotary evaporator with water and warm it to 25 °C.
3. Connect the flask to the rotary evaporator and rotate at 110 rpm under vacuum until the total evaporation of the solvent and the formation of a thin lipid film on the sides of the flask (the process takes approximately 15 min).
4. Remove the residual chloroform using a vacuum pump from 3 h to overnight.
5. Slurry the beads in ultrapure water and degas for at least 2 h (*see* **Note 1**).
6. Resuspend the dried asolectin by pipetting and vortexing in resuspension buffer (0.1 M NaCl supplemented with 2% (w/v) sodium cholate) at a final concentration of 1% (w/v) which means 1 mL of buffer for 10 mg of asolectin.

   Next steps should be held at 4 °C with minimal light exposure of samples (see **Notes 2** and **3***).*
7. Clarify the mixture by sonication at 4 °C for 5 min.
8. Put the mixture into microcentrifuge tube. Centrifuge the tube for 5 min in a table centrifuge, decant the supernatant, and

aliquot it into 500 μL portions into microcentrifuge tubes (2.0 mL).

9. Cut off the narrow end of the 1-mL pipette tip and take 500 μL of Amberlite XAD-2 slurry. Squeeze the water from the tip pushing the tip against a lint-free cloth. As a result, the beads in the tip should be dry.
10. Sonicate the lipid mixture in the Eppendorf tube for 1 min at 4 °C, then quickly (!) add the protein to a final protein concentration of 0.7 mg/mL, and then add the beads prepared in the **step 9**. Wrap the tube in foil (*see* **Note 2**).
11. Repeat **steps 9** and **10** with the lipid mixture in the other Eppendorf tube.
12. Stir on a rotary platform at 4 °C for 2 h.
13. Recover the sample by removing it with a pipette. It may be convenient to centrifuge the sample before pipetting. Repeat **step 9** and put a new portion of beads to the recovered sample.
14. Repeat **steps 12** and **13** three times (*see* **Note 3**).

3.2 Functional Characterization of Retinal Bearing Proteins on Planar Lipid Bilayers

1. Weigh 0.4 mg 1,2-dimyristoyl-sn-glycero-3-ethylphosphopholine and dissolve in 1 mL of n-decane (**solution 1**, *see* **Note 4**). To prepare the membrane-forming solution (**BLM solution**), weigh 2 mg 1,2-di-O-phytanoyl-sn-glycero-3-phosphocholine and dissolve in 0.1 mL of **solution 1**.
2. Wash a Teflon cell with distilled water and dry it. Put and spread a drop of **BLM solution** around a hole in septum and wait until the solution becomes completely dry. Then fill both the compartments of a Teflon cell with a buffer solution containing 10 mM Mes, 10 mM Tris, 100 mM NaCl pH = 6.0, or other required solution. Connect membrane cell to an external measuring circuit via two Ag/AgCl electrodes. In order to avoid artificial photoeffects, separate electrodes from the aqueous compartments of the cell by salt bridge with the same electrolyte solution or place them in light-tight attachments.
3. For electric current measurement, use patch-clamp amplifier (e.g., Warner Instruments, model BC-525C), digitize output signal using an NI-DAQmx (National Instruments), and analyze with a personal computer with the use of WinWCP Strathclyde Electrophysiology Software designed by J. Dempster (University of Strathclyde) or other appropriate program. Connect ground electrode to the front compartment and high-resistance electrode to the rear compartment. Put magnetic stirrers in both the compartments.
4. Form a BLM on the hole in septum by touching the lipid bubble in buffer solution and control the process of BLM formation optically or by means of capacitance measurements

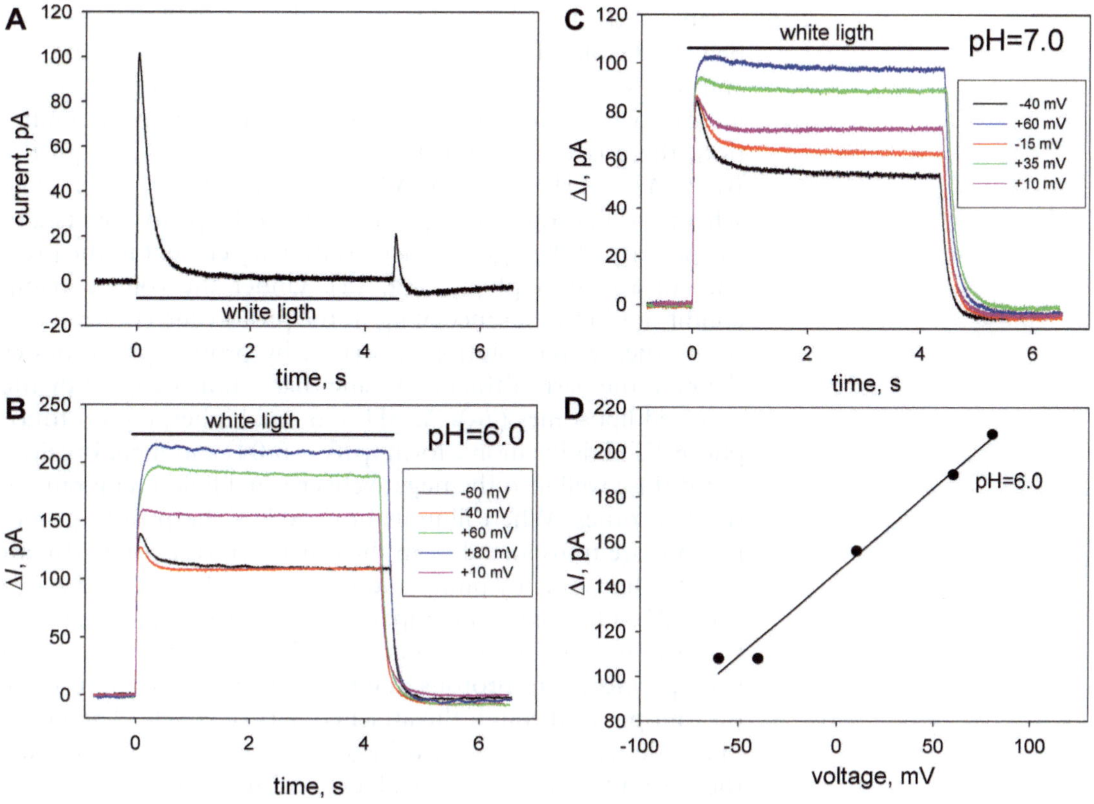

Fig. 2 Photocurrent of bR in the absence (**a**) and in the presence of a protonophore TTFB (**b**, **c**) at different applied voltages. (**d**) voltage dependence of bR steady-state photocurrents at pH = 6.0

(specific capacity range 0.4–0.6 μF/cm^2). Then test the background conductivity of the bilayer lipid membrane (acceptable range $10^{-9} - 10^{-8}$ S/cm^2). The light beam first entered the front compartment. Make sure that when illuminating the lipid membrane, there is no change in the current level.

5. Add 10–20 μL of vesicles with reconstituted bacteriorhodopsin protein to rear compartment and stir the solution. After 10 min, illuminate BLM with a constant light (density of power about 0.8 W/cm^2) produced by a halogen lamp (e.g., Novaflex, World Precision Instruments, Sarasota, FL) and register the transient current at the beginning and end of lighting (Fig. 2a). Continue attachment of the liposomes until the magnitude of the transient current will stop to increase, it takes usually about 1 h. The time constant associated with the fast current transient under long-lasting light excitation is the membrane relaxation time (RC) constant which can vary in the range of several hundred milliseconds. Stationary current at long-lasting illumination is very small or tends to zero at the different applied potentials.

6. Then add 0.1–0.3 μM tetrachloro-2-(trifluoromethyl)-benzimidazole (TTFB) to the front compartment of the cell and stir solution rigorously. The conductance of BLM will increase within 10–20 min to 2–8 μS/cm^2. During illumination, there is a change in the stationary current through the BLM (ΔI). At $|V| \leq 75$ mV $\Delta I = I_L - I_D > 0$ (Fig. 2b, c). The scheme of liposome adsorption on BLM helps to understand the polarity of the light-induced stationary current in the presence of a protonophore (Fig. 1a). Under the voltage clamp conditions in the absence of light, the proton current from one to another water solution mediated by protonophore passes through the part of the membrane that is not occupied by the attached liposomes (I_D). At pH = 6 and higher, the protonophore TTFB is in anionic form (pK_a = 5.04) and therefore does not diffuse well into the negatively charged lipid membrane of the liposomes. When illumination excites the proton pump, protons are transported from the outside in rear compartment into the liposome by pump and then through the contact area (c) of BLM to the front compartment of the cell due to permeability to protons of the contact area (Fig. 1a). The contact area (c) is permeable to protons due to partial redistribution of the protonophore through the attached part of vesicle. Therefore, in a wide range of applied voltages (V), regardless of its value, the current change (ΔI) will be positive. Only at negative potentials at the rear compartment, close to the proton pump potential (E_p), the light-induced current change will decrease to zero.
7. After determining the values of ΔI at different applied voltages, one can plot the dependence of the current change on voltage and approximate it with a linear function (Fig. 2d). $E_p = -V$ at which the current change is zero. By shifting the pH of the aqueous solution in both the compartments, it is easy to determine E_p. The proton pump potential E_p was in the range of 140–200 mV for bacteriorhodopsin (Fig. 2d).
8. The study of channelrhodopsin 2 (ChR2) should be carried out according to the same sequence 2–5. Add 10–20 μL of vesicles with reconstituted ChR2 to the rear compartment and stir the solution. Record the current kinetics upon lighting of the BLM at various applied voltages after about 10–20 min of incubation. The transient current is registered at the beginning and the end of the lighting (Fig. 3a). The amplitude of transient current changes in this case is much smaller than for bacteriorhodopsin. Stationary conductance of the lipid membrane increases slightly in the dark. ΔI is zero at all voltages (Fig. 3a), indicating the attachment of the liposomes to the surface of the BLM without a fusion. Continue the incubation of liposomes for an hour or a little longer.

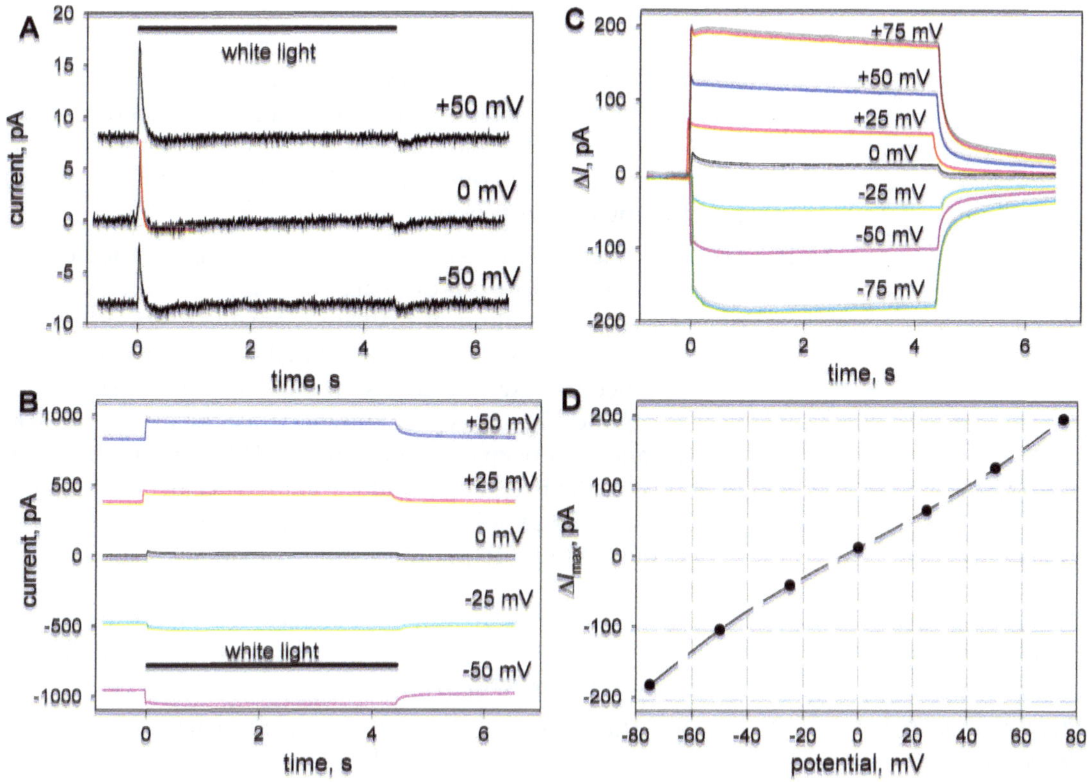

Fig. 3 Photocurrent of ChR2 in the absence (**a**) and in the presence of a protonophore TTFB (**b**, **c**) at different applied voltages. (**d**), voltage dependence of bR steady-state photocurrents

9. Then add 0.1–0.3 μM TTFB to the front compartment of the cell and stir the solution. After increasing the membrane conductance to a stationary level, it is necessary to measure the current kinetics under illumination (J_L) at various applied voltages (Fig. 3b). Note that the membrane may be unstable at voltages above 75 mV. Liposomes with ChR2 under voltage clamp conditions are characterized by a change in the sign of the light-induced stationary current (ΔI) at voltages close to zero (Fig. 3c, d). The slope of the linear dependence of the light-induced current on the voltage was approximately four times larger for ChR2 then for bR. Such a significant light-induced increase in membrane conductance in accordance with the sign of the applied voltage suggests functioning of ChR2 as a light-triggered ion channel. To understand the strong current-voltage dependence of BLM with attached liposomes with ChR2, refer to the scheme in Fig. 1a. Because the anionic form of the protonophore diffuses weakly into negatively charged liposomes, we assume that the proton leakage through the lipid part of the liposomes is very small. So, in the dark, the proton current mediated by TTFB in the lipid membrane is determined by the conductance of the liposome-free regions of

BLM. The electrical resistance of the areas with attached liposomes is a serial connection of the contact zone resistance of the BLM and the liposomes resistance (Fig. 1b). Therefore, when illuminated, the current of protons and monovalent cations through the channel rhodopsin increases significantly. Proton transfer through the contact area of the BLM with liposomes is provided by the protonophore.

10. From the dependence of the current change on the applied voltage (Fig. 3d) one can estimate voltage E_e at which $\Delta I = 0$. At pH = 6.0 E_e is about 4 mV and increases to 25–30 mV at pH = 8.1. These values are not equal to the proton pump voltage since the exact relationship between the resistances of the membrane fragments is unknown. Since the relative proton conductance has been estimated to be ~10^6 times higher than the Na^+ conductivity, the channel conductance decreases as the pH of the aqueous solution increases [20]. At the same time, the protein's ability to pump protons is manifested better.

3.3 Light-Induced Electrogenic Steps of Retinal-Containing Proteins Measured on Phospholipid-Impregnated Collodion Film

3.3.1 Formation of Phospholipid-Impregnated Collodion Film and Experimental Setup

1. Prepare collodion film. Put 0.010–0.15 mL of 1% collodion solution (nitrocellulose) in isoamyl acetate on the surface of distilled water (*see* **Note 5**).
2. Remove the collodion film formed on the surface after a few minutes using a ring with a diameter of ~1 cm, cut from a nitrocellulose membrane filter, bringing the ring under the film from below the surface, so that when removing the ring, the film covers the ring on both sides (*see* **Note 6**).
3. Wait ~3–5 h until the film becomes dry.
4. Soak the film with asolectin solution in *n*-decane (100 mg/mL) and clamp the film between the two symmetrical cylindrical compartments of the cuvette which have a hole of 4 mm in diameter on one side and face each other by these holes. The film should have larger diameter so that it should fill the hole between the compartments of the cuvette (*see* **Note 7**).
5. Fill slowly both compartments of the cuvette with 4 mL of an aqueous solution (25 mM HEPES-TRIS, pH 7.5). Put magnetic stirrers in both the compartments (*see* **Note 8**).
6. Add 20 mM $MgCl_2$ (or $MgSO_4$) to both the compartments of the cuvette.
7. Put 50–100 μL of proteoliposomes to one of the compartments of the cuvette.
8. Incubate the cuvette with proteoliposomes at constant stirring, at room temperature in the dark. The process of the proteoliposomes attachment to the collodion film requires about 2 h.
9. Put two Ag/AgCl electrodes on both the sides of the collodion film. Freshly chlorinated silver wires shielded with thin

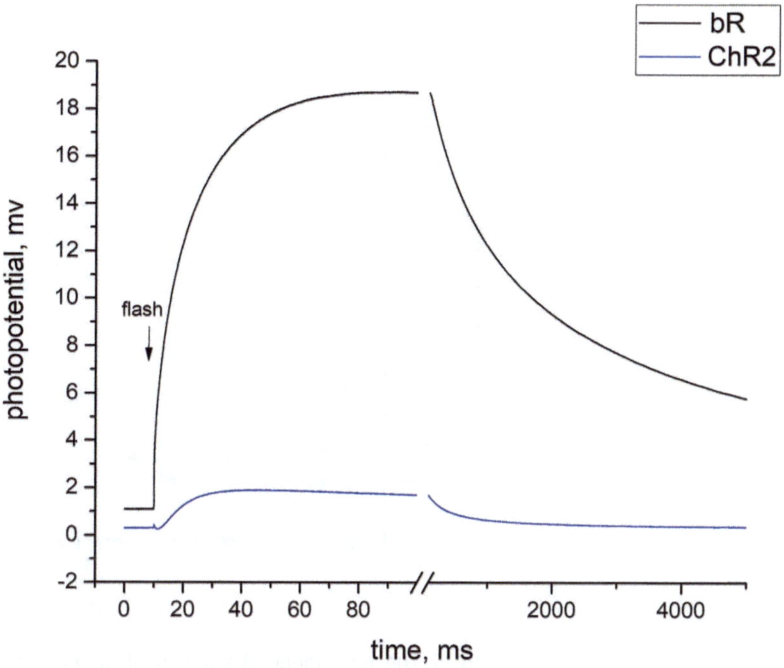

Fig. 4 The light-induced kinetics of bR and ChR2 membrane potential generation in the mode of a single catalytic turnover of the enzyme (photocycle). The moment of light flash by 532 nm laser is indicated by an arrow. The solution was 25 mM HEPES-TRIS (pH 7.5), 20 mM $MgCl_2$

polyethylene should be used to record fast photoelectric response of proteoliposomes attached to the collodion film.

10. Connect the electrodes to an input of an operational amplifier with feedback resistance of 10^{11} Ohm and a gain of 10–100. The output voltage is digitized by an oscilloscope (or its analog) connected to a computer (*see* **Note 9**). For electrical shielding, the cuvette and the amplifier must be placed in a metal casing.
11. For starting the photocycle a pulse of laser light is used (frequency-doubled NdYAG, 532 nm; pulse half-width, 15 ns; pulse energy, 20 mJ; Quantel, Les Ulis, France). The laser beam is fed into the cell compartment through a transparent window at the front compartment of the cuvette, it is focused on the membrane separating the cell compartments.

3.3.2 Electrogenic Steps of bR and ChR2

Figure 4 shows the kinetics of generating the membrane potential of bR and ChR2 in response to a laser flash in the mode of a single photocycle (turnover) of the enzymes. In response to the flash, there is a generation of a membrane potential reaching maximum values in the time scale of tens of milliseconds (the characteristic time of the photocycle), after which there is a passive discharge of

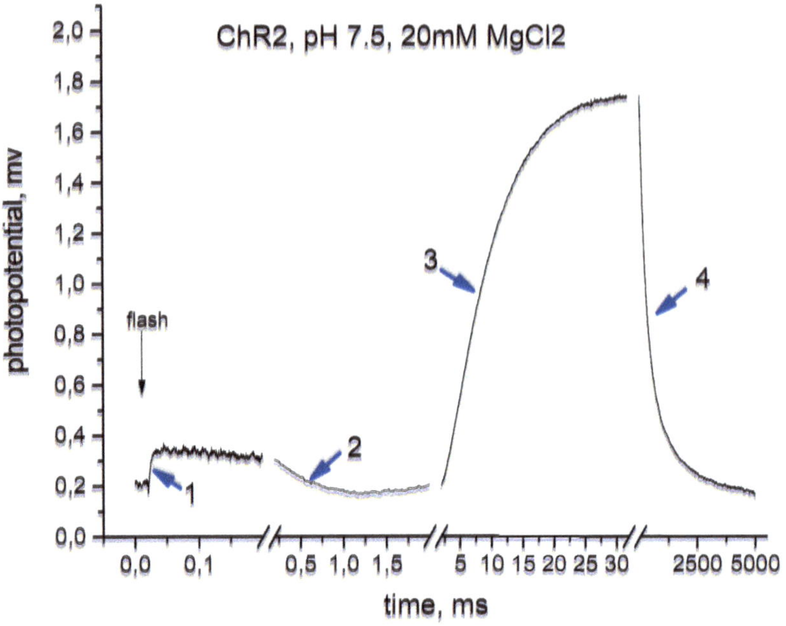

Fig. 5 The light-induced kinetics of generation of the membrane potential of ChR2 in the mode of a single catalytic turnover of the enzyme (photocycle). The moment of the light flash by 532 nm laser is indicated by a black arrow. The blue arrows indicate the kinetic phases 1–4. The solution was 25 mM HEPES-TRIS (pH 7.5), 20 mM $MgCl_2$

the membrane for a few seconds. A laser flash of 532 nm was used to initiate the photocycle. The amplitude of photoelectric response in the case of ChR2 is ~10 times less than that of BR under these conditions which is apparently a result of the shift of the maximum absorption of the optical spectrum of ChR2 in the initial state compared to BR.

Figure 5 shows the kinetics of membrane potential generation of the ChR2 in several time scans, which allows us to distinguish individual kinetic phases in the photoelectric response that reflect the main electrogenic processes in the photocycle. Approximation by the four components provides reliable fit with the following parameters of the electrogenic phases. Phase 1 (~4 μs) reflects the proton transfer from the Schiff base to the primary acceptor and the generation of the M state. The relative amplitude of phase 1 is ~1/10 of the total amplitude of the photoelectric response, which is in agreement with the typical electrogenic distance between the Schiff base and the primary acceptors in retinal proteins [21–25].

Phase 2 (~1.6 ms) has a negative direction, which corresponds to the discharge of the membrane potential. This phase coincides with the appearance and operation of an ion channel in ChR2, through which the membrane potential generated at the previous stage is apparently discharged. This phase is clearly observed in the presence of a 20 mM $MgCl_2$ salt and is inhibited by replacing the buffer with a salt-free one (not shown). Phase 3 (~8.4 ms) reflects the transfer of protons to the base of the Schiff in the mode of operation of the ChR2 proton pump, which is manifested against the background of the channel closed at the previous stage. Phase 4 (~500 ms) reflects passive leakage (discharge) through the membrane. Phase 0—the negative phase (initial drop down just after the flash, not marked in the Fig. 5)—reflects the formation of the early intermediates of the photocycle K and L. The retinal displacement seems to be the main cause of generation of the "negative" phases associated with K intermediate formation as well as with L intermediate formation accompanied by torsion relaxation of retinal [22].

3.4 Fluorescent Tools for Initial Characterization of Proteoliposome Functionality

3.4.1 Oxonol Fluorescence for the Study of Pumping Activity

1. Weigh 0.63 mg Oxonol VI (Mol Probes) and dissolve in 1 mL of DMSO (**Oxonol solution**). Weigh 1 mg valinomycin (Sigma) and dissolve in 1 mL of Ethanol (**Valinomycin solution**).
2. Fill fluorescent cuvette with 1 mL of buffer solution containing 10 mM Tris, 10 mM MES, 100 mM NaCl, pH 6.0, or other required solution. Place the cuvette in a fluorimeter (e.g., Panorama, Lumex, St. Petersburg, Russia) and set excitation at 590 nm and emission at 620 nm (5 nm slits). Set the temperature at 15 °C and wait for 10 min for temperature equilibration. Start recording the kinetics of the fluorescent signal.
3. Add 2 μL of Oxonol solution which produce a high signal of fluorescence after intensive stirring (Fig. 6a).
4. Add 10–20 μL of vesicles with reconstituted bacteriorhodopsin protein and stir solution.
5. Illuminate the cuvette by a green laser module (532-nm) with laser beam coming from above. The beam could be expended to the diameter of the cuvette using a lens. Alternatively illuminate the cell by a halogen lamp (e.g., Novaflex, World Precision Instruments, Sarasota, FL) using a light-guide and a green filter. The Oxonol VI signal started to decrease due to the formation of membrane potential on liposomes because of the bR pumping activity (Fig. 6a). After turning off the illumination, the signal is restored in a minute time scale. Repeat the illumination cycle three times. Similar experiments with ChR2 liposomes showed no pumping activity obviously due to low sensitivity of the method compared to BLM measurements (data not shown).
6. Add 1 μL of Valinomycin solution and stir. The fluorescent signal remains at the nearly same level (*see* **Note 10**).

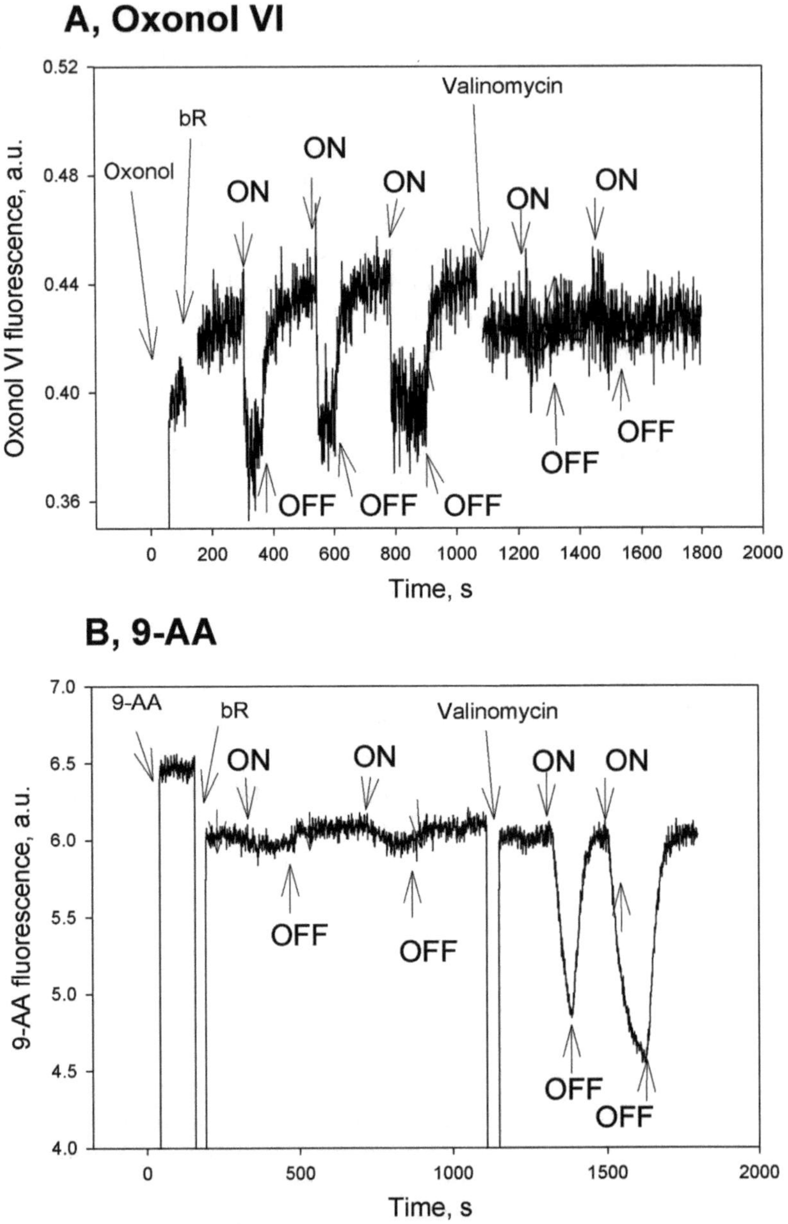

Fig. 6 Proton pumping activity of bR measured by the potential-sensitive probe oxonol VI (4 μM, **a**) or pH-sensitive probe 9-aminoacridine (9-AA, 8 μM, **b**). ON and OFF indicated the start and the end of the illumination with 532-nm laser module. Valinomycin concentration was 1 μg/mL

7. Illuminate the cuvette several times. Note that valinomycin suppressed the light-induced decrease of the fluorescent signal suggesting that valinomycin decreased the magnitude of the membrane potential generation by bR pumps (Fig. 6a).

3.4.2 Use of 9-Aminoacridine for the Characterization of pH Gradient on Proteoliposomes

1. Weigh 0.39 mg 9-aminoacridine and dissolve in 1 mL of Ethanol (**9-AA solution**).
2. Fill fluorescent cuvette with 1 mL of buffer solution containing 10 mM Tris, 10 mM MES, 100 mM NaCl, pH 6.0, or other required solution. Place the cuvette in a fluorimeter (e.g., Panorama, Lumex, St. Petersburg, Russia) and set excitation at 425 nm and emission at 455 nm (5 nm slits). Set the temperature at 15 °C and wait for 10 min for temperature equilibration. Start recording the kinetics of the fluorescent signal.
3. Add 4 μL of 9-AA solution which produce a high signal of fluorescence after intensive stirring (Fig. 6b).
4. Add 10–20 μL of vesicles with reconstituted bacteriorhodopsin protein and stir solution.
5. Illuminate the cuvette by a green laser module. The 9-AA signal started to decrease due to the formation of pH gradient on liposomes because of the bR pumping activity (Fig. 6b). After turning off the illumination, the signal is restored in a minute time scale. Repeat the illumination cycle three times.
6. Add 1 μL of Valinomycin solution and stir. The fluorescent signal remains at the nearly same level.
7. Illuminate the cuvette several times. The decrease in the fluorescent signal caused by illumination increased markedly (Fig. 6b), suggesting the increase in the formation of pH gradient on the membranes of bR liposomes in the presence of valinomycin. Obviously valinomycin induced the transport of potassium/sodium ions which decreased the membrane potential (Fig. 6a) leading to a stimulation of the H+ -pumping activity of bR. Similar experiments with ChR2 liposomes showed no visible pumping activity obviously due to low sensitivity of the method compared to BLM measurements (data not shown).

4 Notes

1. Some detergent-absorbing beads are stored in methanol and should be rinsed with water before using. Beads can be regenerated by rinsing with methanol in most cases, see the instruction manual. The amount of beads depends on the manufacturer. Usually, it is determined by the total weight of detergent, including detergent in the protein sample and sodium cholate in resuspension buffer. To degas, put the beads slurry in the bottle or glass and seal it with parafilm making several small holes. Do not degas the slurry on magnetic stirrer because stirrer bars can be harmful for beads.

2. Light exposure of protein samples and proteoliposomes with embedded rhodopsins should be minimal. We recommend wrapping the sample-containing tubes with foil.
3. Important: **steps 7–14** should be done at 4 °C, preferably in cold room.
4. This cationic lipid facilitates substantially the attachment of negatively charged asolectin proteoliposomes to planar BLM.
5. The area of the water surface used is approximately 50 cm^2 or more.
6. Avoid inhaling isoamyl acetate (work with the fume hood turned on).
7. The measurement is carried out in a Teflon cuvette assembled from two separate compartments with a volume of 4 mL, connected by a hole with a diameter of 4 mm.
8. A resistance of the membrane should be 10^9 Ohm or higher.
9. In this work, an operational amplifier (Burr Brown 3554BM, USA) and a high-speed gage CS8012 digitizer connected to a personal computer were used to amplify the output voltage and digitize the signal, respectively.
10. One should use high concentration of valinomycin (about 1 μM) in sodium buffers while low concentrations (about 10 nM) in potassium buffers.

Acknowledgments

We are grateful to Oleksandr Volkov for producing channelrhodopsin-2 and Christian Baeken for providing us with purple membranes.

The research was supported by the Ministry of Science and Higher Education of the Russian Federation (agreement # 075-00337-20-03, project FSMG -2020-0003). Measurements of electrogenic properties were supported by RFBR grants 19-04-00238 (to TIR) and 18-04-00503a (to SAS).

This work was supported by Russian Science Foundation (RSF) Project 21-64-00018.

References

1. Gushchin I, Gordeliy V (2018) Microbial rhodopsins. In: Subcellular biochemistry, pp 19–56
2. Ernst OP, Lodowski DT, Elstner M et al (2014) Microbial and animal rhodopsins: structures, functions, and molecular mechanisms. Chem Rev 114:126–163
3. Grote M, Engelhard M, Hegemann P (2014) Of ion pumps, sensors and channels—perspectives on microbial rhodopsins between science and history. Biochim Biophys Acta Bioenerg 1837:533–545
4. Drachev LA, Jasaitis AA, Kaulen AD et al (1974) Direct measurement of electric current

generation by cytochrome oxidase, H+-ATPase and bacteriorhodopsin. Nature 249:321–324

5. Bamberg E, Apell HJ, Dencher NA et al (1979) Photocurrents generated by bacteriorhodopsin on planar bilayer membranes. Biophys Struct Mech 5:277–292
6. Drachev LA, Frolov VN, Kaulen AD et al (1976) Reconstitution of biological molecular generators of electric current. Bacteriorhodopsin. J Biol Chem 251:7059–7065
7. Herrmann TR, Rayfield GW (1978) The electrical response to light of bacteriorhodopsin in planar membranes. Biophys J 21:111–125
8. Bamberg E, Dencher NA, Fahr A et al (1981) Transmembranous incorporation of photoelectrically active bacteriorhodopsin in planar lipid bilayers. Proc Natl Acad Sci 78:7502–7506
9. Feldbauer K, Zimmermann D, Pintschovius V et al (2009) Channelrhodopsin-2 is a leaky proton pump. Proc Natl Acad Sci 106: 12317–12322
10. Sineshchekov O, Govorunova EG, Wang J et al (2013) Intramolecular proton transfer in channelrhodopsins. Biophys J 104:807–817
11. Drachev LA, Kaulen AD, Khitrina LV et al (2005) Fast stages of photoelectric processes in biological membranes. Eur J Biochem 117: 461–470
12. Drachev LA, Kalamkarov GR, Kaulen AD et al (1981) Fast stages of photoelectric processes in biological membranes. II Visual rhodopsin. Eur J Biochem 117:471–481
13. Apell H-J, Bersch B (1987) Oxonol VI as an optical indicator for membrane potentials in lipid vesicles. Biochim Biophys Acta Biomembr 903:480–494
14. Bennett AB, Spanswick RM (1983) Optical measurements of ΔpH and $\Delta\psi$ in corn root membrne vesicles: kinetic analysis of Cl– effects on a proton-translocating ATPase. J Membr Biol 71:95–107
15. Antonenko YN, Denisov SS, Silachev DN et al (2016) A long-linker conjugate of fluorescein and triphenylphosphonium as mitochondria-targeted uncoupler and fluorescent neuro- and nephroprotector. Biochim Biophys Acta Gen Subj 1860:2463–2473
16. Sone N, Yoshida M, Hirata H et al (1977) Adenosine triphosphate synthesis by electrochemical proton gradient in vesicles reconstituted from purified adenosine triphosphatase and phospholipids of thermophilic bacterium. J Biol Chem 252:2956–2960
17. Nichols JW, Hill MW, Bangham AD et al (1980) Measurement of net proton-hydroxyl permeability of large unilamellar liposomes with the fluorescent pH probe, 9-aminoacridine. Biochim Biophys Acta Biomembr 596:393–403
18. Volkov O, Kovalev K, Polovinkin V et al (2017) Structural insights into ion conduction by channelrhodopsin 2. Science 358:eaan8862
19. Dencher NA, Heyn MP (1982) Preparation and properties of monomeric bacteriorhodopsin. In: Methods in enzymology, pp 5–10
20. Nagel G, Szellas T, Huhn W et al (2003) Channelrhodopsin-2, a directly light-gated cation-selective membrane channel. Proc Natl Acad Sci 100:13940–13945
21. Gushchin I, Chervakov P, Kuzmichev P et al (2013) Structural insights into the proton pumping by unusual proteorhodopsin from nonmarine bacteria. Proc Natl Acad Sci 110: 12631–12636
22. Kaulen AD (2000) Electrogenic processes and protein conformational changes accompanying the bacteriorhodopsin photocycle. Biochim Biophys Acta Bioenerg 1460:204–219
23. Kalaidzidis IV, Kaulen AD, Radionov AN et al (2001) Photoelectrochemical cycle of bacteriorhodopsin. Biochemistry (Mosc) 66: 1220–1233
24. Siletsky SA, Mamedov MD, Lukashev EP et al (2016) Electrogenic steps of light-driven proton transport in ESR, a retinal protein from Exiguobacterium sibiricum. Biochim Biophys Acta Bioenerg 1857:1741–1750
25. Siletsky SA, Mamedov MD, Lukashev EP et al (2019) Elimination of proton donor strongly affects directionality and efficiency of proton transport in ESR, a light-driven proton pump from Exiguobacterium sibiricum. Biochim Biophys Acta Bioenerg 1860:1–11

Chapter 13

Electrophysiological Characterization of Microbial Rhodopsins by Patch-Clamp Experiments

Thomas Mager

Abstract

Optogenetics is of key importance for progress in basic neuroscience research and the development of innovative future medical treatments. In particular, the use of microbial rhodopsins enables remote control of excitable-cell activity by light. The electrophysiological characterization of microbial rhodopsins is inevitable for the development of variants, which further advance optogenetic applications. Therefore, we provide a detailed description of the application of the patch-clamp method for the electrophysiological characterization of microbial rhodopsins. Here we describe the investigation of light sensitivity, wavelength- and voltage-dependence, photocurrent inactivation, kinetics, and ion selectivity.

Key words Optogenetics, Microbial rhodopsins, Channelrhodopsins, Light-driven ion pumps, Patch-clamp method, Electrophysiological characterization

1 Introduction

Optogenetics combines optics with genetic manipulation to control a specific function of genetically targeted cells using light. It enables the cell-type-specific activation with high spatial and temporal resolution in a reversible manner. Therefore, optogenetics expedites progress in basic research, in particular in cardiology and neurosciences, and is key to the development of future medical applications. Remote control of excitable cell activity with light is primarily accomplished by the use of microbial rhodopsins. In that regard, Channelrhodopsins (ChRs) [1–3], which are light-gated ion channels, as well as light-driven ion pumps [4, 5] are of particular interest. To date there are two ongoing clinical trials for vision restoration, using Channelrhodopsin variants in retinal ganglion cells [6]. Optogenetic hearing restoration is at the preclinical stage. The development of the future optical cochlea implant promises a better speech and music comprehension by improving the frequency resolution [7–9].

Valentin Gordeliy (ed.), *Rhodopsin: Methods and Protocols*, Methods in Molecular Biology, vol. 2501,
https://doi.org/10.1007/978-1-0716-2329-9_13,

The biophysical characterization of microbial rhodopsins, which were discovered in metagenomic studies, led to the identification of variants with advanced properties for optogenetic applications. The use of Chronos allows neuronal photostimulation of fast-spiking neurons with high temporal fidelity due to its fast channel-closing kinetics [10–12]. Chrimson variants have the most red-shifted action spectra of all currently known ChR variants [12, 13]. They are of particular interest because longer-wavelength light penetrates deeper into tissue due to reduced light scattering and also bears a much-reduced risk of phototoxicity. Neuronal silencing can be accomplished by the use of outwardly directed, hyperpolarizing, light-driven ion pumps [4, 5] and anion channelrhodopsins (ACR) [14, 15]. Xenorhodopsin (XeR) activation by yellow light pulses leads to plasma membrane depolarization, due to its inwardly directed proton pumping activity, which is a unique feature among light-driven ion pumps [16, 17], and can therefore be used for cation-gradient independent optogenetic activation [17].

Channelrhodopsins (ChRs) have been optimized by genetic engineering to provide suitable action spectra, kinetics, and ion selectivity [18, 19]. Examples for advanced optogenes are CatCh (calcium-translocating channelrhodopsin) [18], ReaChR [19] (red-activatable channelrhodopsin), and the fast Chrimson variants (f-Chrimson, vf-Chrimson) [8]. In addition to its elevated calcium permeability, CatCh is strongly expressed in neurons and shows a reduced desensitization of the photocurrent. ReaChR has been successfully used for the transcranial optical activation of neurons in living mice, possible because of its increased light sensitivity and red-shifted action spectrum. The fast Chrimson variants (f-Chrimson and vf-Chrimson) combine robust expression, red-light excitability, and ultrafast closing kinetics [8].

The correct application of optogenetic tools requires the detailed knowledge of their biophysical properties. Therefore, the electrophysiological characterization of microbial rhodopsins is of key importance. The patch-clamp method [20] is a powerful technique to study the electrophysiological properties of ion-transporting membrane proteins. Carrying out patch-clamp experiments with microbial rhodopsins in neuroblastoma glioma cells (NG-cells) prior to their further characterization in excitable cells has several advantages. NG cells are neuronal-like cells, which enable the strong overexpression of heterologous proteins already after short expression times. The absence of excitability is advantageous, because spiking interferes with the electrophysiological characterization, e.g., when current–voltage curves are measured. The application of the following protocols enable the investigation of the basic transport properties of microbial rhodopsins by patch-clamp experiments.

2 Materials

2.1 Cell Culture and Transfection

1. CO_2 level- and temperature-controlled incubator. The incubators are used at 37 °C and 5 vol. % CO_2.
2. Cell counter. Use a Neubauer chamber or automated cell counters.
3. Cell culture hood. The use of laminar flow hoods is recommended.
4. Cell culture medium (DMEM$^+$): Dubelcco's Modified Eagle Medium (DMEM) supplemented with 10 vol. % fetal calf serum and 5 vol. % penicillin/streptomycin (10,000 units penicillin and 10 mg/mL streptomycin). Store at 4 °C.
5. 1 mM All-trans-Retinal (ATR) in ethanol (absolute grade). Store at −20 °C.
6. Transfection reagent. Use cationic lipid-based chemical transfectants. Store at 4 °C.
7. Coverslips. Use glass coverslips with a diameter of 10 mm. The use of coated glass coverslips is not necessary.

2.2 Electrophysiological Characterization of Microbial Rhodopsins

1. Patch-Clamp setup with fluorescence microscope. The fluorescence microscope should be equipped with excitation and emission filters for the visualization of the fluorescent fusion-protein expression.
2. Horizontal puller for the fabrication of patch pipettes.
3. Illumination device (*see* **Note 1**): Continuous wave (cw) laser of suitable wavelength (*see* **Note 2**), Nd:YAG-laser in combination with optical parametric oscillator (OPO), multimode light fiber, fiber coupling, fast electromechanical shutter (*see* **Notes 3** and **4**), pulse generator (*see* **Note 5**).
4. Device for the attenuation of the light pulse (*see* **Note 4**). Neutral density filters and variable optical attenuators e.g. based on edge filters with a tilting angle-dependent transmission.
5. Device for solution exchange.
6. Solution heating/cooling device.
7. Standard pipette solution A: 110 mM NaCl, 2 mM $MgCl_2$, 10 mM EGTA, 10 mM HEPES, pH 7.4.

   Bath solution A: 140 mM NaCl, 2 mM $CaCl_2$, 2 mM $MgCl_2$, 10 mM HEPES, pH 7.4.

   Bath Solution B: 140 mM NaCl, 2 mM $MgCl_2$ and 10 mM Tris, pH 9.

   Bath solution C: 140 mM Cation-Cl, 2 mM $MgCl_2$ and 10 mM Tris, pH 9.

   Bath solution D: 90 mM Cation-Cl_2, 2 mM $MgCl_2$ and 10 mM Tris, pH 9.

8. Optical power/energy meter.
9. Software for data analysis.

3 Methods

3.1 Cell Culture and Transfection

1. NG cells are cultured at 37 °C and 5% CO_2 in DMEM$^+$.
2. Seed cells on 24-well plates at a density of ~2 × 10^5 cells/well and a volume of 500 μL DMEM$^+$/well.
3. For heterologous expression of the microbial rhodopsin (*see* **Note 6**), perform a cationic lipid-based transfection of the cells according to the manufacturer's protocol 1 day after seeding.
4. Add 500 μL of fresh DMEM$^+$ and 0.5 μL of ATR (*see* **Note 7**) (at a final concentration of 1 μM) 1 day after the transfection.
5. Transfer the cells to coverslips 2 days after the transfection and at least 2 h prior to the patch-clamp measurements. Resuspend the cells growing in one well in a volume of ~2 mL using a plastic Pasteur pipette (pipette ~50 times) (*see* **Note 8**) and transfer them to three coverslips.

3.2 Electrophysiological Characterization of Microbial Rhodopsins

1. Fabricate patch pipettes with resistances of 2–5 mΩ from thin-walled borosilicate glass on a horizontal puller.
2. If not stated differently, use the standard pipette solution and bath solution A.
3. Identify NG cells, which are heterologously expressing the microbial rhodopsins, by fluorescence microscopy.
4. Place the light fiber directly on top of the fluorescent cell with a micromanipulator (*see* **Note 9**).
5. Perform whole-cell patch-clamp measurements in the voltage-clamp mode. Exclude measurements when the series resistance (R_a) exceeds 10 MΩ. The electrophysiological signals are filtered at 5 kHz and digitized at a frequency of 10 kHz if not stated differently.

3.3 Intensity Dependence of the Photocurrent

1. Perform voltage-clamp measurements at −60 mV.
2. Apply the following illumination protocol: light off: 500 ms, light on: 500 ms, light off: 500 ms (*see* **Note 10**).
3. Apply the illumination protocol at a saturating light intensity.
4. Apply the illumination protocol at lower light intensities. Use neutral density filters and/or a variable optical attenuator for the attenuation of the excitation light (*see* **Note 11**).
5. For rundown control, repeat the illumination protocol at a saturating light intensity in between the application of the lower intensity illumination protocols (*see* **Note 12**).

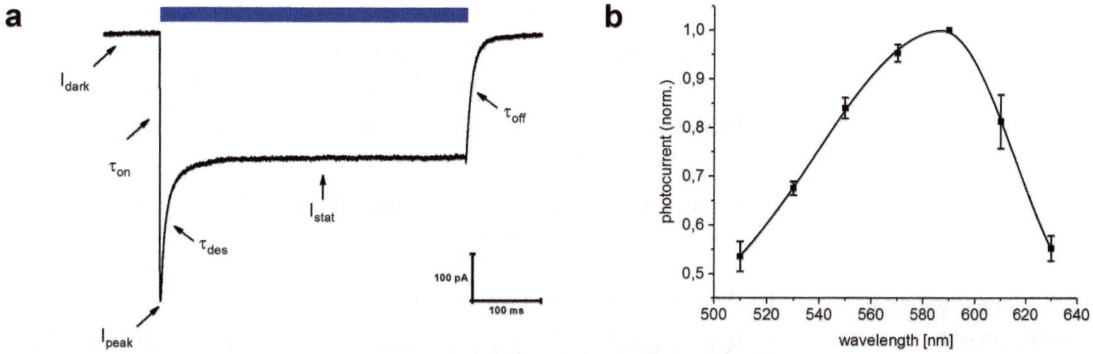

Fig. 1 Whole-cell patch-clamp measurements of Channelrhodopsin photocurrents. (**a**) A typical ChR2 photocurrent evoked by a saturating blue light pulse ($\lambda = 473$ nm, 15 mW/mm^2) with a pulse length of 500 ms. NG cells heterologously expressing ChR2-eYFP were voltage-clamped at a potential of −60 mV. I_{dark}: current measured in the dark, I_{peak}: peak photocurrent, I_{stat}: stationary photocurrent, τ_{on}: time constant of on-kinetics, τ_{des}: time constant of desensitization, τ_{off}: time constant of channel closing. (**b**) Action spectrum of f-Chrimson. Shown are normed peak photocurrents ($n = 3$, error bars show standard deviation) in response to ns light pulses of indicated wavelength. The solid line is a fit to the data using a bi-Gaussian function. The electrophysiological recordings were performed in NG cells heterologously expressing f-Chrimson at a membrane potential of −60 mV. The pulse energies at the different wavelengths were set to equal photon counts of 10^{19} photons/m^2

6. Measure the applied light intensities with an optical power meter.
7. Divide the applied light intensities by the illumination area for the determination of the irradiance.
8. Plot the stationary photocurrents (*see* Fig. 1a) (*see* **Note 13**) at the different irradiance values.
9. Determine the EC50 by a hyperbolic fitting procedure.

3.4 Wavelength Dependence of the Photocurrent: Action Spectrum

1. Perform voltage-clamp measurements at −60 mV.
2. Light pulses of different wavelengths are applied by a Nd:YAG-laser in combination with an optical parametric oscillator (OPO) (*see* **Note 14**).
3. For the light pulse protocol, the pulse energies of the ns-light pulses at the different wavelengths are set to values (*see* **Note 11**) which correspond to equal photon counts (*see* **Note 15**) and which evoke sub-saturating photocurrents (*see* **Note 16**) (in the linear phase of the pulse energy dependence of the photocurrent) by means of an optical energy meter.
4. Apply the illumination protocol close to the optimal wavelength (*see* **Note 17**).
5. Apply the illumination protocol at wavelengths, which are larger or smaller than the optimal wavelength.

6. For rundown control, repeat the illumination protocol close to the optimal wavelength in between the application of the light pulses at the different wavelengths (*see* **Note 12**).
7. Plot the photocurrents (*see* **Note 13**) at the different wavelengths (Fig. 1b).
8. Fit a bi-Gaussian function to the data for the determination of the optimal wavelength.

3.5 Investigation of Peak Current Inactivation and Recovery

1. Perform voltage-clamp measurements at −60 mV.
2. Run a double illumination protocol. Apply two 500 ms light pulses at a saturating light intensity (*see* **Note 18**). Set the interpulse interval to a value, which results in separated photocurrent traces.
3. Apply an additional 500 ms light pulse after different waiting times in darkness, following the double illumination protocols.
4. Calculate the peak current to stationary current ratios of the photocurrents (*see* **Notes 13** and **19**) (*see* Fig. 1a).
5. Plot the peak current to stationary current ratios at the different waiting times in darkness.
6. Determine the time constant of peak current recovery by an exponential fitting procedure.

3.6 Investigation of the Voltage Dependence of the Photocurrent (IV Curve) and Current Density Determination

1. Perform voltage-clamp measurements in the whole-cell configuration. Thereby compensate for pipette capacitance only, do not compensate for whole-cell capacitance.
2. Run the following voltage clamp protocol: Apply voltages ranging from −120 mV to +60 mV with a step size of 20 mV and a holding time at each potential of 1.5 s. Set the holding potential in between the voltage steps to −60 mV.
3. Run the following illumination protocol: light off: 500 ms, light on: 500 ms, light off: 500 ms (*see* **Note 10**).
4. Synchronize the light pulse protocol to the voltage clamp protocol such that the illumination protocol runs at each clamped potential (*see* **Note 5**) and apply the protocols.
5. Plot the stationary photocurrents (*see* **Note 13**) at the different voltages (*IV* curve) (Fig. 2b).
6. For current density determination, at first subtract the values of the stationary currents in the dark from the measured traces at each voltage and plot the resulting current traces (unit of *y*-axes: [pA], unit of *x*-axes: [s]).
7. Integration over time of the transient capacitive currents (*see* Fig. 2a), which are measured in response to the voltage steps from the holding potential (−60 mv) to the various potentials

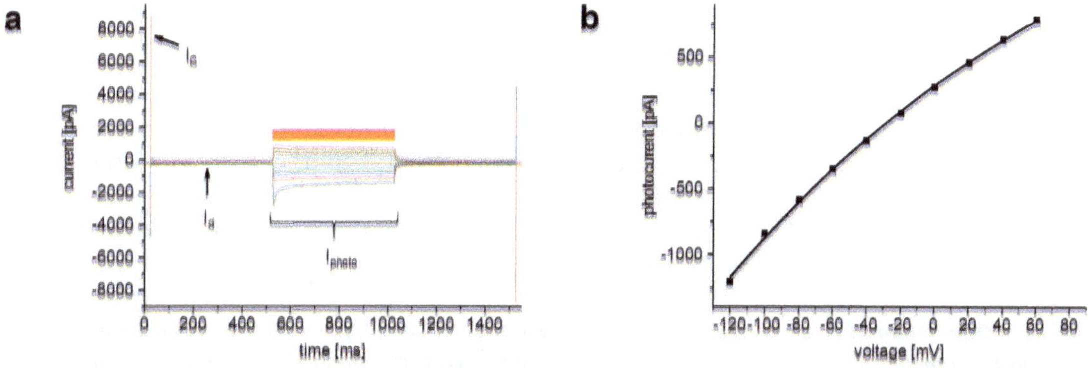

Fig. 2 Voltage dependence of Channelrhodopsin photocurrents. Shown is a representative whole-cell patch-clamp measurement, which was performed in a NG cell heterologously expressing f-Chrimson-eYFP. (**a**) Current traces at voltages ranging from −120 mV to +60 mV. The photocurrents were measured in response to 500 ms light pulses with a wavelength of 594 nm and a saturating light intensity of 23 mW/mm^2. I_C: capacitive currents, I_d: currents at different voltages measured in the dark, I_{photo}: photocurrents at different voltages measured during illumination. (**b**) Current (I) – voltage (V) curve of f-Chrimson-eYFP. The graph shows the stationary photocurrents at the indicated voltages. The data is taken from the measurement shown in (**a**)

of the applied voltage-clamp protocol, will give you the transferred charge for each voltage step.

8. Plot the transferred charge (Q) for each voltage step (ΔU, unit of x-axes: [mV], unit of y-axes: [fC]).
9. Conduct a linear fitting procedure. The linear regression slope corresponds to the capacitance of the cell (C_m, unit: [pF]).
10. Divide the photocurrent at each voltage by the capacitance of the cell for current density determination (unit [pA/pF]).

3.7 Investigation of the Kinetics

1. Perform whole-cell patch-clamp measurements in the voltage-clamp mode. The electrophysiological signals are low pass filtered at 10 kHz and digitized at a frequency of 20 kHz in cases where low pass filtering at 5 KHz and digitization at a frequency of 10 kHz is too slow due to the fast kinetics of the investigated variant.
2. Synchronize light pulse protocols of varying pulse lengths to the voltage-clamp protocol (*see* **Note 5**) and apply the protocols. At first, apply light pulse protocols with pulse lengths of 3 ms and 500 ms at saturating light intensities.
3. Plot the photocurrents (*see* **Note 13**) at the different voltages. The application of the longer light pulse should evoke photocurrents, which clearly show a peak current and a stationary current level (*see* Fig. 2a). For the shorter light pulse, the pulse length should be smaller than the τ_{off} values (*see* **Note 20**), which are determined by an exponential fitting procedure.

Adjust the pulse lengths accordingly if these requirements are not fulfilled.

4. Run the adjusted protocols.
5. Plot the photocurrents at the different voltages. Determine the τ_{on}- (*see* **Note 21**) and τ_{off}-values and, in the case of the longer light pulse, the time constant of the relaxation from peak current to stationary current (τ_{des}) by exponential fits (*see* **Note 22**).

3.8 Determination of the Ion Selectivity

1. In order to assess the permeability of different cations relative to the permeability of sodium ions (P_{cation}/P_{Na}), at first measure the *IV* curve using the standard pipette solution and bath solution B (*see* **Notes 23** and **24**).
2. Exchange bath solution B against bath solution C for monovalent cations or bath solution D for divalent cations and measure the *IV* curve.
3. For drift control, exchange bath solution C or bath solution D against bath solution B and repeat the measurement of the *IV* curve.
4. Plot the photocurrents (*see* **Note 13**) at the different voltages and determine the reversal potentials.
5. Discard data from further analysis when the drift of the reversal potential, which is quantified by the two measurements, which are performed in bath solution B, is larger than 10 mV.
6. Determine the relative permeability (P_{cation}/P_{Na}) from the difference of the reversal potentials of the photocurrents (ΔU_{rev}) measured in bath solution B and bath solution C or bath solution D using the Goldman-Hodgkin-Katz equation (*see* **Note 25**).

4 Notes

1. Alternatively, other illumination devices, e.g., light-emitting diodes (LEDs), mercury arc lamps (HBO lamps), or xenon arc lamps (XBO lamps) can be used. However, in particular for the characterization of microbial rhodopsins of low light sensitivity, the use of lasers is better suited because higher light intensities can be applied.
2. The wavelength of the cw-laser should be close to the maximum of the action spectrum.
3. Alternatively, the duration and the frequency of the light pulses can be controlled by, e.g., acousto-optic modulators.
4. The direct control over intensity, duration, and frequency of the light pulses by the electrical power supply limits the

dynamic range of the light source and increases the pulse to pulse variability, in particular at low light intensities.

5. The most commonly used patch-clamp systems enable software controlled illumination protocols via the application of TTL pulses. This generally allows the exact timing of light-pulse-length and -frequency and the convenient synchronization of the light pulse protocols to the measurement protocols.
6. Coexpression of a fluorescent protein allows a convenient verification of microbial rhodopsin expression. The use of C-terminal fusions or viral 2A peptides for fluorescent protein expression is advantageous over the use of N-terminal fusions in most cases.
7. The addition of ATR increases the photocurrent of microbial rhodopsins in some cases. However, the addition is not mandatory, because NG108–15 cells endogenously produce ATR.
8. Do not use a proteolytic enzyme for the dissociation of the adherent cells and avoid foam formation.
9. This step is omitted when the illumination device is directly coupled into the light path of the microscope or into special pipette holders, designed for optogenetic experiments, which enable the application of light pulses through the patch pipette.
10. The illumination protocol is suitable for most microbial rhodopsins, but requires modifications when slow variants are investigated.
11. The adjustment of the light intensity with sufficient accuracy in a wide dynamic range is not possible with most commercially available variable optical attenuators alone. Therefore, the use of neutral density filters or the combined use of variable optical attenuators and neutral density filters is recommended.
12. Quantify the rundown by the decrease of the photocurrents, which are measured in the repeated runs at identical conditions. Exclude measurements from data analysis when rundown exceeds 10%.
13. Subtract the current measured in the dark from the current measured during illumination in order to determine the photocurrent.
14. Alternatively, the action spectrum can be measured with a XBO or a HBO lamp combined with narrow bandpass filters and neutral density filters. XBO lamps are more suitable than HBO lamps, because the emission profile of HBO lamps contains several strong emission lines in the visible light range.
15. Calculation of pulse energy values which correspond to equal photocounts:

At first, calculate the energy of one photon at the wavelength of the applied light pulse:

$$E = \frac{h \times c}{\lambda}$$

E = Energy of photon expressed in units of joules

h = Planck's constant = 6.626×10^{-34} Js

c = The speed of light = 2.998×10^{8} ms^{-1}

λ = Wavelength of the applied light pulse expressed in units of meters

Multiplication of E with the number of photons/m^2 you want to apply gives you the pulse energy in units of J/m^2. Finally, multiply that value with the illumination area to determine the correct pulse energy for the application of the desired number of photons/m^2.

16. The illumination at energy values, which are saturating or close to saturating, leads to the broadening of the action spectrum.
17. The close to optimal wavelength can be estimated based on prior measurements at different wavelengths.
18. The ratio of the stationary current to the peak current critically depends on the intensity of the light pulse.
19. The peak current to stationary current ratio of the photocurrent measured in response to the light pulse, which is applied after the longest waiting times in darkness should be close to the peak current to stationary current ratio of the photocurrent measured in response to the first pulse of the double illumination protocol.
20. The application of light pulses, which are shorter than the cycle time of the investigated microbial rhodopsin, enables synchronized single-turnover-measurements.
21. The on-kinetics of microbial rhodopsins is usually beyond the time resolution ($\sim R_a \times C_m$) of a whole-cell patch-clamp experiment.
22. Repeat the measurements at different temperatures and varying ionic conditions. An analysis of the data at the different conditions likely gives deeper insights into the ion-transport mechanism.
23. The protocol is not applicable for anion-selective ChRs and light-driven ion pumps.
24. Establishing the whole-cell recording configuration at neutral pH, in bath solution A prior to the measurement in bath

solution B, has a positive effect on the success rate of gigaseal formation and cell health.

25. Use the following equation for monovalent cations:

$$\frac{P_{\text{cation}}}{P_{\text{Na}}} = \frac{[\text{Na}]^{\text{outside}}}{[\text{cation}]^{\text{outside}}} \times e^{\frac{\Delta U_{\text{Rev}} F}{RT}}.$$

For divalent cations, use the following equation:

$$\frac{P_{\text{cation}}}{P_{\text{Na}}} = \frac{[\text{Na}]^{\text{outside}}}{4 \times [\text{cation}]^{\text{outside}}} \times e^{\frac{\Delta U_{\text{Rev}} F}{RT}}.$$

F = Faraday constant = 96,485 Cmol^{-1}
R = Universal Gas Constant = 8.314 $\text{Jmol}^{-1}\ \text{K}^{-1}$
T = Temperature in Kelvin

Acknowledgments

This work was supported by the Cluster of Excellence "Multiscale Bioimaging: from Molecular Machines to Networks of Excitable Cells" (MBExC), University of Göttingen, Germany.

References

1. Nagel G, Ollig D, Fuhrmann M, Kateriya S, Musti AM, Bamberg E, Hegemann P (2002) Channelrhodopsin-1: a light-gated proton channel in green algae. Science 296: 2395–2398
2. Nagel G, Szellas T, Huhn W, Kateriya S, Adeishvili N, Berthold P, Ollig D, Hegemann P, Bamberg E (2003) Channelrhodopsin-2, a directly light-gated cation-selective membrane channel. Proc Natl Acad Sci U S A 100:13940–13945
3. Volkov O, Kovalev K, Polovinkin V, Borshchevskiy V, Bamann C, Astashkin R, Marin E, Popov A, Balandin T, Willbold D, Buldt G, Bamberg E, Gordeliy V (2017) Structural insights into ion conduction by channelrhodopsin 2. Science 358:eaan8862
4. Han X, Boyden ES (2007) Multiple-color optical activation, silencing, and desynchronization of neural activity, with single-spike temporal resolution. PLoS One 2:e299
5. Zhang F, Wang LP, Brauner M, Liewald JF, Kay K, Watzke N, Wood PG, Bamberg E, Nagel G, Gottschalk A, Deisseroth K (2007) Multimodal fast optical interrogation of neural circuitry. Nature 446:633–639
6. Sahel JA, Roska B (2013) Gene therapy for blindness. Annu Rev Neurosci 36:467–488
7. Dieter A, Duque-Afonso CJ, Rankovic V, Jeschke M, Moser T (2019) Near physiological spectral selectivity of cochlear optogenetics. Nat Commun 10:1962
8. Mager T, Lopez de la Morena D, Senn V, Schlotte J, D Errico A, Feldbauer K, Wrobel C, Jung S, Bodensiek K, Rankovic V, Browne L, Huet A, Juttner J, Wood PG, Letzkus JJ, Moser T, Bamberg E (2018) High frequency neural spiking and auditory signaling by ultrafast red-shifted optogenetics. Nat Commun 9:1750
9. Wrobel C, Dieter A, Huet A, Keppeler D, Duque-Afonso CJ, Vogl C, Hoch G, Jeschke M, Moser T (2018) Optogenetic stimulation of cochlear neurons activates the auditory pathway and restores auditory-driven behavior in deaf adult gerbils. Sci Transl Med 10:eaao0540
10. Hight AE, Kozin ED, Darrow K, Lehmann A, Boyden E, Brown MC, Lee DJ (2015) Superior temporal resolution of Chronos versus channelrhodopsin-2 in an optogenetic model of the auditory brainstem implant. Hear Res 322:235–241
11. Keppeler D, Merino RM, Lopez de la Morena D, Bali B, Huet AT, Gehrt A, Wrobel C, Subramanian S, Dombrowski T,

Wolf F, Rankovic V, Neef A, Moser T (2018) Ultrafast optogenetic stimulation of the auditory pathway by targeting-optimized Chronos. EMBO J 37:e99649

12. Klapoetke NC, Murata Y, Kim SS, Pulver SR, Birdsey-Benson A, Cho YK, Morimoto TK, Chuong AS, Carpenter EJ, Tian Z, Wang J, Xie Y, Yan Z, Zhang Y, Chow BY, Surek B, Melkonian M, Jayaraman V, Constantine-Paton M, Wong GK, Boyden ES (2014) Independent optical excitation of distinct neural populations. Nat Methods 11:338–346
13. Oda K, Vierock J, Oishi S, Rodriguez-Rozada-S, Taniguchi R, Yamashita K, Wiegert JS, Nishizawa T, Hegemann P, Nureki O (2018) Crystal structure of the red light-activated channelrhodopsin Chrimson. Nat Commun 9:3949
14. Govorunova EG, Sineshchekov OA, Janz R, Liu X, Spudich JL (2015) Neuroscience. Natural light-gated anion channels: a family of microbial rhodopsins for advanced optogenetics. Science 349:647–650
15. Kim YS, Kato HE, Yamashita K, Ito S, Inoue K, Ramakrishnan C, Fenno LE, Evans KE, Paggi JM, Dror RO, Kandori H, Kobilka BK, Deisseroth K (2018) Crystal structure of the natural anion-conducting channelrhodopsin GtACR1. Nature 561:343–348
16. Inoue K, Ito S, Kato Y, Nomura Y, Shibata M, Uchihashi T, Tsunoda SP, Kandori H (2016) A natural light-driven inward proton pump. Nat Commun 7:13415
17. Shevchenko V, Mager T, Kovalev K, Polovinkin V, Alekseev A, Juettner J, Chizhov I, Bamann C, Vavourakis C, Ghai R, Gushchin I, Borshchevskiy V, Rogachev A, Melnikov I, Popov A, Balandin T, Rodriguez-Valera F, Manstein DJ, Bueldt G, Bamberg E, Gordeliy V (2017) Inward H+ pump xenorhodopsin: mechanism and alternative optogenetic approach. Sci Adv 3:e1603187
18. Kleinlogel S, Feldbauer K, Dempski RE, Fotis H, Wood PG, Bamann C, Bamberg E (2011) Ultra light-sensitive and fast neuronal activation with the ca(2)+–permeable channelrhodopsin CatCh. Nat Neurosci 14:513–518
19. Lin JY, Knutsen PM, Muller A, Kleinfeld D, Tsien RY (2013) ReaChR: a red-shifted variant of channelrhodopsin enables deep transcranial optogenetic excitation. Nat Neurosci 16:1499–1508
20. Sakmann B, Neher E (1984) Patch clamp techniques for studying ionic channels in excitable membranes. Annu Rev Physiol 46:455–472

Chapter 14

Molecular Optimization of Rhodopsin-Based Tools for Neuroscience Applications

Lief E. Fenno, Rivka Levy, and Ofer Yizhar

Abstract

There is no question that genetically encoded tools have revolutionized neuroscience. These include optically modulated tools for writing-in (optogenetics) and reading-out (calcium, voltage, and neurotransmitter indicators) neural activity as well as precision expression of these reagents using virally mediated delivery. With few exceptions, these powerful approaches are derived from naturally occurring molecules that are sourced from diverse organisms that span all kingdoms of life. Successful expression of genetic tools in standard neuroscience model organisms requires optimizing gene structure, taking into account differences in both protein translation and trafficking. Myriad approaches have resolved these two challenges, resulting in order-of-magnitude increases in functional expression. In this chapter, we focus on synthesizing prior experience in successfully enabling the transition of genes across kingdoms with a goal of facilitating the production of the next generation of molecular tools for neuroscience. We then provide a detailed protocol that allows expression and testing of novel genetically encoded tools in mammalian cell lines and primary cultured neurons.

Key words Optogenetics, Neuroscience, Rhodopsin, Bioengineering, Molecular engineering, Neurobiology

1 Introduction

Rhodopsin-based tools have revolutionized neuroscience in the last two decades. Fueled by the discovery and application of channelrhodopsin on the one hand [1, 2] and by advances in viral vector technology [3–6] and powerful transgenic targeting techniques [7–9] on the other, rhodopsins and light-delivery technology have made their way into the heart of systems neuroscience, a field that traditionally relied on electrical currents for both recording and stimulating neurons in the brain. The advent of optogenetics [10–12] has spurred new interest in the discovery of rhodopsin proteins with new functional properties [13–15] and in the design of new molecular variants of the existing rhodopsins [16, 17] to diversify the optogenetic toolbox in the spectral, kinetic, and ionic

Valentin Gordeliy (ed.), *Rhodopsin: Methods and Protocols*, Methods in Molecular Biology, vol. 2501,
https://doi.org/10.1007/978-1-0716-2329-9_14,

selectivity domains. Beyond rhodopsins, optogenetics employs diverse light-sensitive proteins from several families, including LOV-domain proteins, light-activated enzymes, and receptors [18, 19].

Optogenetics relies on the expression of light-sensitive proteins in the membrane or cytoplasm of targeted cells. The genes encoding these proteins are almost always derived from simpler organisms than those in which they are applied as tools. For example, channelrhodopsin, a blue-light sensitive cation channel commonly applied in mammalian animal models to drive neuronal activity, originates from the unicellular algae *Chlamydomonas rheinhardtii.* Other tools, including several light-driven ion pumps used for neuronal silencing through membrane hyperpolarization, arise from diverse prokaryotes, while diverse light-modulated protein systems from plants are able to control a wide range of protein–protein interactions [20, 21]. A critical challenge in successfully employing a newly discovered, or engineered, light-actuated protein as an optogenetic tool is the expression of the gene encoding it in the model organism in which it is to be applied, its transcription, translation, and proper trafficking to the appropriate cellular compartment to enable the desired optogenetic effector function.

1.1 Adapting Microbial Genes for Modulation of Neural Function

It is remarkable that a protein evolved for energy production in archaea or light-driven phototaxis in algae can successfully confer millisecond light-gated control of neural activity in mammalian neurons and modulate animal behavior. The universality that underlies this process, namely the storage of information within DNA, transcription and processing into messenger RNA ("mRNA"), and translation of this mRNA into a protein, is known as the *central dogma* of molecular biology. With only a few exceptions, this represents a highly conserved process within all known lifeforms. While the birds-eye view presents a simple and conserved process, subtle—yet critical—differences exist in each one of these steps, both within and across kingdoms of life. This variation must be considered to achieve successful expression of genes encoding heterologous light-actuated proteins in mammalian neurons. The most frequently encountered differences can be classified into three categories: open reading frame, untranslated sequences, and protein trafficking.

1.1.1 Open Reading Frame

The open reading frame (ORF) is the portion of the gene that will be translated into protein and is denoted on the 5′ (upstream) terminus with the start codon and the 3′ (downstream) terminus with the stop codon, where a codon is one of the 64 possible three-base DNA sequences [22, 23].

The start codon, typically ATG in mammals, is critical in both being recognized as the location that protein translation should begin as well as setting the *reading frame*. Because codons are read in three base sets, there are three potential reading frames for any given sequence of DNA, and the start codon, in conjunction with non-translated DNA elements, denotes which of these options should be used for protein translation (for detailed review, see [22]). There are three typically used stop codons in mammals, TAA, TAG, and TGA. Beyond the start and stop codons, the remaining 60 codons represent 20 amino acids, with some redundancy (where more than one codon denotes a single amino acid). During the process of translation, individual codons are recognized by unique tRNA molecules and the pools of these molecules vary by organism. Changing the codon bias to match that typically found in a specific target organism is called *codon optimization* and has been found to improve expression levels of heterologously expressed genes as it allows the utilization of more ubiquitous tRNAs in the target [24–26]. This process is now largely automated by commercial entities that offer gene synthesis. In addition to codon optimization, bespoke gene synthesis also enables evaluation and elimination of secondary structures that may play a role in regulating prokaryotic gene transcription and impair function in animal cells.

1.1.2 Untranslated Functional DNA Elements

In a mammalian expression cassette, the ORF is flanked on the 5′ end by the *promoter* and *Kozak* sequences and on the 3′ end by the *poly(A) signal*, all of which differ from prokaryotes and are critical to proper expression in eukaryotes. Promoters are DNA sequences that determine which cells will express a gene that is downstream of the promoter and are the major regulatory element in mammalian gene transcription. In neuroscience, a handful of promoter sequences are commonly used (EF1a, CaMKIIa, hSyn), although a complete accounting of their function across neuron subtypes has not been described. The Kozak sequence [27] is unique to eukaryotes; it is a six base pair (bp) sequence that directs protein translation from mRNA (e.g., denotes which potential ATG sequence should be used as the translation starting point). It is critical that the Kozak sequence is directly 5′ of the ATG start codon of the opsin gene. Last, the poly(A) signal contains a consensus sequence that is recognized during mRNA processing and directs the addition of a *poly(A) tail*, literally a string of adenosine RNA bases, onto the 3′ terminus of the RNA. The poly(A) tail improves mRNA stability by inhibiting degradation and is part of the chaperone system that exports mRNA from the nucleus. See Table 1 for a list of DNA elements contributing to opsin expression.

1.1.3 Protein Trafficking

After mRNA is exported from the nucleus, it is translated to protein by ribosomes. At this point, the newly synthesized protein is near the nucleus and must be transported to its final destination. In the case of membrane-resident optogenetic tools, this requires

Table 1
A list of DNA elements contributing to opsin expression

Category	Item	Description	References
ORF	Start Codon	ATG	
	Stop Codon	TAA, TAG, TGA	
	Codon Optimization	The process of optimizing a gene for expression in a specific species by applying a unique codon bias. Typically completed at time of gene synthesis with commercial algorithms	
Untranslated elements	Kozak	Eukaryotic consensus sequence directing translational initiation at the correct ATG	[27, 28]
	Promoters	Determine the level of gene expression and which cells will express the gene	[29]
	Poly(A) signal	Directs attachment of a poly(A) "tail" onto the transcribed mRNA during RNA processing, important for stability and nuclear export. SV40 and hGH are commonly used	[30]
Trafficking motifs	Kir2.1 ER export and TS motifs	Sequences that improve membrane trafficking and functional expression of optogenetic tools	[24, 31]
Plasmids	addgene.org	Addgene is a worldwide, nonprofit repository of plasmids and stocks a wide range of optogenetic expression plasmids at minimal cost	

trafficking to the cell membrane (although this destination could also be other sub-cellular compartments, such as the mitochondria [32] or synaptic vesicles [33]). Substantial effort in optimizing optogenetic tool trafficking has improved functional expression dramatically, mainly due to an increase in the efficiency and selectivity of membrane transport. Non-optimized optogenetic proteins are prone to aggregation, accumulation within the endoplasmic reticulum ("ER"), and poor membrane expression [24]. The addition of a 7aa ER export motif from Kir2.1 and a 18aa membrane trafficking signal from the nicotinic acetylcholine receptor abolished ER aggregation and improved photocurrent magnitude [24]. The addition of a Kir2.1 membrane trafficking signal was found to further enhance membrane expression, with concomitant improvement in photocurrent [31]. Additional modifications can be made (Table 1) to restrict the trafficking of an optogenetic protein to specific cellular sub-compartments, including the soma and proximal dendrites (by adding a soma-targeting motif from the Kv2.1 channel [34] or a N-terminal domain from the kainate receptor KA2 [35]), the axon initial segment (by adding an AIS-targeting motif derived from sodium channels (Na(v)II-III) [36]), the dendritic compartment (through the addition of a melanophilin myosin-binding domain [37]), or the axonal compartment (by the addition of a neurexin targeting motif [38]). Beyond their effects on protein localization, these modifications can lead to dramatic changes in the overall concentration of the optogenetic tool in the membrane and therefore influence its efficacy.

Trafficking varies widely between prokaryotes and eukaryotes, resulting from substantial differences in cellular organization, membrane composition, and the process of protein translation across these kingdoms. Differences also exist between simpler eukaryotes and mammalian cells. While the mechanisms of membrane protein trafficking are not always known in specific organisms from which a protein of interest is derived, the approaches above are a useful starting point. Each protein is unique, and there is no single process that is universally applicable. This necessitates a trial-and-error process that benefits from a standardized pipeline approach for testing and validating new rhodopsin sequences in heterologous systems.

1.2 Functional Evaluation of Novel Optogenetic Tools In Vitro

After designing and cloning an expression vector containing a novel optogenetic tool that has been optimized for mammalian cells, the next step is to assess whether it properly traffics to the membrane, and whether its expression in any way impacts the health of expressing cells. The most common approach is to fuse the optogenetic tool with a fluorescent protein (e.g., yellow fluorescent protein, "YFP") scale and purify high-quality, endotoxin-free DNA stocks, and transfect the DNA into cultured cells, such as the immortalized human HEK293 cell line or neurons grown in culture. As most mature neurons are post-mitotic, cultured neurons must be derived from primary sources, such as the cortex or hippocampus of neonatal rodents. HEK293 cells are readily transduced and tolerate commercially available lipid-based transfection reagents (e.g., FuGENE, Lipofectamine), while primary neuronal cultures should be transfected using calcium-phosphate precipitation. After transfection, cells may be analyzed either alive or after fixation using fluorescence microscopy. Most commonly, this is done with confocal imaging in order to evaluate optical sections with subcellular resolution and closely examine membrane localization. Promising optogenetic tools are then functionally evaluated, typically with single-cell patch-clamp electrophysiology; such functional evaluation is beyond the scope of this chapter.

1.2.1 Co-expression of Optogenetic Tools with Fluorescent Proteins

Fluorescent proteins ("FPs") were originally found in diverse organisms, with green fluorescent protein (GFP) from jellyfish being the most well-known [39]. Like other fluorescent materials, these proteins absorb a photon of a particular energy (wavelength) and emit a photon of a lower energy (red-shifted wavelength). There now exists an entire rainbow of fluorescent proteins, which, when tagged to optogenetic tools or other proteins, provide a visual readout of cellular expression or activity [40]. For the purposes of evaluating cell health and opsin trafficking, the FP should be directly fused to the opsin. The FP should be positioned on the intracellular terminus of the 7-transmembrane opsin, which is typically the C-terminus. Short linker sequences [41] between the opsin and FP decrease the likelihood of interference, and this fusion strategy is generally well-tolerated.

In experimental use, it may be more useful to have the opsin and FP generated from the same expression cassette (e.g., with a single virus), but not to have the two proteins fused together. This is useful in two-photon microscopy and for resolving structures of expressing neurons during histology, as the FP fills the cell cytosol, while the opsin remains trafficked to the membrane [42, 43]. As payload restrictions of commonly used viral delivery options (e.g., AAV) make it difficult to have two separate sets of promoters, ORFs, and poly(A) signals, bi-cistronic expression strategies are use and these are able to generate two separate proteins from a single mRNA. The two most common options are *Internal Ribosomal Entry Sites* ("IRES") [44] and *2A* [45, 46]. These sequences are both from viruses and work in slightly different ways: the IRES sequence generates a secondary structure that acts to independently recruit ribosomes to transcribe the second ORF, while p2a either creates a secondary structure that causes the ribosome to drop the peptide being translated [45] or catalyzes its own cleavage [46]. 2A sequences are generally favored due to their smaller size compared to most IRES sequences. Of note, the use of these sequences for bi-cistronic expression may result in lower expression levels of the second ORF [45, 47].

1.2.2 Preparation and Transfection of High-Quality DNA for Cultured Cell Transfection

After optimizing and cloning a novel optogenetic gene into a mammalian expression vector and considering whether and how to affix a fluorescent protein, concentrated, high-quality DNA must be generated for transfection. Mini-prep format DNA preparations are generally insufficient for neuronal transfection, although they may be used for screening in HEK293 cells. We typically use endotoxin-free maxi-prep kits from Qiagen, although endotoxin-free kits from other suppliers are also available. Endotoxins are large lipopolysaccharides produced by *E. coli* and related bacteria that negatively impact cell health and decrease transfection efficiency. Commercial endotoxin-free DNA preparation kits effectively remove these molecules from plasmid DNA preparations at scale. DNA concentration and quality must be quantified prior to transfection.

Primary neurons do not tolerate lipid-based commercial transfection reagents, and calcium phosphate transfection is recommended. DNA is mixed with high-concentration calcium and phosphate ("$CaPO_4$"), causing the DNA to precipitate [48]. This mixture is then applied to cells, dusting them with particles that are visible under a typical light microscope. $CaPO_4$ typically transduces fewer cells than commercial transfection reagents, but neurons tolerate $CaPO_4$-mediated gene expression while remaining healthy enough to be examined by confocal microscopy and electrophysiology. $CaPO_4$ transfection is exquisitely sensitive to protocol parameters [49], and may be tailored to improve neuron transfection rates [50, 51].

1.2.3 Cell Health and Overexpression

Overexpression of a foreign membrane protein in mammalian cells can result in detrimental effects on cell health. High-level, long-term expression of ChR2 has been reported to cause abnormalities when delivered by in utero electroporation or high titer virus [52], while high-titer delivery of recombinases—commonly used to activate optogenetic tools in a cell type-specific manner—may be lethal [53]. Therefore, introduction of a novel optogenetic tool requires a thorough characterization of the health of cells expressing the protein. In many cases, a moderate expression level can be tolerated well while an increase in expression, through increased viral titers or a strong promoter such as the cytomegalovirus (CMV) [54] or CAG [55] promoters can lead to impaired cell health.

1.3 Optimization of Prokaryotic Rhodopsin Genes for Mammalian Neuron Expression

While cloning protocols to create modified DNA fragments are outside the scope of this chapter (and many options exist [56]), general considerations for the design elements required for successful heterologous expression of novel optogenetic tool transgenes is critical. We suggest approaching the design of such genes using a similar categorization as described above: focusing first on the ORF, then the non-translated elements, trafficking motifs, and finally additional functional elements. It is useful to examine and use as source material plasmids that have been previously published; most of these are available from Addgene (www.addgene.com), a non-profit plasmid repository that is used by laboratories worldwide. There is an ever-growing array of software options for designing plasmids, including both free (ApE, benchling.com) and commercial (SnapGene, VectorNTI) options. After designing a transgene and deciding on a cloning strategy, synthesizing the sequence from scratch is generally not cost-prohibitive, and there are many companies able to complete such a task within days to weeks (Eurofins, GenScript, Twist, Atum).

1.4 Protocol: Preparation of Dissociated Primary Neuronal Cultures

This protocol provides the essential information for preparation of high-quality, dissociated neuronal cultures from the hippocampus of newborn rats. Cells are prepared on glass coverslips to facilitate downstream experiments (electrophysiology, histology) and dividing cell growth is arrested using FUDR to prevent glial overgrowth.

2 Materials

2.1 Materials for Primary Neuron Culture

2.1.1 Neurobasal Media (NB+)

- 250 mL Neurobasal Medium (Gibco).
- 10 mL B-27 Supplement (Gibco).
- 2.5 mL Glutamax-I Supplement (Gibco).
- 3.12 mL Fetal Bovine Serum (FBS; Gibco).

Combine all and sterile-filter into 250 mL bottle. Incubate before use for several hours in a 37 °C water bath. Media can be stored at 4 °C when not in use and is good for 1 month. Can be scaled.

2.1.2 Serum Media (SM)

- 200 mL Minimum Essential Medium, with Earle's salts, without L-glutamine (Gibco).
- 10 mL Fetal Bovine Serum (Gibco).
- 3 mL Hi-Glucose/MEM (Gibco).
- 0.2 mL Mito+ Serum Extender (BD Biosciences).

Combine all and sterile-filter into 250 mL bottle. Incubate before use for several hours in a 37 °C water bath. Media can be stored at 4 °C when not in use and is good for 1 month. Can be scaled.

2.1.3 Hi-Glucose/MEM

- 50 mL Minimum Essential Medium w/ Earle's salts w/o L-glutamine.
- 12.75 g Glucose.

Dissolve Glucose in Minimum Essential Medium. Sterile-filter into a 50 mL tube. Media can be stored at 4 °C when not in use and is good for 6 months.

2.1.4 Mito+ Serum Extender

- 1 bottle Mito+ Serum Extender.
- 5 mL sterile, double distilled water.

Dissolve Mito+ Serum Extender in 5 mL double-distilled water. Prepare 0.25 mL aliquots and store at −20 °C.

2.1.5 5-Fluoro-2′-Deoxyuridine (FUDR)

- 100 mg 5-Fluoro-2′-Deoxyuridine ("FUDR," 1 bottle).
- 60 mg Uridine.
- 12 mL MEM (Minimum Essential Medium, with Earle's salts, without L-glutamine).

Weigh 60 mg Uridine into a 50 mL conical tube. In the hood, add MEM to a concentration of 5 mg/mL. Invert several times to dissolve and mix. Add 10 mL of 5 mg/mL Uridine in MEM to 100 mg bottle of FUDR. Wait until FUDR is dissolved (usually, a few minutes). Sterile-filter solution into 50 mL conical tube.

Prepare 0.5 mL aliquots and store at −20 °C. Avoid multiple freeze-thaw cycles.

2.1.6 Dissection Solution

- 1 L double-distilled water.
- 9.4 g NaCl (160.8 mM).
- 0.37 g KCl (4.96 mM).
- 0.27 g $MgSO_4$-$7H_2O$ (1.1 mM).
- 0.43 g $CaCl_2$ (3.87 mM).
- 1.2 g HEPES (5.03 mM).
- 1.0 g Glucose (5.55 mM).

Combine all in 1 L flask and mix until dissolved using magnetic stir bar. Adjust pH to 7.4 using NaOH. Sterile-filter solution into 500 mL bottles and store at 4 °C.

2.1.7 Enzymatic Digestion Solution

Enzymatic Digestion Solution must be prepared at least 1 h before dissection to allow for it to equilibrate in an incubator prior to use. Ten milliliters of Enzymatic Digestion Solution is sufficient for 8 hippocampi (from 4 brains).

- 10 mL Hippocampal Dissection Solution.
- 10 μL Cysteine (200 mg/mL) kept in aliquots in −20 °C.
- 100 μL $CaCl_2$ (100 mM).
- 100 μL EDTA (50 mM).
- 15 μL NaOH (1N).
- 100 μL DNAse I (Add DNAse just before sterile filtering).
- Lot-dependent volume of Papain for 10 mL of solution:

$$\text{Volume of Papain (10 mL)} = \frac{100,000}{X \times \Upsilon}$$

  - X = # u/mgP (# on left side of bottle).
  - Υ = # mgP/mL (# on right side of bottle).

(Note: Vortex Papain before adding to solution.)

Combine all components except DNAse I and store in incubator. Add DNAse I immediately prior to dissection, then filter sterilize.

2.1.8 Inactivating Solution

- 25 mg Bovine Albumin.
- 25 mg Trypsin Inhibitor.
- 10 mL Serum Media.

Combine ingredients, swirl to mix, and store in hood until use. Filter while hippocampi are digesting.

2.1.9 Matrigel

- Matrigel (BD Biosciences).

Thaw manufacturer stock overnight on ice at 4 ° C. Aliquot 300 μL into 1.5 mL tubes; maintain on ice while aliquoting. Store at −80 ° C.

2.2 Materials for Preparing Tissue Cultures Plates for Neuron Culture

- 24-well tissue culture plates.
- #1 glass coverslips (Menzel Glaser) 12 mm diameter.
- 16.5 mL HCl.
- 183 mL double-distilled water.
- 96% EtOH.
- Matrigel (BD Biosciences)—light sensitive!
- Neurobasal medium (Gibco).

2.3 Tools for Hippocampal Dissection

Maintaining a sterile environment is critical. All equipment must be sterilized in 70% EtOH. Spray the dissection area with 70% ethanol and wipe down before starting.

2.3.1 Decapitation and Brain Removal: Equipment Needed (Fig. 1)

- Straight scissors with large finger loop (e.g., F.S.T 14002-14).
- Straight microscissors with 8 mm cutting edge (e.g., F.S.T 15024-10).
- 45° angled tip microdissection forceps (e.g., Dumont #7).
- Serrated and angled microdissection forceps (e.g., F.S.T 11009-13).
- Spray bottle with 70% EtOH.
- Stereoscope for visualizing tissue dissection.

2.3.2 Hippocampus Dissection: Equipment Needed (Fig. 2)

- 2× micro-dissection forceps (e.g., Dumont #511254-20).
- Angled up micro-dissection scissors with 5 mm cutting edge (*e.g., F.S.T 15005-08*).

2.3.3 Workspace: Equipment Needed

- 150 mm petri dish.
- 3× 100 mm petri dish.
- 35 mm petri dish.
- Ice bucket with ice.
- Plastic disposal container for animal tissues.
- P0-1 rat pups (2 hippocampi sufficient for a single 24-well plate).
- 0.22 μm Steriflip filters (EMD Millipore).
- 40 μm tube-top cell strainer (Falcon).
- Trypan Blue (0.4% liquid; Sigma).

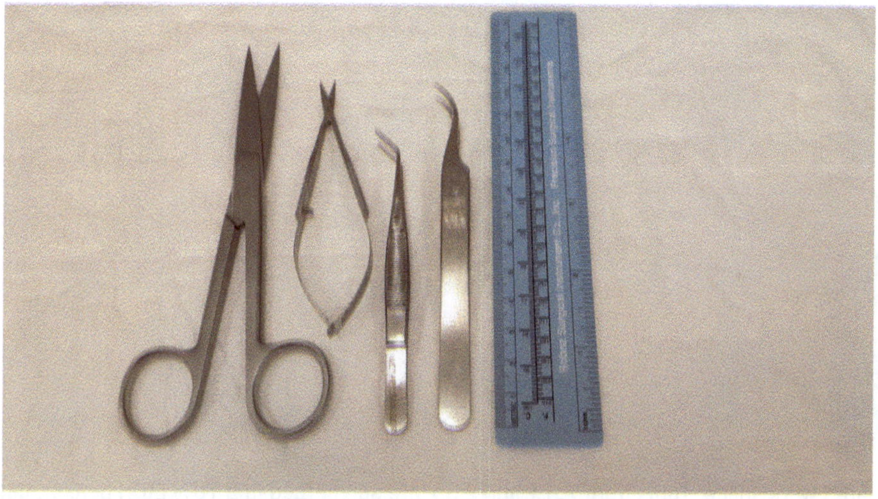

Fig. 1 Dedicated tools for decapitation, cutting of skull, and brain extraction. Ruler for scale

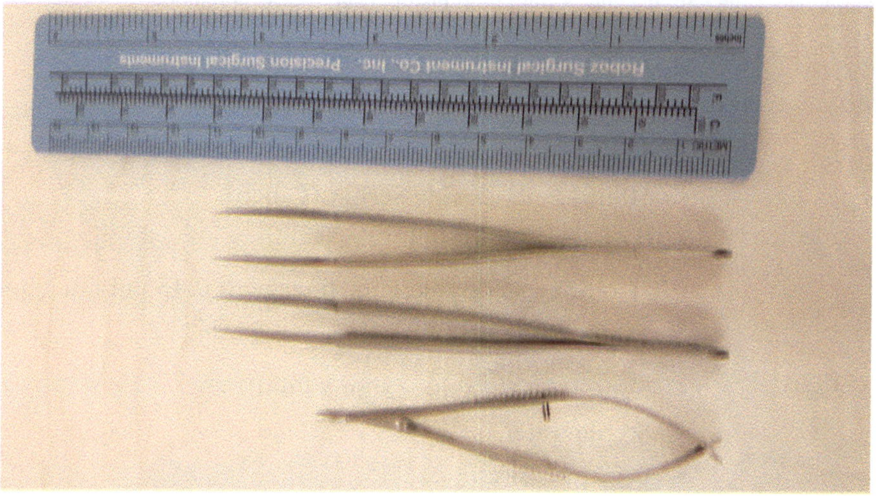

Fig. 2 Dedicated tools for isolating hippocampi from whole brains. Ruler for scale

2.4 Materials for Tissue Culture Preparations (Fig. 3)

1. For efficiency and sterile purposes, arrange dissection equipment in the order used. Ensure EtOH has evaporated.
2. Maintain Dissection Media on ice throughout procedure.
3. Fill 150 mm and one of the 100 mm petri dishes with ice.
4. Add ice-cold Dissection Media to 100 mm petri dish and place it on the closed lid of the ice-filled 150 mm petri dish.
5. Add ice-cold Dissection Media to 35 mm petri dish and place it on closed lid of the ice-filled 100 mm petri dish.

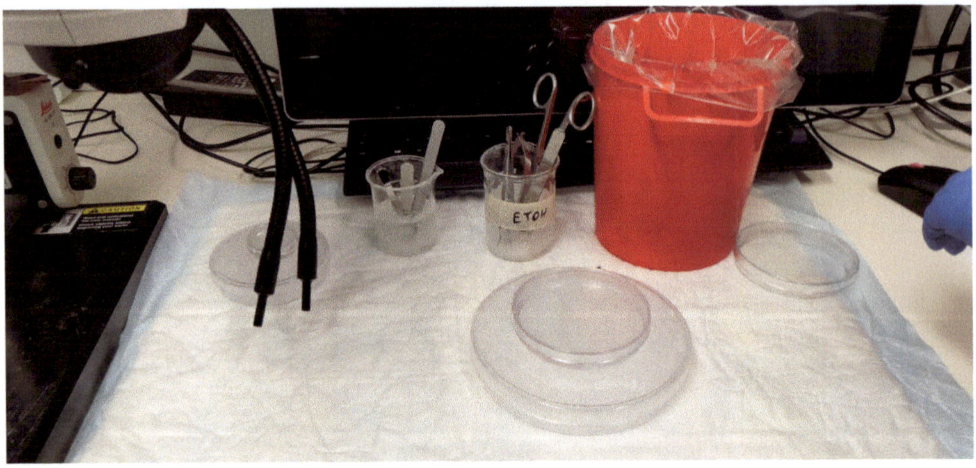

Fig. 3 Prepared work space. Tools from 2.3.1 and 2.3.2 are submerged in a 70% EtOH beaker, padded at the bottom with Kimwipes to protect sharp edges. Two petri dishes containing ice are prepared for handling live tissue in dissection solution. A high-quality adjustable light source is essential for proper visualization of brains and hippocampi under the stereoscope. A third petri dish is used for decapitation

2.4.1 HEK293 Cell Culture Media

- 500 mL DMEM (Gibco).
- 5 mL Penicillin-Streptomycin (Gibco).
- 5 mL Glutamax (Gibco).
- 50 mL Fetal Bovine Serum (Gibco).
- 0.45 μm 500 mL filter flask.

Combine all ingredients, filter with 0.45 μm filter, and store at 4 °C for up to 3 months.

2.4.2 Poly-D-Lysine Solution

- 5 mg Poly-D-Lysine (Sigma P6407).
- Sterile water.
- 0.22 μm Filter.

Dissolve Poly-D-Lysine in 100 mL of water and filter to make a solution of 50 μg/mL. Note that pre-coated coverslips may be purchased (e.g., Neuvitro GG121.5PDL). 20× stocks may be prepared and stored at −20 °C.

2.4.3 Tissue Culture Plate Preparation

- 24-well tissue culture plates.
- 12 mm #1 glass coverslips (Menzel Glaser).
- 100% EtOH.
- Forceps.
- Kimwipe (or other napkin).
- Culture Media.
- Poly-D-Lysine Solution.

- HEK293 cells (ATCC CRL-1573).
- Trypsin (such as TrypLE Express, Gibco).
- All-Trans Retinal (Sigma R2500).

2.5 Materials for Calcium-Phosphate Transfection of Primary Neuron Cultures

- Concentrated (>500 ng/μL), endotoxin-free DNA.
- 2 M $CaCl_2$ (Gibco).
- Sterile H_2O.
- 2× HEPES-buffered saline (pH 7.05).! *pH is critical for efficient transfection.*
- Minimal Essential Media (MEM) (Gibco).

3 Methods

3.1 Preparation of Coverslips and Culture Plates

While dissociated hippocampal neurons can easily grow on the plastic substrate of commercially available tissue culture plates, they do not readily adhere to glass coverslips. Growing neurons on glass coverslip is advantageous since the coverslips can be easily moved between solutions, for example, to perform electrophysiological recordings or for immunohistochemistry and imaging. Glass coverslips should be pretreated with acid, then cleaned with ethanol and flamed to provide a thin film of carbon to which Matrigel will be applied. Neurons are then plated on the Matrigel-coated coverslips, to which they adhere easily. Coverslips can be treated with 8% HCl and then stored in EtOH for months. The final step involving flame treatment and Matrigel coating should ideally be performed on the same day of the neuronal cell preparation.

*! **step 1** should be completed in a chemical fume hood wearing appropriate gloves.*

3.1.1 Preparing Coverslips

1. Combine 16.5 mL HCl with 183 mL double-distilled water in a 250 mL glass bottle using a disposable serological pipette to make a 8% solution of HCl. Cap bottle and gently swirl to ensure mixing. Do not shake.
2. Carefully remove coverslips from packaging and place 300 coverslips (from 3 packages) into 10 cm petri dish. Cover with 8% HCl solution and gently swirl. Incubate coverslips in acid overnight on a shaker set on low speed to ensure coverslips are thoroughly washed.
3. The following day, dispose of acid solution according to institutional protocol. Wash coverslips with 96% EtOH three times to remove all remaining acid. Store coverslips in petri dish and cover in 96% ethanol. Seal petri dish with parafilm to prevent evaporation of EtOH. Before use wash again with 96% EtOH.

*! **step 4** must be completed with as much sterility as possible.*

*! **step 4** requires an open flame. Plan the work to prevent the possibility of igniting the EtOH reservoir and ensure that equipment for extinguishing fires is nearby.*

! Glass coverslips will shatter if held at the flame for too long or are ignited with an excess of EtOH still on the glass.

4. One day prior to dissection, prepare 24-well tissue culture plates (typically, one 24-well plate per 2 hippocampi). Use 70% EtOH-sterilized forceps to pick out a single coverslip from the petri dish stock. Gently touch the rim of the coverslip onto a Kimwipe to wick excess EtOH, briefly run each side of the coverslip through a flame to dry, then place into a well of the 24-well plate. Keep the lid of the 24-well plate on except to place coverslips in wells.
5. Once coverslips are dried and placed into tissue culture plates, they are ready for glass coverslip coating. Keep in sterile environment until use.

3.1.2 Coating of Glass Coverslips with Matrigel

! Do not thaw Matrigel prior to use.

! Keep Matrigel cold throughout Protocol 3.1.2.

1. Prepare 8.7 mL of NB (**not NB+**) medium and place on ice.
2. Add 1 mL of the ice-cold NB medium from **step 1** to 300 μL frozen Matrigel aliquot kept on ice and pipette until Matrigel dissolves.
3. Combine dissolved 300 μL Matrigel and 1 mL NB medium with remaining 7.7 mL NB medium.

   ! If Matrigel gels, wrap 15 mL tube in foil and place on a shaker for 20 min at 4 ° C to prevent precipitation and dissolve evenly in NB.
4. Wrap 15 mL tube in foil to protect from light and store at 2–8 °C. Matrigel should be prepared at least one day prior to use and can be stored for up to 1 month. Do not use if solution develops clumps or strands.
5. On the day of dissection, add 50 μL of cold Matrigel solution to each glass coverslip. Matrigel should coat the entire coverslip, but should also remain on the glass as a drop. Take care not to shake or bump 24-well plate after adding Matrigel solution.
6. Place plates in incubator after addition of Matrigel solution. The solution will be removed immediately prior to **step 3.4.2** (cell plating).

3.2 Tissue Dissection

1. **Brain Extraction.** Decapitate P0-1 rats using large straight scissors (Fig. 1) in an empty 100 mm petri dish. Spray heads with 70% EtOH. Carefully insert one scissor blade into the rear of the skull and snip along the central suture, from the base

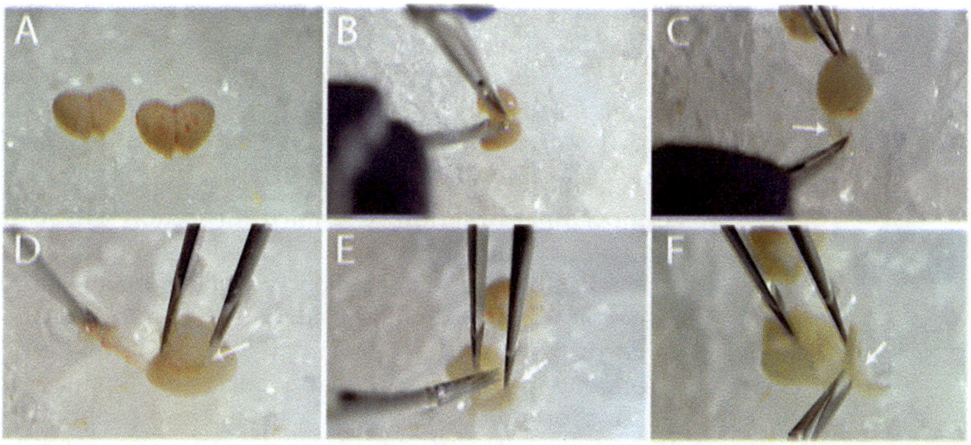

Fig. 4 Dissection of hippocampal tissue from neonatal brain. (**a**) Two brains in dissection media shortly after extraction. (**b**) Hemispheres are separated along the midline using two pairs of forceps. (**c**) Removal of meninges from external side of hemisphere (cortical side). (**d**) Removal of meninges from the inner aspect of the hemisphere. Note that the hippocampus is visible from this side, folded into the tissue below the cortex (arrow). (**e**) Separation of the hippocampus (arrow) from the hemisphere by applying gentle force with the forceps. (**f**) The hippocampus (arrow) is cut away from the surrounding tissue using a pair of angled micro-dissection scissors

toward the nose, being careful not to cut into the brain. Create lateral skull incisions bilaterally at the base and nose, then retract both sides of the skull laterally with serrated forceps to expose the brain. Remove the cerebellum with angled serrated forceps, then remove the brain carefully with the same forceps (a sterile, metal spatula may also be used; slide under the brain and lift) and place tissue in the 100 mm dish containing ice-cold Dissection Solution (Fig. 4a). Keep on ice.

2. **Hemisphere separation**. With micro-dissecting forceps (Fig. 2), separate the hemispheres carefully through the midline of the brain by closing the forceps along the midline toward the center (Fig. 4b).

   ! Take care to avoid damaging the hippocampi with instruments.

3. **Removal of meninges.** The meninges are tissue layers that protect the brain and separate it from the skull. After separating the hemispheres, the meninges will have been interrupted and this makes it easy to remove them. Use micro-dissecting forceps to peel away the meninges, starting from the edge opposite that of the hippocampus (olfactory bulb) and on the rounded, outer side of the hemisphere (Fig. 4c). Flip the hemisphere over so that the hippocampus is facing up. Continue to remove the meninges on this side of the hemisphere carefully (Fig. 4d), as they are attached to a fissure

in the hippocampus itself. Gently remove the remaining meninges and discard.

! The meninges are more readily removed when the dissection solution is cold.

! The banana-shaped hippocampi should be visible after hemisphere separation (arrow in Fig. 4d*).*

4. **Hippocampal isolation.** Use micro-dissection scissors to clip both ends of the hippocampus and free it from the overlying cortex. Gently fold out the hippocampus using angled micro-dissection scissors, carefully guiding it so that it detaches from the cortex (Fig. 4e). Take care not to include non-hippocampal tissue. Cut away any tissue that connects the hippocampus to the rest of the hemisphere with micro-dissection scissors (Fig. 4f) and then transfer it to the 35 mm petri dish with ice-cold Dissection Solution using micro-dissection forceps. Repeat these steps for four brains at a time. If needed, keep a first set of hippocampi in a 4 °C refrigerator and prepare a new set of petri dishes with fresh cold dissection media.
5. **Tissue digestion.** Add DNAse and papain to Enzymatic Digestion Solution, invert to mix, and sterile-filter. We use a Steriflip 0.22 μm filter.
6. Transfer all hippocampi to the 50 mL tube containing the Enzymatic Digestion Solution using a 10 mL pipette (let tissue settle to the bottom of the pipette prior to dispensing into the tube), set tube horizontally in a 37 °C incubator. Digest in the incubator for 25 min.
7. During this incubation period, sterile-filter the inactivating solution. A Steriflip 0.22 μm filter is recommended.
8. After 25-min digestion, remove as much of the Enzymatic Digestion Solution as possible using a 10 mL pipette, taking care not to remove any tissue. Add 10 mL Inactivation Solution to the tissue and let stand for 2 min at room temperature in a tissue culture hood.
9. Remove as much of Inactivation Solution as possible using a 10 mL pipette, taking care not to remove any tissue.

3.3 Single-Cell Dissociation

1. Add 1 mL Serum Media to 50 mL tube containing tissue. Triturate tissue by pipetting up and down 10 times using a 1 mL pipette tip.

   ! Minimize air bubbles by keeping the tip near the bottom of the tube. Do not withdraw or eject all of the media.

   ! Triturate gently to ensure maximum cell viability and a high cell count. Do not over-triturate.
2. Bring total volume to 5 mL using Serum Media and transfer entire contents to new 50 mL tube through a 40 μm tube-top cell strainer.

3. Centrifuge 5 min at 300 rcf on tabletop centrifuge at room temperature.
4. Remove 24-well plates from the incubator and remove Matrigel from each coverslip using vacuum aspiration. Hold cover slip in well using a pipette tip while applying suction.
5. After centrifuging, carefully aspirate supernatant using a pipette. Do not aspirate or disturb the cell pellet. Add 1 mL Serum Media and use 1 mL pipette to gently resuspend pellet until homogenous (usually 3–5 times).
6. Add additional 9 mL of Serum Media, mix by gently using 10 mL pipette.

3.4 Cell Viability Analysis and Plating

1. Place 30 μL of the cell suspension into a 0.5 mL tube (or other small tube) and add 10 μL of Trypan Blue. Trypan Blue stains non-viable cells.
2. Quantify viable cell concentration with a hematocytometer using a light microscope under 20× magnification. Discard 30 μL sample after quantification.

   Resuspend cells with Serum Media to desired density and add 130 μL per coverslip. Our typical plating density is 60,000–75,000 per 130 μL.

   ! Maintain the 130 μL cell suspension as a drop on the glass coverslip.
3. Place 24-well plates in 37 °C incubator for at least 2 h, but no more than 3 h.
4. Gently add 1 mL of NB+ Media to each well. Do not change media at any point.

3.5 Glial Inhibition of Dissociated Hippocampal Cultures

! FUDR should be added at 4 days in vitro *(DIV),* e.g., *for cultures created on Thursday, add FUDR on Monday.*

! Each well requires 5 μL of FUDR stock (50 μL after 1:10 dilution); scale recipes for total number of wells.

1. Thaw FUDR stock at room temperature.
2. Add 0.5 mL FUDR stock to 4.5 mL NB+ media (1:10 dilution).
3. Add 50 μL of diluted FUDR per well without touching media.

! Only take one plate out of the incubator at a time.

! In the event of a possible contamination of the dispenser tip, replace it with a sterile tip.

! Neural culture media does not contain antibiotics; if concerned about contamination, skip well.

3.6 Transfection of Hippocampal Neurons with Optogenetic Constructs

$CaPO_4$-mediated transfection of cell cultures is an inexpensive and effective method to express heterologous proteins [48]. Take care to ensure that the pH of the HEPES is correct, as small changes in this variable decrease transfection efficiency. Optimization of the protocol for neuronal cell lines has been previously described [49–51], and it is possible that you will find improvements in transfection efficiency in your own hands by experimenting with protocol parameters.

3.7 Calcium Phosphate Transfection

! Note that neuron media does not contain antibiotics. Take care to prevent contamination, including frequent 70% EtOH spray of hands and tools, working in a biological hood, and changing pipette tips.

- Pre-warm MEM in 37 °C water bath.
- In a sterile hood, mix 2 μg of DNA with 1.875 μL 2 M $CaCl_2$ (final Ca^{2+} concentration will be 250 mM). Bring volume to 15 μL total with sterile H_2O and mix well.
- Add 15 μL of 2× HBS and mix briefly (3–5 s) at low speed with a Vortex and incubate for 20 min at room temperature.
- While DNA-Calcium-Phosphate mixture is incubating, remove *and keep* neuron media in a tube, replace with minimal amount (~300 μL) pre-warmed MEM. Keep reserved neuron media at 37 °C or in the tissue culture incubator.
- Add DNA-$CaPO_4$ mixture drop-wise to the center of each well and distribute by gently swirling plate. Do not apply to the well wall.
- Incubate for 1 h in tissue culture incubator at 37 °C.

  !Optional: Examine plate under light microscope and look for precipitates.
- Aspirate MEM + DNA-$CaPO_4$ from well, rinse 3× with 1× pre-warmed MEM, and replace with reserved neuron media.
- Expression is typically apparent within 48–72 h.

3.8 Preparation of HEK293 Cells for Optogenetic Tool Expression and Testing

HEK293 cells were derived from human tissue and are immortalized. They are useful for determining optogenetic tool ion conductance properties using electrophysiology and are easy to maintain in culture.

3.8.1 Coverslip Preparation

! step 1 must be completed with as much sterility as possible.

! step 1 requires an open flame and a reservoir of EtOH. Plan the work to prevent the possibility of igniting the EtOH reservoir and ensure that equipment for extinguishing fires is nearby.

! Glass coverslips will shatter if held at the flame for too long or are ignited with an excess of EtOH still on the glass.

1. Prepare coverslips, 24-well plate, 100% EtOH, forceps, napkin, and open flame.
2. Dip coverslip in 100% EtOH using forceps, wick excess off using Kimwipe, and light a fire to sterilize. There should not be more than 1–2 s of fire before the EtOH is exhausted.
3. Place coverslip in 24-well plate and repeat until desired number of wells are prepared.

3.8.2 Poly-D-Lysine Coating

1. In a sterile environment (e.g., tissue culture hood), coat each coverslip with Poly-D-Lysine Solution (approximately 150 μL).
2. Incubate at room temperature for at least 1 h.
3. Aspirate solution and rinse 2× with sterile water.
4. Allow to dry completely.

3.8.3 HEK Cell Plating

! This protocol assumes growing cultures of HEK293 cells.

! HEK293 cells are typically classified as BSL-2, use appropriate precautions and PPE.

1. Thaw, maintain, and passage HEK293 cells as per manufacturer protocol. Cell quality and growth rate decrease when cells are allowed to reach confluence.
2. Pre-warm Growth Media to 37 °C.
3. Calculate surface area of experimental wells (e.g., an entire 24-well plate typically has ~46 cm^2 of total surface area).
4. Note approximate confluence of stock flask.
5. Aspirate media from stock flask and add just enough trypsin to coat flask (e.g., ~7 mL for a 225 cm^2 flask).
6. Allow to digest at room temperature, tapping flask with palm every minute. Watch for cell detachment.
7. When cells detach, add Growth Media and pipette using 10 mL serological pipette, washing flask bottom and triturating cells to create single-cell suspension.
8. (a) Calculate stock volume (mL) needed for *total* surface area of all target wells. Target confluence is typically 10%:

$$\left[\frac{\text{volume trypsin} + \text{Growth Media in stock flask mL}}{\left(\frac{\%\text{confluence of stock flask}}{\%\text{target confluence}}\right) * (\text{surface area stock flask cm}^2)}\right] * (\text{total surface area wells cm}^2)$$

   (b) A hematocytometer may also be used to target a cell density of 1.25×10^5 cells per mL [57].
9. Dilute calculated stock cell volume into 0.5 mL of Growth Media per target well (e.g., 20 mL for 40 wells). Mix well by inverting 5–6 times and distribute 0.5 mL per well to Poly-D-

Lysine coated coverslips and place at 37 °C incubator with 5% CO_2.

10. 24 h later, transfect using commercially available reagents (Lipofectamine, FuGENE, etc.).

    ! Note that Growth Media does NOT contain retinal that is necessary for opsin function, and this must be supplemented at 1 μM prior to optogenetic experiments.

References

1. Nagel G, Szellas T, Huhn W et al (2003) Channelrhodopsin-2, a directly light-gated cation-selective membrane channel. Proc Natl Acad Sci U S A 100:13940–13945
2. Boyden ES, Zhang F, Bamberg E et al (2005) Millisecond-timescale, genetically targeted optical control of neural activity. Nat Neurosci 8:1263–1268
3. Tervo DGR, Hwang BY, Viswanathan S et al (2016) A designer AAV variant permits efficient retrograde access to projection neurons. Neuron 92:372–382
4. Chan KY, Jang MJ, Yoo BB et al (2017) Engineered AAVs for efficient noninvasive gene delivery to the central and peripheral nervous systems. Nat Neurosci 20:1172–1179
5. Nathanson JL, Yanagawa Y, Obata K et al (2009) Preferential labeling of inhibitory and excitatory cortical neurons by endogenous tropism of adeno-associated virus and lentivirus vectors. Neuroscience 161:441–450
6. Grimm D, Lee JS, Wang L et al (2008) In vitro and in vivo gene therapy vector evolution via multispecies interbreeding and retargeting of adeno-associated viruses. J Virol 82: 5887–5911
7. Sohal VS, Zhang F, Yizhar O et al (2009) Parvalbumin neurons and gamma rhythms enhance cortical circuit performance. Nature 459:698–702
8. Atasoy D, Aponte Y, Su HH et al (2008) A FLEX switch targets Channelrhodopsin-2 to multiple cell types for imaging and long-range circuit mapping. J Neurosci 28:7025–7030
9. Fenno LE, Mattis J, Ramakrishnan C et al (2014) Targeting cells with single vectors using multiple-feature Boolean logic. Nat Methods 11:763–772
10. Zhang F, Wang LP, Boyden ES et al (2006) Channelrhodopsin-2 and optical control of excitable cells. Nat Methods 3:785–792
11. Deisseroth K, Feng G, Majewska AK et al (2006) Next-generation optical technologies for illuminating genetically targeted brain circuits. J Neurosci 26:10380–10386
12. Fenno L, Yizhar O, Deisseroth K (2011) The development and application of optogenetics. Annu Rev Neurosci 34:389–412
13. Zhang F, Prigge M, Beyriere F et al (2008) Red-shifted optogenetic excitation: a tool for fast neural control derived from Volvox carteri. Nat Neurosci 11:631–633
14. Zhang F, Wang LP, Brauner M et al (2007) Multimodal fast optical interrogation of neural circuitry. Nature 446:633–639
15. Marshel JH, Kim YS, Machado TA et al (2019) Cortical layer-specific critical dynamics triggering perception. Science 80:365
16. Yizhar O, Fenno LE, Prigge M et al (2011) Neocortical excitation/inhibition balance in information processing and social dysfunction. Nature 477:171–178
17. Berndt A, Lee SY, Wietek J et al (2016) Structural foundations of optogenetics: determinants of channelrhodopsin ion selectivity. Proc Natl Acad Sci U S A 113:822–829
18. Baker BJ, Mutoh H, Dimitrov D et al (2008) Genetically encoded fluorescent sensors of membrane potential. Brain Cell Biol 36(1–4):53–67
19. Kim CK, Adhikari A, Deisseroth K (2017) Integration of optogenetics with complementary methodologies in systems neuroscience. Nat Rev Neurosci 18:222–235
20. Wang W, Wildes CP, Pattarabanjird T et al (2017) A light- a nd calcium-gated transcription factor for imaging and manipulating activated neurons. Nat Biotechnol 35:864–871
21. Levskaya A, Weiner OD, Lim WA et al (2009) Spatiotemporal control of cell signalling using a light-switchable protein interaction. Nature 461:997–1001
22. Kozak M (2005) Regulation of translation via mRNA structure in prokaryotes and eukaryotes. Gene 361:13–37

23. Ikemura T (1985) Codon usage and tRNA content in unicellular and multicellular organisms. Mol Biol Evol 2:13–34
24. Gradinaru V, Thompson KR, Deisseroth K (2008) eNpHR: a Natronomonas halorhodopsin enhanced for optogenetic applications. Brain Cell Biol 36:129–139
25. Wu Y, Wang C, Sun H et al (2009) High-efficient FLPo deleter mice in C57BL/6J background. PLoS One 4:e8054
26. Kim EJ, Jacobs MW, Ito-Cole T et al (2016) Improved monosynaptic neural circuit tracing using engineered rabies virus glycoproteins. Cell Rep 15:692–699
27. Kozak M (1987) An analysis of 5′-noncoding sequences from 699 vertebrate messenger rNAS. Nucleic Acids Res 15:8125–8148
28. Kozak M (1986) Point mutations define a sequence flanking the AUG initiator codon that modulates translation by eukaryotic ribosomes. Cell 44(2):283–292
29. Yizhar O, Fenno LE, Davidson TJ et al (2011) Optogenetics in neural systems. Neuron 71: 9–34
30. Fitzgerald M, Shenk T (1981) The sequence 5′-AAUAAA-3′ forms part of the recognition site for polyadenylation of late SV40 mRNAs. Cell 24:251–260
31. Gradinaru V, Zhang F, Ramakrishnan C et al (2010) Molecular and cellular approaches for diversifying and extending optogenetics. Cell 141:154–165
32. Tkatch T, Greotti E, Baranauskas G et al (2017) Optogenetic control of mitochondrial metabolism & Ca2+ signaling by mitochondria-targeted opsins. Proc Natl Acad Sci U S A 114:E5167–E5176
33. Rost BR, Schneider F, Grauel MK et al (2015) Optogenetic acidification of synaptic vesicles and lysosomes. Nat Neurosci 18:1845–1852
34. Baker CA, Elyada YM, Parra A et al (2016) Cellular resolution circuit mapping with temporal-focused excitation of soma-targeted channelrhodopsin. elife 5:e14193
35. Shemesh OA, Tanese D, Zampini V et al (2017) Temporally precise single-cell-resolution optogenetics. Nat Neurosci 20: 1796–1806
36. Wu C, Ivanova E, Cui J et al (2011) Action potential generation at an axon initial segment-like process in the axonless retinal AII amacrine cell. J Neurosci 31:14654–14659
37. Lewis TL, Mao T, Svoboda K et al (2009) Myosin-dependent targeting of transmembrane proteins to neuronal dendrites. Nat Neurosci 12:568–576
38. Stachniak TJ, Ghosh A, Sternson SM (2014) Chemogenetic synaptic silencing of neural circuits localizes a hypothalamus→midbrain pathway for feeding behavior. Neuron 82:797–808
39. Tsien RY (1998) The green fluorescent protein. Annu Rev Biochem 67:509–544
40. Specht EA, Braselmann E, Palmer AE (2017) A critical and comparative review of fluorescent tools for live-cell imaging. Annu Rev Physiol 79:93–117
41. Chen X, Zaro JL, Shen WC (2013) Fusion protein linkers: property, design and functionality. Adv Drug Deliv Rev 65(10):1357–1369
42. Packer AM, Peterka DS, Hirtz JJ et al (2012) Two-photon optogenetics of dendritic spines and neural circuits. Nat Methods 9:1202–1205
43. Prakash R, Yizhar O, Grewe B et al (2012) Two-photon optogenetic toolbox for fast inhibition, excitation and bistable modulation. Nat Methods 9:1171–1179
44. Pelletier J, Sonenberg N (1988) Internal initiation of translation of eukaryotic mRNA directed by a sequence derived from poliovirus RNA. Nature 334:320–325
45. Donnelly MLL, Luke G, Mehrotra A et al (2001) Analysis of the aphthovirus 2A/2B polyprotein "cleavage" mechanism indicates not a proteolytic reaction, but a novel translational effect: a putative ribosomal "skip". J Gen Virol 82:1013–1025
46. Donnelly MLL, Hughes LE, Luke G et al (2001) The "cleavage" activities of foot-and-mouth disease virus 2A site-directed mutants and naturally occurring "2A-like" sequences. J Gen Virol 82:1027–1041
47. Mizuguchi H, Xu Z, Ishii-Watabe A et al (2000) IRES-dependent second gene expression is significantly lower than cap-dependent first gene expression in a Bicistronic vector. Mol Ther 1:376–382
48. Graham FL, van der Eb AJ (1973) A new technique for the assay of infectivity of human adenovirus 5 DNA. Virology 52:456–467
49. Jordan M, Schallhorn A, Wurm FM (1996) Transfecting mammalian cells: optimization of critical parameters affecting calcium-phosphate precipitate formation. Nucleic Acids Res 24: 596–601
50. Jiang M, Chen G (2006) High Ca2+-phosphate transfection efficiency in low-density neuronal cultures. Nat Protoc 1: 695–700
51. Köhrmann M, Haubensak W, Hemraj I et al (1999) Fast, convenient, and effective method to transiently transfect primary hippocampal neurons. J Neurosci Res 58:831–835

52. Miyashita T, Shao YR, Chung J et al (2013) Long-term channelrhodopsin-2 (ChR2) expression can induce abnormal axonal morphology and targeting in cerebral cortex. Front Neural Circuits 7:8

53. Fenno LE, Ramakrishnan C, Kim YS et al (2020) Comprehensive dual- and triple-feature intersectional single-vector delivery of diverse functional payloads to cells of behaving mammals. J Clean Prod 107:836–853.e11

54. Saito T, Nakatsuji N (2001) Efficient gene transfer into the embryonic mouse brain using in vivo electroporation. Dev Biol 240:237–246

55. Hitoshi N, Ken-ichi Y, Jun-ichi M (1991) Efficient selection for high-expression transfectants with a novel eukaryotic vector. Gene 108:193–199

56. Green MR, Sambrook J (2012) Molecular cloning: a laboratory manual. Cold Spring Harbor Laboratory Press, Cold Spring Harbor

57. Wietek J, Wiegert JS, Adeishvili N et al (2014) Conversion of channelrhodopsin into a light-gated chloride channel. Science 344:409–412

Chapter 15

Optogenetic Studies of Mitochondria

Kai Chen, Patrick Ernst, Xiaoguang Margaret Liu, and Lufang Zhou

Abstract

While optogenetic approaches have been widely used for remote control of cell membrane excitability and intracellular signaling pathways, their application in mitochondrial study has been limited, largely due to the challenge of effectively and specifically expressing heterologous light-gated rhodopsin channels in the mitochondria. Here, we describe the methods for expressing functional channelrhodopsin 2 (ChR2) proteins in the mitochondrial inner membrane with an unusually long mitochondrial leading sequence and characterizing optogenetic-mediated mitochondrial membrane potential ($\Delta\Psi_m$) depolarization. We then illustrate how this next-generation optogenetic approach can be used to study the effect of $\Delta\Psi_m$ on mitochondrial functions such as mitophagy, programed cell death, and preconditioning-mediated cytoprotection. We anticipate that this innovative technology will enable new insights into the mechanisms by which changes in $\Delta\Psi_m$ differentially impacts mitochondrial and cellular functions.

Key words Mitochondria, Optogenetics, Mitochondrial protein import, Apoptosis, Mitophagy, Preconditioning

1 Introduction

Mitochondria lie on the crossroad of cellular metabolic and signaling pathways [1]. Beyond ATP production, mitochondria also play a pivotal role in regulating key processes maintaining cellular functions such as cell division, reactive oxygen species (ROS) production, Ca^{2+} handling, and programmed cell death [2]. Not surprisingly, defects in mitochondrial functions have been implicated in a variety of human maladies such as heart failure [3–5], diabetes [6], and cancer [7].

The normal functioning of mitochondria relies on maintaining the inner membrane potential ($\Delta\Psi_m$). A profound loss of $\Delta\Psi_m$ not only impairs energy production but can also initiate a cascade of events that eventually lead to cell death [8]. Importantly, there is evidence that reversible, partial $\Delta\Psi_m$ depolarization can protect cells against stress-induced injury [9, 10]. Thus, precise manipulation of $\Delta\Psi_m$ would allow for a better understanding of the

Valentin Gordeliy (ed.), *Rhodopsin: Methods and Protocols*, Methods in Molecular Biology, vol. 2501,
https://doi.org/10.1007/978-1-0716-2329-9_15,

mechanisms by which mitochondria affect cellular functions and fate, and for the development of mitochondrial-targeted therapies. Yet, putting this theory into practice is challenging due to the lack of an appropriate method to specifically and precisely control $\Delta\Psi_m$ in an experimental setting. Current strategies to manipulate mitochondria are mainly via pharmacological intervention, such as by using chemicals uncouplers (e.g., FCCP and CCCP) [9, 10] or permeability transition pore activators (e.g., Atr and polyP) [11, 12]. However, those pharmacological reagents cannot perturb $\Delta\Psi_m$ with high spatial resolution and often induce nonspecific side effects. While laser flashing can trigger local $\Delta\Psi_m$ depolarization via the photooxidation effect of fluorescent dyes such as tetramethylrhodamine methyl ester (TMRM) [13–17], this approach requires high-concentration (i.e., >100 nM) TMRM loading and high-energy laser illumination, which may cause unpredictable outcomes.

Optogenetics is an emerging technology that utilizes genetically encoded, light-sensitive rhodopsin proteins (e.g., Channelrhodopsin 2 [ChR2]) [18, 19] expressed in the cellular membrane to remotely control the activity of the cells or animals [20, 21]. Due to high spatiotemporal resolution, optogenetics has evolved as a powerful tool in neuronal [22], cardiac [23–26], and skeletal muscle [27] research. Despite these advancements, application of this technique in mitochondrial study has been limited until recently [28–31]. One challenge to developing a mitochondrial-based optogenetics is effective and specific expression of the heterologous light-gated channel proteins in the highly folded inner mitochondrial membrane (IMM).

Here, we describe the methods for expressing functional ChR2 in the IMM by using the unusually long (140 aa) mitochondrial leading sequence (MLS) of the ATP binding cassette subfamily B (*a.k.a.* ABCB) [32], and confirming high-level mitochondrial ChR2 expression by live cell confocal imaging and immunostaining. We then narrate the procedure to illuminate cells expressing mitochondrial ChR2 with blue light (475 nm) to induce $\Delta\Psi_m$ depolarization in a controllable manner. Finally, we illustrate how this new-generation optogenetic technology can be used for the study of mitochondrial functions such as mitophagy, programmed cell death, and preconditioning-mediated cytoprotection.

It is anticipated that the methods discussed herein will allow for targeting expression of other rhodopsin proteins such as the light-driven inward H^+ pump [33] and Na^+ pump [34] in the mitochondria. The capability of inducing controllable $\Delta\Psi_m$ depolarization will expand mitochondrial study and enable new insights into how changes in $\Delta\Psi_m$ impact cellular signaling and health.

2 Materials

2.1 Plasmid Construction

1. pCMV-myc-mito-GFP vector (Life Technologies, Catalog Number: V82220) (*see* **Note 1**).
2. MLS of renal outer medullary potassium channel (ROMK) (Gene Art, Life technologies) (*see* **Note 2**).
3. ABCB MLS (Gene Art, Life technologies). For simplification, ABCB is used to denote ABCB MLS hereinafter.
4. pcDNA-CMV-YFP vector (Addgene, Plasmid #13033).
5. pLenti-EF1a-hChR2(H134R)-EYFP-WPRE vector (Addgene, Plasmid Number: 20942). For simplification, ChR2 is used to denote hChR2(H134R) hereinafter.
6. XbaI restriction enzyme.
7. BamHI-HF restriction enzyme.
8. PCR primers.
   (a) ABCB-F: 5′- GCTTGGTACCGAGCTCGGATCCGC CACCATGCGCGCC-3′.
   (b) ABCB-R: 5′- TAGTCCATGAATTCAACTGCAGCTGA CAGTCTCCCGC-3′.
   (c) ChR2-eYFP-F: 5′- AGCTGCAGTTGAATTCATGGAC TATGGCGGCGCTT-3′.
   (d) ChR2-eYFP-R: 5′- CACTATAGAATAGGGCCCTCTA GATTACACCTCGTTCTCGTAGCA-3′.
9. Q5 High-Fidelity DNA polymerase.
10. Deoxynucleotide (dNTP) solution mix.
11. Ultrapure agarose.
12. TAE buffer.
13. PCR purification kit.
14. Gel extraction kit.
15. Ampicillin antibiotic.
16. LB agar medium.
17. LB medium.

2.2 Cell Culture, Plasmid Transfection, and ChR2 Expression

1. 15 mm #1.5 roundglass coverslip.
2. Hela and H9C2 cells.
3. Dulbecco's modified Eagle medium.
4. Fetal bovine serum.
5. L-Glutamine.
6. Penicillin-Streptomycin antibiotics.

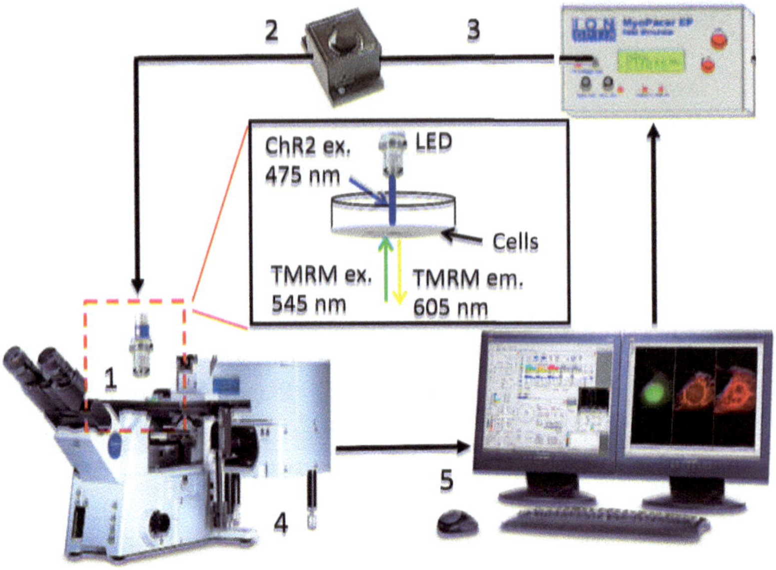

Fig. 1 The scheme of setup for simultaneous optical illumination and confocal imaging. 1: Recording chamber containing cells; 2: 475 nm LED driver connected with a 10× objective lens; 3: Stimulator; 4: Confocal microscope, including 545 nm laser, excitation filter, emission filter and PMT detector; 5: Computer for image acquisition and stimulator control

7. Lipofectamine 3000 transfection reagent.
8. Mitotracker Deep red.

2.3 Optical Illumination and Confocal Imaging (Fig. 1)

1. 470 nm fiber-coupled LED (ThorLabs, Catalog Number: M470F3).
2. T-Cube LED driver (ThorLabs, Catalog Number: LEDD1B).
3. Ø1000 μm, 0.39 NA, SMA-SMA fiber patch cable (ThorLabs, Catalog Number: M35L01).
4. 10× objective lens.
5. Culture Dish Incubator (Warner Instruments, Catalog Number: DH-35iL).
6. Myopacer 100 field stimulator (IonOptix, Westwood, MA).
7. Olympus FV1000 confocal microscope.
8. Computer for image acquisition and stimulator control.
9. ImageJ software (Wayne Rasband, National Institutes of Health, Bethesda, MD).

2.4 Mitochondrial Membrane Potential Measurement and Cell Death Assay

1. TMRM.
2. MitoView 633.
3. Olympus FV1000 confocal microscope.

4. Trypsin-EDTA.
5. Trypan blue.
6. Hemocytometer.
7. LDH assay kit.
8. AlamarBlue assay kit.
9. Pan-caspase inhibitor, Z-VAD-FMK.
10. Necroptosis inhibitor, 7-Cl-O-Nec-1.
11. Tecan GENios microplate reader.
12. Nanodrop 2000 spectrometer.

2.5 Immunocytochemistry

1. Fixation solution: 4% Formaldehyde in phosphate-buffered saline (PBS), pH 7.4.
2. Permeabilization buffer: PBS containing 0.2% Triton X-100.
3. Blocking buffer: 10% goat serum and 0.3% Triton® X-100 in PBS.
4. Diluent solution: 1% bovine serum albumin, 1% goat serum, 0.3% Triton X-100 in PBS.
5. Anti-ChR2 mouse monoclonal antibody (Progen, Catalog Number: 610180).
6. Anti-LC3 mouse monoclonal antibody.
7. Goat anti-mouse IgG secondary antibody, Alexa Fluor 555.
8. Goat anti-mouse IgG2b secondary antibody, Alexa Fluor 647.
9. Glass coverslip (*see* Subheading 2.2, **item 1**).
10. Microscope slide.
11. Antifade mounting medium.
12. Nail polish.

2.6 Mitochondrial Autophagy Assay

1. mCherry-Parkin plasmid DNA (Addgene, Plasmid Number: 23956).
2. pcDNA-CMV-ABCB-ChR2-YFP plasmid DNA.
3. Neon transfection system (Invitrogen, Catalog Number: MPK10025).
4. LysoTracker Deep Red.
5. 35 mm glass-bottom dish.
6. 470 nm fiber-coupled LED (*see* Subheading 2.3, **items 1–3**).

3 Methods

3.1 Targeting Express ChR2 in Mitochondrion

3.1.1 Construct pcDNA-ABCB-ChR2-eYFP Plasmid

1. Amplify ABCB MLS gene fragment by PCR using ABCB-F and ABCB-R primers with Q5 High-Fidelity DNA Polymerase.
2. Amplify ChR2-eYFP gene fragment by PCR using pLenti-EF1a-hChR2(H134R)-EYFP-WPRE vector as a template using ChR2-eYFP-F and ChR2-eYFP-R primers, with Q5 High-Fidelity DNA Polymerase.
3. Run a portion of both products on a gel to confirm fragment size and purify the remaining products (*see* **Note 3**).
4. Digest the backbone vector (pcDNA-CMV-YFP) overnight at 37 °C with the restriction enzymes BamHI-HF and XbaI. Inactivate the digest at 65 °C.
5. Run the digested vector on a gel. Excise the piece of gel containing the backbone fragment (5–5.5 kb) and extract the backbone DNA from the gel.
6. Ligate all three fragments together using the Gibson ligation kit following the manufacturer's instruction.
7. Transform the ligation product into *E. coli*, and then spread the *E. coli* onto an agar plate containing ampicillin.
8. Pick individual *E. coli* colonies and grow in liquid culture for 20 h.
9. Screen the colonies using PCR (*see* **Note 4**).
10. Isolate plasmid DNA from the liquid cultures and verify the construction via restriction enzyme digestion.
11. Perform Sanger sequencing to confirm the sequence of the constructed pcDNA-ABCB-ChR2-eYFP (termed as ABCB-ChR2-eYFP hereinafter) plasmid.

3.1.2 Determine Mitochondrial ChR2-eYFP Expression by Confocal Microscopy

1. Culture cells (e.g., H9C2 or HeLa) on a 35 mm glass-bottom dish.
2. Transfect cells with ABCB-ChR2-eYFP plasmid using lipofectamine 3000.
3. Forty-eight hours later, stain cells with MitoTracker Deep Red (100 nM) for 30 min, then wash cells 2–3 times to remove excess dye.
4. Mount the culture dish on the stage of a confocal microscope.
5. Image eYFP and MitoTracker by the 515 nm argon laser and 635 nm laser diode lines, respectively.
6. Analyze images using ImageJ software. YFP and MitoTracker fluorescence colocalization indicates mitochondrial expression of ChR2.

3.1.3 Confirm Mitochondrial ChR2-YFP Expression in H9C2 Cells by Immunocytochemistry

1. Culture cells (e.g., H9C2 or HeLa) on 15 mm glass coverslips.
2. Transfect cells with ABCB-ChR2-eYFP plasmid using lipofectamine 3000.
3. Forty-eight hours later, stain cells with MitoTracker Deep Red (100 nM) for 30 min at 37 °C. Wash cells three times with PBS to remove excess dye.
4. Incubate cells in fixation solution for 10 min at room temperature, and then wash three times with ice-cold PBS.
5. Incubate sample for 10 min with permeabilization buffer to permeabilize the cells (*see* **Note 5**). Wash cells in PBS three times for 5 min.
6. Incubate cells with blocking buffer for 2 h at room temperature to block nonspecific binding of antibody.
7. Incubate cells with anti-ChR2 primary antibody (1:200 dilution) for overnight at 4 °C.
8. After a washing step, block sample again with blocking buffer (30 min at room temperature), and incubate with Alexa Fluor 555 secondary antibody (1:200 dilution) for 1 h at room temperature.
9. After a final wash step, mount coverslip on slide, and image MitoTracker and Alexa Fluor 555, respectively, using a confocal microscope.
10. Analyze MitoTracker and Alexa Fluor 555 colocalization using ImageJ software.

3.2 Light Illumination Causes Targeted Mitochondrial Depolarization in ABCB-ChR2-eYFP Expressing Cells

1. Transfect H9C2 cells with ABCB-ChR2-eYFP plasmid using lipofectamine 3000.
2. Stain cells with mitochondrial membrane potential fluorescent dye TMRM (20 nM) for 15 min. Wash 2–3 times to remove excess dye.
3. Mount coverslip in a microincubation system on the stage of a confocal microscope. Select 3–5 cells within the field of view and illuminate them with 470 nm light emitting diode (LED) pulses (5 mW/mm^2) using the optical stimulation setup shown in Fig. 1.
4. Use un-illuminated cells within the same field of view as control.
5. Image TMRM in both the unilluminated and illuminated cells by a confocal microscope, before and after illumination.
6. Use ImageJ software to analyze TMRM fluorescence and determine relative $\Delta\Psi_m$. Compare results between groups.

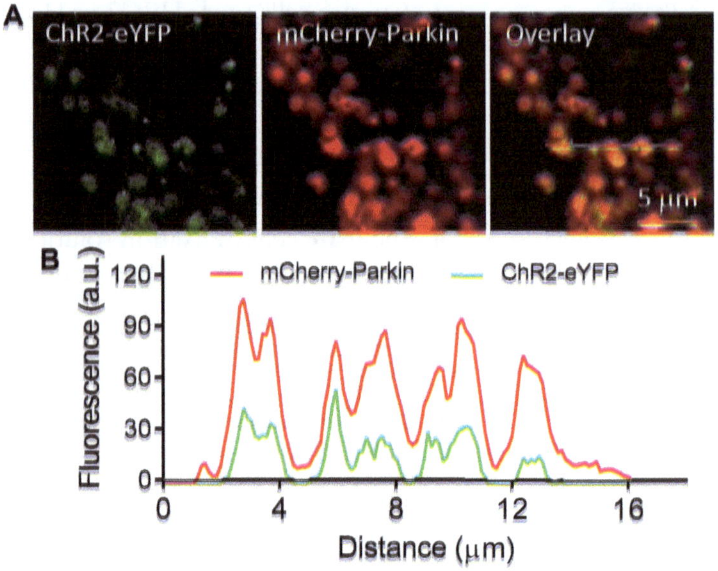

Fig. 2 (**a**) Representative confocal images show overlay of mCherry (red) and YFP (green) in blue light (475 nm)-illuminated HeLa cells co-expressing ABCB-ChR2-eYFP and mCherry-Parkin, indicating translocation of Parkin to mitochondria. (**b**) Line scanning (denoted by the yellow line in the overlay panel in **a**) shows relative position of mCherry-Parkin (red line) and ChR2-eYFP (green line) proteins, indicating accumulation of Parkin surrounding mitochondria

3.3 Optogenetic-Mediated Mitochondrial Autophagy (Mitophagy)

3.3.1 Measure Parkin Translocation to Mitochondrion

1. Co-transfect HeLa cells with ABCB-ChR2-eYFP plasmid and mCherry-Parkin plasmid (*see* **Note 6**) using the Neon Transfection System (*see* **Note 7**).
2. Seed the cells in a 35 mm glass-bottom dish and culture for 2 days.
3. Illuminate cells with LED light at 0.5 mW/mm^2 for 24 h (*see* **Note 8**).
4. Image YFP and mCherry fluorescence via a confocal microscope at 0, 4, 8, and 24 h of light illumination.
5. Analyze images using ImageJ software. Parkin translocation to the mitochondria indicates mitochondrial depolarization and induction of Parkin-dependent mitophagy (see example in Fig. 2).

3.3.2 Analyze Colocalization of Mitochondria and Lysosomes

1. Co-transfect HeLa cells with ABCB-ChR2-eYFP plasmid and mCherry-Parkin plasmid (*see* **Note 6**), and culture cells in an incubator at 37 °C for 48 h.
2. Illuminate the transfected cells with LED light at 0.5 mW/mm^2 for 24 h.
3. Stain cells with LysoTracker Deep Red (50 nM) at 37 °C for 30 min. Wash cells three times to remove excess dye.

4. Image YFP and LysoTracker Deep Red using a confocal microscope.
5. Analyze colocalization of mitochondria (indicated by YFP) and lysosomes to assess activation of mitophagy.
6. Use the mock-transfected and the un-illuminated ABCB-ChR2-eYFP-expressing cell cultures as controls.

3.3.3 Measure Mitochondrial Autophagy by Autophagosome Marker LC3

1. Co-transfect HeLa cells with ABCB-ChR2-eYFP plasmid and mCherry-Parkin plasmid (*see* **Note 6**), and culture cells in an incubator at 37 °C for 48 h.
2. Illuminate cells with LED light at 0.5 mW/mm^2 for 24 h.
3. Incubate cells with fixation buffer for 10 min at room temperature, and then wash three times with ice-cold PBS.
4. Incubate cells with blocking solution to permeabilize cells and block nonspecific binding of antibody. Wash sample three times with PBS.
5. Incubate cells with the LC3 primary antibody (diluted at a ratio of 1:200) overnight at 4 °C.
6. After a washing step, block sample again with blocking buffer (30 min at room temperature), and incubate with AlexaFluor 647 secondary antibody (diluted at a ratio of 1:200) for 1 h at room temperature.
7. Mount coverslip on slides, and image YFP and AlexaFluor 647 using a confocal microscope.
8. Analyze LC3 aggregation and colocalization with mitochondrion (indicated by YFP) to determine activation of mitophagy.
9. Use mock-transfected cells and un-illuminated ABCB-ChR2-eYFP-expressing cells as controls.

3.4 Optogenetic-Mediated Cell Death

3.4.1 Assess Light Illumination-Mediated Cytotoxicity in ABCB-ChR2-eYFP-Expressing Cells

1. Prepare control (e.g., non-transfected and mock-transfected) and ABCB-ChR2-eYFP-transfected HeLa cells in a 24-well plate.
2. Illuminate cells by (0.5 mW/mm^2) LED light from underneath (*see* **Note 9**) for 24 h (*see* **Note 10**).

   If using Trypan Blue method.
3. Detach cells with 0.25% Trypsin-EDTA (*see* **Note 11**).
4. Stain cells with Trypan Blue and count with a hemocytometer.
5. Calculate cell viability and compare among groups.

   If using lactate dehydrogenase (LDH) assay.
7. Remove 200 mL cell culture medium and add to a 96-well plate at the end of light illumination.
8. Add 100 mL of the reaction mixture to each well and incubate the plate for 30 min at room temperature, protected from light.

9. Measure the absorbance at 490 and 680 nm using a microplate reader.
10. Calculate cell viability following the manufacturer's instruction.

If using alamarBlue assay.

11. Add 10 mL alamarBlue reagent to cells in 100 mL culture medium, incubate cells for 4 h in an incubator at 37 °C, protected from light.
12. Measure the absorbance at 570 nm using a spectrometer.
13. Calculate cell viability following the manufacturer's instruction.

3.4.2 Determine Time Dependence of Optogenetic-Mediated Cell Death

1. Prepare mock-transfected and ABCB-ChR2-eYFP-transfected HeLa cells in a 24-well plate.
2. Illuminate both groups of cells with LED light (0.5 mW/mm^2) for 2, 4, 8, 12, and 24 h, respectively (*see* **Note 12**).
3. Detach the cells with 0.25% Trypsin-EDTA at each time point.
4. Stain cells with Trypan Blue, count with a hemocytometer, and calculate cell viability.

3.4.3 Determine Light Irradiance Dependence of Optogenetic-Mediated Cell Death

1. Prepare mock-transfected and ABCB-ChR2-eYFP-transfected HeLa cells in a 96-well plate.
2. Illuminate cells with LED light at 1, 3, 5, and 7 mW/mm^2, respectively, for 6 h.
3. Stain the cells with Hoechst 33342 and propidium iodide (PI) simultaneously.
4. Image cells with a fluorescent microscope and calculate cell death (*see* **Note 13**).

3.4.4 Explore the Mechanistic Pathway (Apoptosis vs. Necrosis) Underlying Optogenetic-Mediated Cell Death

1. Prepare mock-transfected and ABCB-ChR2-eYFP-transfected HeLa cells in 24-well plate.
2. Illuminate cells with LED light at 0.5 mW/mm^2 for 24 h with no chemical inhibitor, the pan-caspase inhibitor Z-VAD-FMK (20 μM), or the necroptosis inhibitor 7-Cl-O-Nec-1 (100 μM).
3. Detach the cells with 0.25% Trypsin-EDTA.
4. Stain cells with Trypan Blue, count with a hemocytometer, and calculate cell viability.

3.5 Optogenetic-Mediated Preconditioning and Cytoprotection

1. Transfect Hela cells with ABCB-ChR2-eYFP plasmid.
2. Forty-eight hours post transfection, split cells into three groups and treat with: (a) no light, (b) 6 h of 4 mW/mm^2 LED illumination, and (c) 2 h of 0.2 mW/mm^2 LED illumination followed by 6 h of 4 mW/mm^2 LED illumination, respectively.

3. At the end of treatment, stain cells with MitoView 633 (25 nM) at 37 °C for 15 min.
4. Image MitoView using a confocal microscope to measure $\Delta\Psi_m$.
5. Detach the remaining cell cultures with 0.25% Trypsin-EDTA.
6. Stain cells with Trypan Blue, count with a hemocytometer, and calculate cell viability.

4 Notes

1. The MLS of cytochrome c oxidase subunit VIII (*a.k.a.* mito) has been widely used for importing protein to mitochondria. Thus, we tested its capability to transport ChR2 to mitochondria in the beginning. However, our data revealed that neither mito itself nor its duplicate and triplicate repeats successfully led to mitochondrial ChR2 expression. It is worth mentioning that Tkatch et al. reported that tandem repeats of the *mito* sequence could target ChR2 to mitochondria in a recent paper [31].
2. ROMK is a candidate MLS of mitochondrial K_{ATP} channel located in the IMM. Our data showed that similar to mito, ROMK MLS could not effectively translocate ChR2 to mitochondria.
3. When doing a multi-fragment ligation using Gibson Assembly, do not digest the PCR products as this will remove the fragment overlaps necessary for the ligation.
4. We find that this method allows for early-stage screening of the colonies (i.e. before isolating plasmid DNA). The procedure is the following: (a) Dilute 10 μL from each liquid culture with 40 μL ddH_2O. (b) Boil the samples at 98 °C for 10 min followed by 5 min centrifugation at maximum speed (>10,000 × *g*). (c) Use 1–2 μL of the supernatant as template DNA for the PCR, being careful not to disturb the pellet of cell debris at the bottom of the tube.
5. The optimal percentage of Triton X-100 should be determined for each cell type.
6. Wild type HeLa cells do not express endogenous Parkin. To examine optogenetic-induced Parkin translocation and the canonical PINK1/Parkin-mediated mitophagy, HeLa cells are transfected with mCherry-Parkin plasmid to overexpress Parkin.
7. We find co-transfection to be more successful using electroporation than using Lipofectamine 3000.
8. For all long-term (i.e., on the order of hours), photostimulation 90 ms pulses at a frequency of 4 Hz can be used.

9. Illuminating cells from underneath is best, especially if using well-plates. Illuminating cells from above will result in much more internal reflection of the light within the well-plate before the light reaches the cells.
10. If illuminating cells in well-plates, it can be difficult lining up the cells of interest with the LED light. To make it easier, one can use lab tape around the well(s) of interest on the bottom of the plate and then adjust the plate until maximum light is transmitted.
11. Some dead cells will likely detach and may be lost while replacing culture medium with Trypsin, giving an inaccurate assessment of cell death. This can be addressed by saving the culture medium from each sample and using it to neutralize the Trypsin and resuspend the cells after they have detached.
12. It is best to have samples for each time-point in separate dishes/plates, so removing them to measure cell death does not disrupt the photostimulation of the remaining groups.
13. Hoechst 33342 will stain the nuclei of all cells, while PI will stain the nuclei of dead cells. Cell death can be calculated by the number of PI+ (red) nuclei divided by the number of Hoechst+ (blue) nuclei.

Acknowledgments

This work was supported by National Institute of Health (NIH) R01 HL121206 (to L. Z.) and American Heart Association (AHA) Predoctoral Fellowship 18PRE34060188 (to P. E.).

References

1. Honda HM, Korge P, Weiss JN (2005) Mitochondria and ischemia/reperfusion injury. Ann N Y Acad Sci 1047:248–258. https://doi.org/10.1196/annals.1341.022
2. Bohovych I, Khalimonchuk O (2016) Sending out an SOS: mitochondria as a signaling hub. Front Cell Dev Biol 4:109. https://doi.org/10.3389/fcell.2016.00109
3. Brown DA, Perry JB, Allen ME, Sabbah HN, Stauffer BL, Shaikh SR, Cleland JG, Colucci WS, Butler J, Voors AA, Anker SD, Pitt B, Pieske B, Filippatos G, Greene SJ, Gheorghiade M (2017) Expert consensus document: mitochondrial function as a therapeutic target in heart failure. Nat Rev Cardiol 14(4): 238–250. https://doi.org/10.1038/nrcardio.2016.203
4. Kim S, Song J, Ernst P, Latimer MN, Ha CM, Goh KY, Ma W, Rajasekaran NS, Zhang J, Liu X, Prabhu SD, Qin G, Wende AR, Young ME, Zhou L (2020) MitoQ regulates redox-related noncoding RNAs to preserve mitochondrial network integrity in pressure-overload heart failure. Am J Phys Heart Circ Phys 318(3):H682–H695. https://doi.org/10.1152/ajpheart.00617.2019
5. Goh KY, He L, Song J, Jinno M, Rogers AJ, Sethu P, Halade GV, Rajasekaran NS, Liu X, Prabhu SD, Darley-Usmar V, Wende AR, Zhou L (2019) Mitoquinone ameliorates pressure overload-induced cardiac fibrosis and left ventricular dysfunction in mice. Redox Biol 21: 101100. https://doi.org/10.1016/j.redox.2019.101100
6. Duchen MR, Szabadkai G (2010) Roles of mitochondria in human disease. Essays Biochem 47:115–137. https://doi.org/10.1042/bse0470115

7. Ralph SJ, Neuzil J (2009) Mitochondria as targets for cancer therapy. Mol Nutr Food Res 53(1):9–28. https://doi.org/10.1002/mnfr.200800044
8. Desagher S, Martinou JC (2000) Mitochondria as the central control point of apoptosis. Trends Cell Biol 10(9):369–377
9. Weisova P, Anilkumar U, Ryan C, Concannon CG, Prehn JH, Ward MW (2012) 'Mild mitochondrial uncoupling' induced protection against neuronal excitotoxicity requires AMPK activity. Biochim Biophys Acta 1817(5):744–753. https://doi.org/10.1016/j.bbabio.2012.01.016
10. Brennan JP, Southworth R, Medina RA, Davidson SM, Duchen MR, Shattock MJ (2006) Mitochondrial uncoupling, with low concentration FCCP, induces ROS-dependent cardioprotection independent of KATP channel activation. Cardiovasc Res 72(2):313–321. https://doi.org/10.1016/j.cardiores.2006.07.019
11. Hausenloy DJ, Maddock HL, Baxter GF, Yellon DM (2002) Inhibiting mitochondrial permeability transition pore opening: a new paradigm for myocardial preconditioning? Cardiovasc Res 55(3):534–543
12. Solesio ME, Elustondo PA, Zakharian E, Pavlov EV (2016) Inorganic polyphosphate (polyP) as an activator and structural component of the mitochondrial permeability transition pore. Biochem Soc Trans 44(1):7–12. https://doi.org/10.1042/BST20150206
13. Aon MA, Cortassa S, Marban E, O'Rourke B (2003) Synchronized whole cell oscillations in mitochondrial metabolism triggered by a local release of reactive oxygen species in cardiac myocytes. J Biol Chem 278(45): 44735–44744. https://doi.org/10.1074/jbc.M302673200
14. Brady NR, Elmore SP, van Beek JJ, Krab K, Courtoy PJ, Hue L, Westerhoff HV (2004) Coordinated behavior of mitochondria in both space and time: a reactive oxygen species-activated wave of mitochondrial depolarization. Biophys J 87(3):2022–2034
15. Zhou L, Aon MA, Liu T, O'Rourke B (2011) Dynamic modulation of Ca(2+) sparks by mitochondrial oscillations in isolated guinea pig cardiomyocytes under oxidative stress. J Mol Cell Cardiol 51(5):632–639. https://doi.org/10.1016/j.yjmcc.2011.05.007
16. Goh KY, Qu J, Hong H, Liu T, Dell'Italia LJ, Wu Y, O'Rourke B, Zhou L (2016) Impaired mitochondrial network excitability in failing guinea-pig cardiomyocytes. Cardiovasc Res 109(1):79–89. https://doi.org/10.1093/cvr/cvv230
17. Zhou L, O'Rourke B (2012) Cardiac mitochondrial network excitability: insights from computational analysis. Am J Phys Heart Circ Phys 302(11):H2178–H2189. https://doi.org/10.1152/ajpheart.01073.2011
18. Nagel G, Szellas T, Huhn W, Kateriya S, Adeishvili N, Berthold P, Ollig D, Hegemann P, Bamberg E (2003) Channelrhodopsin-2, a directly light-gated cation-selective membrane channel. Proc Natl Acad Sci U S A 100(24):13940–13945. https://doi.org/10.1073/pnas.1936192100
19. Hegemann P, Gartner W, Uhl R (1991) All-trans retinal constitutes the functional chromophore in Chlamydomonas rhodopsin. Biophys J 60(6):1477–1489
20. Boyden ES (2011) A history of optogenetics: the development of tools for controlling brain circuits with light. F1000 Biol Rep 3:11. https://doi.org/10.3410/B3-11
21. Deisseroth K (2011) Optogenetics. Nat Methods 8(1):26–29. https://doi.org/10.1038/nmeth.f.324
22. Zhang F, Wang LP, Boyden ES, Deisseroth K (2006) Channelrhodopsin-2 and optical control of excitable cells. Nat Methods 3(10): 785–792. https://doi.org/10.1038/nmeth936
23. Arrenberg AB, Stainier DY, Baier H, Huisken J (2010) Optogenetic control of cardiac function. Science 330(6006):971–974. https://doi.org/10.1126/science.1195929
24. Bruegmann T, Malan D, Hesse M, Beiert T, Fuegemann CJ, Fleischmann BK, Sasse P (2010) Optogenetic control of heart muscle in vitro and in vivo. Nat Methods 7(11): 897–900. https://doi.org/10.1038/nmeth.1512
25. Entcheva E (2013) Cardiac optogenetics. Am J Physiol Heart Circ Physiol 304(9): H1179–H1191. https://doi.org/10.1152/ajpheart.00432.2012
26. Li Q, Ni RR, Hong H, Goh KY, Rossi M, Fast VG, Zhou L (2017) Electrophysiological properties and viability of neonatal rat ventricular myocyte cultures with inducible ChR2 expression. Sci Rep 7(1):1531. https://doi.org/10.1038/s41598-017-01723-2
27. Asano T, Ishizua T, Yawo H (2012) Optically controlled contraction of photosensitive skeletal muscle cells. Biotechnol Bioeng 109(1): 199–204. https://doi.org/10.1002/bit.23285
28. Berry BJ, Trewin AJ, Milliken AS, Baldzizhar A, Amitrano AM, Kim M, Wojtovich AP (2019) Controlling the mitochondrial protonmotive force with light to impact

cellular stress resistance. bioRxiv:742536. https://doi.org/10.1101/742536

29. Ernst P, Xu N, Qu J, Chen H, Goldberg MS, Darley-Usmar V, Zhang JJ, O'Rourke B, Liu X, Zhou L (2019) Precisely control mitochondrial membrane potential with light to manipulate cell fate decisions. Biophys J 17(4):631–645. https://doi.org/10.1101/469668
30. Hara KY, Wada T, Kino K, Asahi T, Sawamura N (2013) Construction of photoenergetic mitochondria in cultured mammalian cells. Sci Rep 3:1635. https://doi.org/10.1038/srep01635
31. Tkatch T, Greotti E, Baranauskas G, Pendin D, Roy S, Nita LI, Wettmarshausen J, Prigge M, Yizhar O, Shirihai OS, Fishman D, Hershfinkel M, Fleidervish IA, Perocchi F, Pozzan T, Sekler I (2017) Optogenetic control of mitochondrial metabolism and Ca2+ signaling by mitochondria-targeted opsins. Proc Natl Acad Sci 114(26):E5167–E5176. https://doi.org/10.1073/pnas.1703623114
32. Graf SA, Haigh SE, Corson ED, Shirihai OS (2004) Targeting, import, and dimerization of a mammalian mitochondrial ATP binding cassette (ABC) transporter, ABCB10 (ABC-me). J Biol Chem 279(41):42954–42963. https://doi.org/10.1074/jbc.M405040200
33. Shevchenko V, Mager T, Kovalev K, Polovinkin V, Alekseev A, Juettner J, Chizhov I, Bamann C, Vavourakis C, Ghai R, Gushchin I, Borshchevskiy V, Rogachev A, Melnikov I, Popov A, Balandin T, Rodriguez-Valera F, Manstein DJ, Bueldt G, Bamberg E, Gordeliy V (2017) Inward H(+) pump xenorhodopsin: mechanism and alternative optogenetic approach. Sci Adv 3(9):e1603187. https://doi.org/10.1126/sciadv.1603187
34. Kovalev K, Polovinkin V, Gushchin I, Alekseev A, Shevchenko V, Borshchevskiy V, Astashkin R, Balandin T, Bratanov D, Vaganova S, Popov A, Chupin V, Buldt G, Bamberg E, Gordeliy V (2019) Structure and mechanisms of sodium-pumping KR2 rhodopsin. Sci Adv 5(4):eaav2671. https://doi.org/10.1126/sciadv.aav2671

Chapter 16

In Vivo and In Vitro Characterization of Cyclase and Phosphodiesterase Rhodopsins

Yuehui Tian, Shiqiang Gao, and Georg Nagel

Abstract

Rhodopsins with enzymatic activity were found in microbes, in 2004 hypothetically from sequence data and since 2014 by experimental proof. So far three different types are known: light-activated guanylyl cyclase opsins (Cyclop) in fungi, light-inhibited two-component guanylyl cyclase opsins (2c-Cyclop) in green algae, and rhodopsin phosphodiesterases (RhoPDE) in choanoflagellates. They are integral membrane proteins with eight transmembrane helices (TM), different to the other microbial (type I) rhodopsins with 7 TM. Therefore, we propose a classification as type Ib rhodopsins for opsins with 8 TM and type Ia for the ones with 7 TM. To characterize those rhodopsins or their mutants, the expression in *Xenopus laevis* oocytes proved to be an efficient strategy. Functional analysis was initially performed "in oocyte" (in vivo), but more detailed characterization can be obtained with an in vitro assay. In this chapter, we describe procedures how to extract membranes from oocytes after cRNA microinjection and heterologous protein expression. Enzymatic activity of these membranes is then analyzed under different illumination conditions. In addition, fluorescent labeling of the rhodopsins is employed to quantify the expression level and the absolute activity of designed mutants. We discuss strengths and pitfalls, associated with this expression system, and strategies for selecting potentially useful optogenetic tools.

Key words *Xenopus laevis* oocytes, Membrane extraction, cAMP, cGMP, In vitro assay, Enzyme rhodopsins, Cyclic nucleotides, Optogenetics

1 Introduction

The unfertilized oocytes from African clawed frog *Xenopus laevis* have been widely used as a heterologous expression system for about half a century [1]. The system can be easily manipulated by directly injecting cRNA (complementary RNA), transcribed from synthetic or engineered linear DNA with targeted exogenous gene fragments [2]. This enables us to study characteristics of heterologous proteins in detail after only 1–3 days post injection. In most cases, an injection of the exogenous cRNA at a range of 1–40 ng can readily produce detectable target proteins in each oocyte, thus being an economical approach. The diameter of *Xenopus* oocytes is

Valentin Gordeliy (ed.), *Rhodopsin: Methods and Protocols*, Methods in Molecular Biology, vol. 2501,
https://doi.org/10.1007/978-1-0716-2329-9_16,

normally >1 mm; hence, the injection process is simple, and electrophysiological experiments are easily performed, both under a basic binocular microscope. The first light-sensitive protein, expressed in oocytes, was a type II rhodopsin: bovine rhodopsin [3] followed by characterization of a type I rhodopsin: bacteriorhodopsin [4]. Later followed the discovery of the light-gated cation channels and type I rhodopsins—Channelrhodopsin 1 and 2 (ChR1, ChR2)—by expressing genes from the green alga *Chlamydomonas reinhardtii* [5, 6].

The first light-sensitive enzymes, characterized by expression in oocytes, were the flavoproteins PACα and PACβ [7]. These soluble proteins with adenylyl cyclase activity could be functionally expressed and partially characterized in vivo (in oocyte), by light activation of co-expressed cAMP-sensitive ion channels.

The first enzyme rhodopsin, identified from the fungus *Blastocladiella emersonii*, was demonstrated in vivo (in the fungus) to be a possibly light-activated guanylyl cyclase, named *Be*GC1 [8]. A direct proof for the light-activated guanylyl cyclase activity was obtained by expression in *Xenopus* oocytes [9, 10]. Extracting the membrane from oocytes combined with in vitro assays provided an efficient way to study in detail the parameters of this novel rhodopsin which was then named *B. emersonii* Cyclase opsin or *Be*Cyclop [9]. The in vitro assays yielded an action spectrum with a maximal activity at around 530 nm, a photocycle time of about 320 ms (at 20 °C), and a ratio of activity in light to activity in dark (L/D) of 5000 [9]. In this study also the first proof of cytosolic N- and C-terminal ends was given, by fusing two split YFP fragments at both ends and successful fluorescence in oocytes. The best explanation for the cytosolic N-terminus is an opsin architecture with 8 TM helices [9].

Already in 2004 light-sensitive adenylyl or guanylyl cyclases were proposed to exist from the analysis of sequences, found in the *C. reinhardtii* genome [11]. But only in 2018 the first such rhodopsins from *Volvox carteri* and *C. reinhardtii* were experimentally shown to be light-inhibited guanylyl cyclases [12]. Also for these rhodopsins, an 8 TM architecture was shown, and as they transmit the light signal to the guanylyl cyclase via a two-component system, they were named 2c-Cyclop (two-component cyclase opsin). Recently, the first bi-stable two-component cyclase opsin, named switch-Cyclop1, was successfully engineered with light-switchable activity [13]. In 2018 a light-sensitive rhodopsin phosphodiesterase (RhoPDE), again with an 8 TM architecture, was characterized after expression in oocytes [14]. The RhoPDE from *Salpingoeca rosetta* was not only characterized in HEK293 cells [15, 16] and *Xenopus* oocytes [14] but also verified to contain an 8 TM topology in a recent structure study [17]. Notably, other RhoPDEs were still identified from genomes of some other protists [18], and RhoPDE homologs from *Choanoeca flexa* sp. nov. were proved to play an important role under dark and illuminated conditions in inverting the curvature of cup-shaped colonies via a rhodopsin-cGMP signaling pathway [19]. Those RhoPDEs were recently characterized in HEK293 cells, with *Cf*Rh-PDE1, *Cf*Rh-PDE4, and *Mr*Rh-PDE exhibiting light-enhanced cGMP hydrolysis activity [20]. To engineer and improve light-gated RhoPDEs with more new strategies [21], in vitro characterization using *Xenopus* oocytes provides an efficient way to achieve useful optogenetic tools.

In vitro assays have several advantages: (1) The reaction conditions can be controlled, as membrane fragments can be directly mixed with reaction buffer including the substrates GTP, ATP, cGMP, or cAMP as required, and the mixture can be illuminated at defined intensity by optical fibers, connected to LEDs or lasers with various wavelengths. (2) The concentrations of cAMP or cGMP can be measured by immunoassay, and the properties of different enzyme rhodopsins, like sensitivity or action spectrum, can be analyzed in detail. (3) The protein expression levels can be directly manipulated by adjusting the amount of cRNA injected to oocytes. This allows us to fit the enzyme activity to a measurable range with controllable reaction time or membrane amount. (4) The wild-type of enzyme rhodopsins and corresponding mutants, fused with a YFP tag, can be easily detected and compared by membrane extraction from the same batch of oocytes. This is valuable in selecting candidate mutants for further application. (5) The co-expression in oocytes gives more opportunities to study two or more components in various combinations. For example, the light-activated guanylyl cyclase *Be*Cyclop can be combined with cGMP-sensitive CNG cation channels, thus demonstrating its activity in vivo [9, 10]. (6) As mentioned above, the oocyte system also enables us to examine the topology of transmembrane domains of enzyme rhodopsins by the BiFC (Bimolecular Fluorescence Complementation) experiment with split YFP [9]. So far, all studied enzyme rhodopsins show a cytosolic N-terminus and 8 TM (transmembrane) helices instead of an extracellular N-terminus and 7 TM, typical for the ion transporting and sensory rhodopsins.

In this chapter, we describe how to handle the oocytes for heterologous expression of enzyme rhodopsins. We depict how to extract the oocyte membrane fractions and operate in vitro reactions under the dark and illuminated conditions. This easily handled system paves the way to screen the natural enzyme rhodopsins or their mutants. They will generate useful optogenetic tools with efficient light/dark switchable control, thus regulating second messenger cGMP or cAMP for future application at spatiotemporal resolutions in different cell types, tissues, or living organisms.

2 Materials

2.1 Xenopus *Oocytes Preparation and Incubation*

1. Female *X. laevis* with unfertilized oocytes.
2. Sterilized scalpels, surgical scissors, and stainless steel point tweezer.
3. Tricaine methane-sulfonate (MS-222) for anesthesia (1 g/L).
4. Surgical monofilament non-absorbable nylon sutures.
5. Collagenase (5 mg/mL) for digestion, ND96 buffer without $CaCl_2$.
6. ND96 incubation buffer: 96 mM NaCl, 2 mM KCl, 1 mM $CaCl_2$, 1 mM $MgCl_2$, 5 mM HEPES and pH 7.4, 50 μg/mL gentamycin (*see* **Note 1**).
7. Bright field microscope for sorting and injecting oocytes.
8. 18 °C incubator for oocyte expression (mostly 1–3 days) and 4 °C refrigerator for longer incubation without or after expression.

2.2 cRNA Preparation and Microinjection

1. AmpliCap-MaxT7 High Yield Message Maker Kit or similar.
2. DEPC H_2O or RNase-free H_2O, 5 M ammonium acetate (CH_3COONH_4).
3. Linearized plasmid DNA.
4. Refrigerated benchtop centrifuge.
5. 37 °C incubator for in vitro transcription.
6. Micro-volume spectrophotometer.
7. Mineral oil with Sudan red.
8. Plastic petri dishes, 60 × 15 mm, 94 × 16 mm.
9. Vertical puller for glass capillaries (e.g., PP-83; Narishige).
10. Injection (glass) capillaries.
11. Drummond nanoinjector or similar.
12. ND96 buffer with 1–10 μM all-trans-retinal (*see* **Note 2**).

2.3 Fluorescence Check and Storage of Oocytes

1. 473 nm blue laser (Changchun New Industries Optoelectronics Tech).
2. 530 nm long pass optical filters to check YFP emission fluorescence.
3. Fluorescent microscope.
4. Eppendorf tubes, 1.5 mL.
5. 10-mL rubber head glass dropper with tip opening of ~2 mm diameter.

6. Liquid N_2.
7. Storage of samples at −80 °C freezer.

2.4 Extraction Buffer and Fluorescence Measurement from Oocytes

1. Extraction buffer: 100 mM NaCl, 75 mM Tris-HCl, 5 mM DTT, with or without 5 mM $MgCl_2$ as designed, 5% glycerol, and pH 7.4 (*see* **Note 3**).
2. Falcon conical centrifuge tubes 15, 50 mL.
3. Pipettors and pipet tips 20 μL, 200 μL, 1 mL.
4. Scientific Vortex G-560 E or similar.
5. Promega Quantus™ Fluorometer or similar.
6. YFP protein standard (1 mg/mL).

2.5 In Vitro Reactions and cAMP/cGMP Immunoassay

1. Reaction buffer: 100 mM NaCl, 75 mM Tris-HCl, 5 mM DTT, with or without 5 mM $MgCl_2$ as designed, pH 7.4 (*see* **Note 4**).
2. Microcentrifuge tubes, black, 1.5 mL.
3. Circulating water bath, countdown timer.
4. Stop solution, 0.1 M HCl.
5. DetectX® direct (or similar) cyclic GMP or cAMP enzyme immunoassay kit.
6. Costar 96-well EIA/RIA strip well plate.
7. 96-well plate shaker, microplate mixer, 0–1500 rpm.
8. Multiskan ascent (or similar) 96 plate reader, absorbance at 450 nm.

2.6 Adjustment of Illumination Conditions

1. 532, 596, 635 nm, or other lasers (e.g., Changchun New Industries Optoelectronics Tech).
2. LEDs with different wavelengths such as 380, 505, 635 nm.
3. White light source, e.g., PhotoFluor II (89North), narrow bandwidth interference filters (e.g., from Edmund Optics) at different wavelengths including 365, 422, 460, 497, 517, 541, 563, 568, 580, 600, 641, 658, 714 nm.
4. Spectrometer (e.g., Ocean Optics).
5. Laser check power meter (Coherent Technologies or similar), or Plus 2 optical power and energy meter (LaserPoint 281 s.r.l, Italy).
6. Optical fibers, rubber plug.
7. Multichannel power supplies 0–3 V.

3 Methods

3.1 Preparation of X. laevis Oocytes

1. Female *X. laevis* are raised for the harvest of unfertilized oocytes. Before surgical processes, sterilize instruments such as scalpels, scissors, and tweezers in advance.
2. Prepare several liters of ice-cold water with the concentration of 1 g/L tricaine methane-sulfonate (MS-222) to anesthetize a frog for 30 min. Manipulate the surgical process on ice covered with tricaine solution-soaked tissue papers. Place the frog with abdomen on the top side and operate a 1 cm long incision by scalpel and scissor under the muscle in the abdomen skin. From the ovary of the frog, strip out a certain number of oocytes and cut by scissor.
3. Put the oocytes in ND96 buffer (plus 50 μg/mL gentamycin) without $CaCl_2$ in 50-mL falcon tube. Sew the incision by surgical monofilament non-absorbable nylon sutures. Prepare fresh water for the recovery of anesthetized frog.
4. Wash the oocytes by ND96 buffer without $CaCl_2$ two times and digest for 1–1.5 h with 5 mg/mL collagenase (*see* **Note 5**). Then add the ND96 buffer (with 1 mM $CaCl_2$) and wash oocytes until they are clean enough to keep in culture dishes. Put the oocytes in a 16–18 °C incubator for next steps.

3.2 In Vitro Transcription and cRNA Microinjection

1. Linearize pGEMHE vectors with targeted DNA fragments by the specific NheI restriction enzyme. Do in vitro transcription, based on the instructions from manufacturer, and mix with a few aliquots of linearized DNA with total amount of 0.6–1 μg. Adjust the final mixed volume to 10 μL and incubate the mixture at 37 °C for 3 h.
2. After incubation, add 12 μL 5 M ammonium acetate (CH_3COONH_4) to the mixture and freeze at −20 °C for 1 h. Collect the pellet after centrifugation at 4 °C, 20,000 × *g* for 1 h and wash the pellet with 50 μL 70% ethanol (prepared in DEPC H_2O or RNase-free H_2O). After centrifugation for 1 h, discard the supernatant and place the tube with pellet in a fuming cupboard to dry.
3. Add 10–15 μL RNase-free H_2O to dissolve the pellet and mix thoroughly on ice. Detect the RNA concentration by Nano-Photometer. Then store the samples at −20 °C for further use (*see* **Note 6**).
4. Prepare injection capillaries via a vertical Puller. Under a microscope with an ocular lens wide field scale line, break each tip of the capillaries to obtain a sharp end with an opening of 30–40 μm diameter. Fill the capillaries with red-colored mineral oil by a syringe needle, and avoid the air bubble inside.

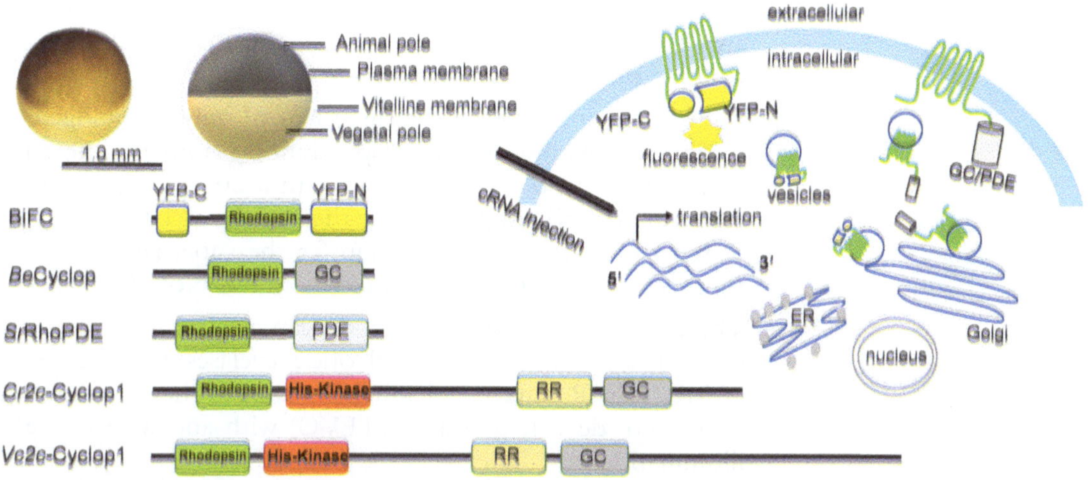

Fig. 1 Scheme, depicting expression of different enzyme rhodopsins in *X. laevis* oocytes. Enzyme rhodopsins *Be*Cyclop, *Sr*RhoPDE, *Cr*2c-Cyclop1, and *Vc*2c-Cyclop1 are shown with various domains indicated by different colored boxes. *GC* guanylyl cyclase, *PDE* phosphodiesterase, *RR* response regulator. The BiFC construct has the YFP-C (155-238 aa of YFP) in the N-terminal and YFP-N (1-154 aa of YFP) in the C-terminal, and the Rhodopsin is fused in between. The processes of cRNA injection and heterologous expression in oocytes are shown in the right cartoon. After the cRNA translation, membrane proteins are trafficking to the plasma membrane in the vesicles generated from ER and Golgi. All enzyme rhodopsins showed eight transmembrane (8 TM) topology, and both the N- and the C-terminals locate in the cytosolic side of different membranes

5. Put the oocytes in a grooved plastic chamber. Fill cRNA into the capillary and inject into *Xenopus* oocytes by Drummond Nanoinjector or similar. Adjust the amount of injected RNA to 1–40 ng in total. Regulate the injected volume at a range of 1–50 nL for each oocyte. For co-expression systems, design and mix individual RNA samples in certain ratios (*see* **Note 7**).
6. After cRNA microinjection, incubate oocytes with ND96 buffer in a 16–18 °C incubator for 2–3 days. The process of heterologous expression is depicted in Fig. 1. Add ATR (all trans retinal) directly in ND96 buffer to final 1–10 μM under red light 635 nm illumination in a dark room. After incubation, remove the incubated buffer completely and freeze oocytes with liquid N_2 and store at −80 °C for further tests (*see* **Note 8**).

3.3 In Vivo Assay with Xenopus Oocytes

1. After cRNA injection, put the oocytes in a plastic petri dish, covered with aluminum foil to make sure the incubation proceeds under a complete dark condition. Change the fresh ND96 buffer with ATR each day under red light illumination in a dark room, and meanwhile remove the bad oocytes.
2. After 2–3 days incubation, start the illumination treatment of oocytes in a closed container, covered by a metal plate with LED light. Record the illumination time and measure the light

power in a certain distance, the same as that of between oocytes and LED. Light treatment can also be done within fiber-coupled tubes after transferring oocytes inside.

3. Aliquot the oocytes evenly in three 1.5-mL tubes, homogenize the oocytes within 200 μL stop solution (sample dilute containing 0.1 M HCl) by pipetting with a 200-μL pipette after removing the ND96. Centrifuge the samples at 10,000 × *g* for 5 min to remove sediment. Transfer the supernatant to new tubes for cGMP or cAMP measurement (*see* **Note 9**).
4. The cAMP and cGMP production can also be monitored by co-expressing light-gated nucleotide cyclases together with cAMP/cGMP-gated channels and the following two-electrode voltage clamp (TEVC) with and without light illumination.

3.4 BiFC (Bimolecular Fluorescence Complementation) Experiment

1. Perform sequence alignments among enzyme rhodopsins and 7 TM channelrhodopsins in online servers such as TMHMM, JPred4 [22–24]. An additional helix (named TM-0) of enzyme rhodopsins is predicted.
2. In each construct for BiFC test, fuse the N-terminal YFP fragment 1–154 aa to the C-terminal end of the rhodopsin domain, and fuse the C-terminal YFP fragment 155–238 aa to the N-terminus of the rhodopsin domain (*see* **Note 10**) (Fig. 1).
3. After in vitro transcription, inject 30 ng cRNA into each oocyte for 3 days expression. Adjust the excitation and emission wavelengths to 515 and 535 nm, respectively. Detect YFP fluorescence imaging of oocytes in a confocal microscope (Leica DM6000).

3.5 Extraction of the Crude Membranes from Xenopus Oocytes

To characterize enzyme rhodopsins, extract the membranes of oocytes after the heterologous expression for in vitro assay (Fig. 2). The extraction processes are depicted as follows:

1. Collect a certain number of oocytes in a 1.5-mL tube with 500 μL extraction buffer and homogenize by pipette (*see* **Note 11**). Then vortex the samples for 30 s and put them on ice for 30 s, with 2–3 times repetition for thorough homogenization.
2. Centrifuge the homogenous sample at 500 × *g* for 15 min at 4 °C to remove the cell debris. Then transfer the supernatant to a new tube and centrifuge at 30,000 × *g* for another 15 min. Discard the supernatant and gently wash the pellet with 500 μL extraction buffer. Repeat the washing and centrifugation steps for two times, and remove the lipid on the top of supernatant.

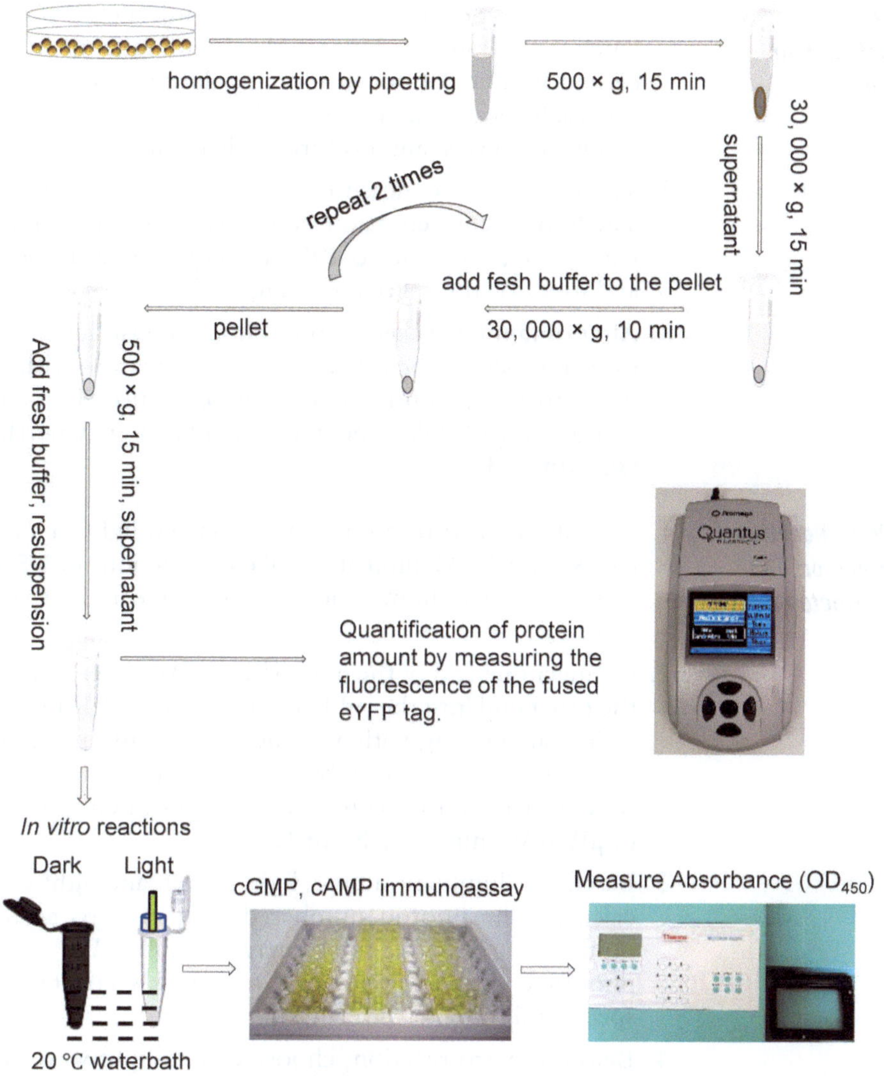

Fig. 2 Crude membrane extraction and in vitro reactions with membranes from *X. laevis* oocytes. The upper part shows the procedure of crude membrane extraction from oocytes. The target protein amount in the crude membrane extract is quantified by a fluorometer. After in vitro reactions in the dark or in the light conditions, cGMP or cAMP concentrations are measured by ELISA kits

3. After washing steps, discard the supernatant and suspend the pellet with a certain volume of extraction buffer. For example, suspend enriched membrane from 1 oocyte with 2–10 μL extraction buffer for quantitative in vitro assay afterwards.
4. Centrifuge the suspended membrane at 500 × *g* for 15 min to remove the large fragments. Transfer the membrane fraction in the supernatant to a new tube for next use (*see* **Note 12**).

3.6 Protein Quantification with YFP Fusion

1. Prepare the oocytes without cRNA injection from the same batch as a control. Extract the membrane fraction from a certain number of oocytes for the fluorescence measurement. For each measurement, select 1–10 oocytes, based on actual expression levels, and do three individual tests for each group.
2. The samples were measured with the Quantus™ Fluorometer. The fluorescence emission value was measured at 510–580 nm with excitation of up to 495 nm. YFP standard was measured and used for the calibration (Fig. 2).
3. After obtaining cyclase or phosphodiesterase activity, the turnover number can be calculated in combination with the quantified protein amount based on fluorescence. This enables the comparison of wild-type and mutants of enzyme rhodopsins (*see* **Note 13**).

3.7 In Vitro Reaction of Membrane or Cytosol Extracted from Oocytes

1. Operate in vitro reaction tests under red light in a dark room (*see* **Note 14**). Manipulate total dark reactions in 1.5-mL black tubes, while the illumination tests are done in transparent tubes (Fig. 2).
2. Connect one end of the fiber with LED or laser sources, and fix the other end into the rubber plug on top of 1.5-mL tubes. Cut a circular opening with a diameter of 4 mm in the bottom of tube. Detect light power by a power meter from the bottom of tube. Calculate the corresponding light intensities to the unit of μW mW/mm^2 (*see* **Note 15**).
3. Based on illumination time, light power, and light wavelength, calculate the photon number during in vitro assay. Photon number = E_{total}/E, total energy $E_{total} = W * s$ (J); single photon energy $E = hc/\lambda$; h = Planck's constant, c = the speed of light, λ = wavelength.
4. Before in vitro reaction, choose certain light wavelengths and set the proper light intensities. For membrane samples, the reaction mixture contains membrane fraction and reaction buffer in a volume ratio of 1:9. For different enzyme rhodopsins, change the substrates as required. For example, use 0.25 mM ATP and 0.2 mM GTP to test the activity of 2c-Cyclops. Employ the final substrate 1–100 μM cGMP-Na or cAMP-Na individually to test RhoPDEs activity. Operate reactions in a 20 °C water bath and record the reaction time under light and in the dark.

3.8 cGMP and cAMP Measurement

1. Measure cGMP or cAMP concentrations by an immunoassay kit (Fig. 2). Mix 50 μL diluted samples with 25 μL cGMP/cAMP-peroxidase conjugate and 25 μL monoclonal antibody specific for either cyclic nucleotide in Costar 96-well plate. Incubate the plate in a 96-well plate shaker at 150 rpm for 2 h.

2. After incubation, wash the plate four times with washing buffer and dry it on tissue papers. Add 100 μL substrate and incubate the plate for another 30 min in quiescence. Then add 50 μL stop solution to the plate wells and detect absorbance at 450 nm wavelength by a 96/384 plate reader.
3. Dilute the samples for the accurate measurement according to preliminary experiments. The standard curves of our assay kit range from 0.5 to 32 nM for cGMP and 0.56 to 150 nM for cAMP. Record the OD450 values of samples from the plate reader and read out the concentrations from the standard curve (*see* **Note 16**).

4 Notes

1. The fresh ND96 incubation buffer should be changed each day, and bad oocytes are removed from petri dishes in order to keep oocytes in a good state. After 2–3 days incubation, oocytes with YFP-tagged proteins are selected, based on the fluorescence. Remove oocytes with weak or no expression. Then the oocytes with obvious expression can be further used in the in vitro assay.
2. After RNA injection, all-trans-retinal (ATR) is added into the ND96 buffer to final range of 1–10 μM. This should be operated in a dark room with red light (e.g., 635 nm) in case of changing the configuration of ATR. The quality of ATR should be checked every month to make sure the absorbance spectrum still peaks at 380 nm. This will ensure enzyme rhodopsins can be formed from opsins and are functional.
3. In extraction and reaction buffer, fresh DTT should be added to final 5 mM to preserve enzyme stability. Based on our data, the activity of enzyme rhodopsins can be dramatically reduced at a lower concentration of NaCl. Normally, 100–300 mM NaCl gives the best light-regulated efficiency.
4. In reaction buffer, additional substrates contain 2 mM GTP for *Be*Cyclop, both 0.25 mM ATP and 0.2 mM GTP for 2c-Cyclop, 1–100 μM cGMP-Na or cAMP-Na for RhoPDE. The fresh ATP or GTP stock solution should be added right before the reaction, and the storage time at −20 °C should not exceed half a year, to avoid the reduction of enzyme activity or light regulation. Sometimes, the concentrations of substrates cGMP or cAMP should be regulated in a reasonable range to observe the light regulation. For example, a fully saturating concentration of 1 mM cGMP makes the dark activity of *Sr*RhoPDE similar to its light activity. However, when the concentrations were lower, such as 100 μM or even below

10 μM, the light regulation was clearly observed. This allows us to distinguish different substrate affinities of *Sr*RhoPDE under light and dark conditions [14].

5. The quality of oocytes should be confirmed after surgical processes in *X. laevis*. The collagenase digestion time should be controlled in the range of 1–1.5 h to remove follicle cells and blood vessels, releasing individual oocytes. This stage needs to be checked in time, because shorter or longer time for the digestion can influence the quality of oocytes for further micro-injection and heterologous expression. Do not add $CaCl_2$ solution in this process to avoid influencing the collagenase digestion efficiency. After digestion, $CaCl_2$ stock solution should be added to final 1 mM in the incubation ND96 buffer. Good oocytes should be sorted out to a new petri dish in the same day. This will keep oocytes longer time during incubation.
6. In in vitro transcription, cRNA quality should be carefully confirmed. Linearized DNA should be dissolved by RNase-free H_2O or DEPC H_2O after digestion. This will avoid newly generated cRNA to be degraded. When comparing wild-type with its mutants, we often choose to use the same batch of transcribed cRNA samples. The originally transcribed cRNA samples can be divided into two tubes, one for long-term storage and the other for immediate use.
7. The heterologous protein expression levels are largely depending on the amount of cRNA injection. But the total cRNA injected should not exceed 40 ng for each oocyte. The cRNA injection needs to be kept of the same volume when comparing enzyme rhodopsins, and the total volume should be no more than 50 nL. The co-expression system needs to mix RNA samples in a designed ratio for one-time injection.
8. All the ND96 buffer has to be aspirated out before oocytes are frozen in liquid nitrogen. This will prevent degradation of heterologous proteins and reduction of enzyme activities.
9. The "in vivo" method has the advantage of being fast and easy to check for enzyme activity. However, due to the long time incubation in the dark for protein expression, the cAMP and cGMP produced in the dark will be accumulated and further modulated by endogenous PDE. In this case, the dark activity cannot be reliably quantified. Furthermore, the protein amount cannot be accurately quantified with this method.
10. The BiFC assay is employed to visualize interaction between two proteins or protein domains [25]. Here we chose to use two truncated YFP fragments at residue 155 (YFP-N 1-154 and YFP-C155-238), which were recommended [26]. The two split YFP parts fused to both termini of the enzyme rhodopsin domain produced bright fluorescence signals in

Xenopus oocytes, indicating that enzyme rhodopsins such as *Be*Cyclop, *Cr*2c-Cyclop1, *Vc*2c-Cyclop1, COP5 (HKR) and *Sr*RhoPDE share similar 8 TM topology [9, 12, 14].

11. In the extraction process, oocytes can be directly homogenized by pipette instead of sonication. Short cycles of vortexing and keeping on ice tend to keep enzymatic activity stable.
12. After extraction processes, the fluorescence from a certain number of oocyte membranes should be measured immediately. This enables us to calculate how much membrane and incubation time would be used in in vitro reactions from the preliminary data.
13. To select useful enzyme rhodopsins as optogenetic tools, some mutants would be engineered in order to enhance their light-regulated efficiencies. YFP tag is normally fused with wild type or mutants of enzyme rhodopsins for the detailed comparison between the fluorescence and enzyme activities. This enables us to find out which mutant becomes more promising for the future application. It is notable that the N- or C-terminal YFP fusion to enzyme rhodopsins can yield quite different results, for example, YFP::RhoPDE exhibited a better expression level as well as light-regulation than RhoPDE::YFP [14].
14. In vitro tests should be operated in a dark room with red light (e.g. 635 nm). This will exclude the influence of blue/green photons in white light. Transparent tubes can be used when using only one wavelength of light. However, black tubes have to be employed when multiplexing light sources are used to avoid interactive influence at the same time.
15. The light wavelengths of LEDs should be confirmed by a spectrometer. Before in vitro tests, light intensities should be measured at least three times to confirm the stability. The fibers have to be fixed stably between light source with screws and tubes with a suitable rubber plug. After in vitro reaction, the intensity should be measured again to confirm the stability of the light source. Other types of optical fibers or light sources should be chosen if the intensity is not strong enough.
16. To make the measurement more accurate, the OD450 values of samples should locate in the middle range of the standard curve. In order to avoid variabilities, cGMP or cAMP standards should be prepared together with samples for each measurement.

References

1. Gurdon JB, Lane CD, Woodland HR et al (1971) Use of frog eggs and oocytes for the study of messenger RNA and its translation in living cells. Nature 233:177–182
2. Wagner CA, Friedrich B, Setiawan I et al (2000) The use of Xenopus laevis oocytes for the functional characterization of heterologously expressed membrane proteins. Cell Physiol Biochem 10:1–12

3. Knox BE, Khorana H, Nasi E (1993) Light-induced currents in Xenopus oocytes expressing bovine rhodopsin. J Physiol 466:157–172
4. Nagel G, Möckel B, Büldt G et al (1995) Functional expression of bacteriorhodopsin in oocytes allows direct measurement of voltage dependence of light induced H+ pumping. FEBS Lett 377:263–266
5. Nagel G, Ollig D, Fuhrmann M et al (2002) Channelrhodopsin-1: a light-gated proton channel in green algae. Science 296: 2395–2398
6. Nagel G, Szellas T, Huhn W et al (2003) Channelrhodopsin-2, a directly light-gated cation-selective membrane channel. Proc Natl Acad Sci 100:13940–13945
7. Schröder-Lang S, Schwärzel M, Seifert R et al (2007) Fast manipulation of cellular cAMP level by light in vivo. Nat Methods 4:39–42
8. Avelar GM, Schumacher RI, Zaini PA et al (2014) A rhodopsin-guanylyl cyclase gene fusion functions in visual perception in a fungus. Curr Biol 24:1234–1240
9. Gao S, Nagpal J, Schneider MW et al (2015) Optogenetic manipulation of cGMP in cells and animals by the tightly light-regulated guanylyl-cyclase opsin CyclOp. Nat Commun 6: 1–12
10. Scheib U, Stehfest K, Gee CE et al (2015) The rhodopsin–guanylyl cyclase of the aquatic fungus Blastocladiella emersonii enables fast optical control of cGMP signaling. Sci Signal 8:rs8
11. Kateriya S, Nagel G, Bamberg E et al (2004) "Vision" in single-celled algae. Physiology 19: 133–137
12. Tian Y, Gao S, von der Heyde EL et al (2018) Two-component cyclase opsins of green algae are ATP-dependent and light-inhibited guanylyl cyclases. BMC Biol 16:1–18
13. Tian Y, Nagel G, Gao S (2021) An engineered membrane-bound guanylyl cyclase with light-switchable activity. BMC Biol 19:1–9
14. Tian Y, Gao S, Yang S et al (2018) A novel rhodopsin phosphodiesterase from Salpingoeca rosetta shows light-enhanced substrate affinity. Biochem J 475:1121–1128
15. Yoshida K, Tsunoda SP, Brown LS et al (2017) A unique choanoflagellate enzyme rhodopsin exhibits light-dependent cyclic nucleotide phosphodiesterase activity. J Biol Chem 292: 7531–7541
16. Lamarche LB, Kumar RP, Trieu MM et al (2017) Purification and characterization of RhoPDE, a retinylidene/phosphodiesterase fusion protein and potential optogenetic tool from the choanoflagellate Salpingoeca rosetta. Biochemistry 56:5812–5822
17. Ikuta T, Shihoya W, Sugiura M et al (2020) Structural insights into the mechanism of rhodopsin phosphodiesterase. Nat Commun 11: 1–12
18. Richter DJ, Fozouni P, Eisen MB et al (2018) Gene family innovation, conservation and loss on the animal stem lineage. elife 7:e34226
19. Brunet T, Larson BT, Linden TA et al (2019) Light-regulated collective contractility in a multicellular choanoflagellate. Science 366: 326–334
20. Sugiura M, Tsunoda SP, Hibi M et al (2020) Molecular properties of new enzyme rhodopsins with phosphodiesterase activity. ACS Omega 5:10602–10609
21. Tian Y, Yang S, Gao S (2020) Advances, perspectives and potential engineering strategies of light-gated phosphodiesterases for optogenetic applications. Int J Mol Sci 21:7544
22. Möller S, Croning MD, Apweiler R (2001) Evaluation of methods for the prediction of membrane spanning regions. Bioinformatics 17:646–653
23. Krogh A, Larsson B, Von Heijne G et al (2001) Predicting transmembrane protein topology with a hidden Markov model: application to complete genomes. J Mol Biol 305:567–580
24. Drozdetskiy A, Cole C, Procter J et al (2015) JPred4: a protein secondary structure prediction server. Nucleic Acids Res 43:W389–W394
25. Kerppola TK (2006) Design and implementation of bimolecular fluorescence complementation (BiFC) assays for the visualization of protein interactions in living cells. Nat Protoc 1:1278–1286
26. Kerppola TK (2008) Bimolecular fluorescence complementation (BiFC) analysis as a probe of protein interactions in living cells. Annu Rev Biophys 37:465–487

Chapter 17

Optogenetic Control of Human Stem Cell-Derived Neurons

Rouhollah Habibey, Johannes Striebel, Kritika Sharma, and Volker Busskamp

Abstract

Spontaneous and optogenetically evoked activities of human induced pluripotent stem cell (hiPSC)-derived neurons can be assessed by patch clamp and multi-electrode array (MEA) electrophysiology. Optogenetic activation of these human neurons facilitates the characterization of their functional properties at the single neuron and circuit level. Here we showcase the preparation of hiPSC-derived neurons expressing optogenetic actuators, in vitro optogenetic stimulation and simultaneous functional recordings using patch clamp and MEA electrophysiology.

Key words Induced pluripotent stem cells (iPSCs), Optogenetic stimulation, Patch clamp, Multi-electrode array (MEA) electrophysiology, Long-term iPSC-derived neuronal culture, Banker culture, Neuron-astrocyte co-culture

1 Introduction

Human-induced pluripotent stem cells (hiPSCs) are generated by reprogramming mature fibroblasts into pluripotent stem cells using the so-called Yamanaka factors (Oct3/4, Sox2, Klf4, c-Myc) or combination of other transcription factors like Oct4, Sox-2, Nanog, and Lin-28 [1]. Human iPSCs can be derived from patients or healthy subjects. Due to their high proliferation rates, hiPSCs offer an unlimited resource for programming somatic cell types including neurons. Using hiPSCs also reduces ethical concerns as opposed to embryonic stem cells (ESCs) [2, 3].

Stem cell-derived neurons are currently used in both basic and biomedical research, for instance, for modeling brain circuitry and neurological diseases in vitro. These cells are also used for cell replacement therapeutic approaches [4, 5]. To this end, a deep understanding of biophysical and functional properties of these iPSCs-derived neurons at the single cell and network level is

Valentin Gordeliy (ed.), *Rhodopsin: Methods and Protocols*, Methods in Molecular Biology, vol. 2501, https://doi.org/10.1007/978-1-0716-2329-9_17,

important. Neuronal activity can be recorded by patch clamp and multi-electrode array (MEA) electrophysiology methods [6–8].

Combining electrophysiology recordings with optogenetic tools provides a further flexible platform for simultaneous activation/inhibition and recording of single neurons or networks. Rhodopsin-based optogenetic actuators are light-gated ion channels or light-driven ion pumps that either activate neurons by depolarization or inhibit neuronal activity by inducing hyperpolarization [8, 9]. Optogenetic tools offer high temporal and spatial precision on controlling neuronal activity and targeting specific neuronal subtypes. There is a steady expansion of the optogenetic toolbox providing a versatile resource for a broad range of applications in the network neuroscience [10, 11]. For instance, the introduction of excitatory and inhibitory actuators into neurons in vivo or in vitro enables functional connectivity studies to reveal the architecture of neuronal networks [12, 13]. In addition, optogenetic tools can be exploited for tracking the functional maturation of the hiPSC-derived neurons [14–17]. Here, we demonstrate detailed and stepwise guide for in vitro recordings of hiPSC-derived neurons and networks expressing optogenetic actuators by patch clamp and MEA electrophysiology.

2 Materials

2.1 In Vitro Expression of Optogenetic Actuators

The expression of optogenetic actuators in hiPSC-derived neuronal cells in vitro is achieved by inserting the corresponding genes via viral vectors or non-viral methods such as the piggyBac system. All technical aspects for a robust optogenetic delivery to human stem cell-derived neurons have been previously described in detail [18]. Viral particles, for example, lentiviruses transferring the genetic information for a desired optogenetic actuator, are added to the culturing media leading to their integration into the host genome. After an incubation period of 24–48 h, the optogenetic actuator is expressed (*see* **Note 1**). Whereas viral vectors efficiently transduce stem cell-derived neurons [3], the piggyBac system without the need of an biosafety level 2 facility [18] can also be used to generate transgenic hiPSC lines conditionally expressing optogenetic actuators. A continuous expression of optogenetic tools in hiPSCs is not stable over many passages [6].

2.2 Long-Term Neuronal Culture

1. **Human iPSC cell line** (e.g., iNGN cell line; this is a hiPSC line in which the two neurogenic transcription factors Neurogenin-1 and Neurogenin-2 are driven by the doxycycline-inducible TetON promoter, leading to postmitotic neurons in 4 days post induction (dpi) [1, 18]).

2. **Optogenetic actuator**: Lentiviral particles including the genes for the desired optogenetic actuators have been already prepared or purchased and are ready to use [18] (*see* **Notes 1** and **2**).
3. **Matrigel-coated plates:** Thaw Matrigel stock solution on ice and aliquot it based on dilution factor provided on each bottle. We use Matrigel hESC-Qualified Matrix (Corning) and prepare 12 mL of coating solution each time. Coating solution includes 12 mL Dulbecco's Modified Eagle Medium (DMEM) + GlutaMAX I + HEPES, one aliquot of Matrigel, and 1% Penicillin-Streptomycin solution. 1 mL of the coating solution is added to each well of a six-well plate and incubated for 1 h at room temperature. Coated plates can be stored at 4 °C for 2 weeks (Fig. 4).
4. **PBS without calcium and magnesium:** Prepare aliquots of 1× PBS without calcium and magnesium at 50 mL falcon tubes and store at room temperature.
5. **mTeSR1 medium:** 400 mL mTeSR1 Basal medium, 100 mL mTeSR1 5× Supplement, 1% Penicillin-Streptomycin solution, and if needed, 3.3 μg/mL ROCK Inhibitor. Sterile-filter (0.45 μm) and store at 4 °C for a maximum of 2 weeks.
6. **TrypLE Express:** Aliquot and store TrypLE Express (Thermo Fisher Scientific) in 15 mL falcon tubes at room temperature.
7. **Doxycycline:** Dissolve 10 mg doxycycline hycalate powder in 20 mL PBS without calcium and magnesium, sterile-filter using 0.22 μm filters and store aliquots at −20 °C. After thawing the doxycycline solution store it in 4 °C. Doxycycline either stored or upon usage should be protected from light.
8. **mTeSR + doxycycline:** For differentiation, add 1 μL of 0.5 mg/mL doxycycline solution to 1 mL mTesR1.
9. **Poly-D-Lysine (PDL):** Prepare aliquots of a 0.1 mg/mL PDL solution.
10. **Laminin**: 50 μg/mL laminin. *See* Subheadings 3.3 and 4.
11. **Complete BrainPhys medium**: 10 mL of BrainPhys Neuronal Medium, 200 μL NeuroCult SM1 Neuronal Supplement, 100 μL N2 Supplement-A (Stemcell Technologies), 20 μL of 10 μg/mL Recombinant Human Brain-Derived Neurotrophic Factor (BDNF, Peprotech), 20 μL of 10 μg/mL Recombinant Human Glial-Derived Neurotrophic Factor (GDNF, Peprotech), 50 μL of 40 mM ascorbic acid with final concentration 200 nM, and 1% Penicillin-Streptomycin. Complete BrainPhys medium is stored at 4 °C for a maximum of 2 weeks. Aliquots of N2 supplement-A and NeuroCult SM1 Neuronal Supplement are stored at −20 °C and aliquots of GDNF, BDNF, and ascorbic acid are stored at −80 °C.

12. **Ara-C**: 11 mg Cytosine β-d-arabinofuranoside hydrochloride powder is dissolved in 15 mL ddH_2O to reach 2.5 mM concentration. Sterile-filtered Ara-C solution is aliquoted and stored at −20 °C. At 4 dpi, 2 μL of Ara-C is added to 1 mL of complete BrainPhys medium (Fig. 4).
13. **Rat primary astrocytes**: Rat primary astrocytes (Gibco) are expanded, frozen, and thawed for several experiments. Each aliquot is propagated to generate many secondary batches sufficient for multiple experiments (Fig. 4).
14. **Astrocyte medium**: 10 mL DMEM + 4.5 g/L Glucose + Pyruvate, 100 μL N2 Supplement-A, 10% Fetal Bovine Serum, and 1% Penicillin-Streptomycin (*see* **Note 3**).
15. **Paraffin-dot coverslips**: 18 mm coverslips are prepared with three paraffin dots on their borders (*see* **Note 4**) to work as a spacer between astrocyte culture on coverslip and neuronal culture on MEA substrate (Fig. 4).
16. **Accutase**: Accutase is thawed and stored as 1 mL aliquots at −20 °C.

2.3 Patch Clamp Recording

2.3.1 Extracellular and Intracellular Solutions

1. **Extracellular solution** is prepared with carbogen (95% O_2–5% CO_2) bubbling or without bubbling but using HEPES to buffer the pH. Composition of extracellular solution with carbogen bubbling includes: 126 mM NaCl, 3 mM KCl, 2 mM $MgSO_4$, 2 mM $CaCl_2$, 1.25 mM NaH_2PO_4, 26.4 mM $NaHCO_3$, 10 mM glucose. Composition of extracellular solution with HEPES buffer includes: 130 mM NaCl, 20 mM HEPES, 10 mM glucose, 5 mM KCl, 2.5 mM $CaCl_2$, 1 mM $MgCl_2$, and NaOH to adjust the pH to 7.4 . The osmolarity of extracellular solution is kept between 305 and 315 mosm. This solution can be kept for 1 week at 4 °C.
2. **Intracellular solution** for patch pipette includes: 125 mM K-Gluconate, 2 mM $CaCl_2$, 2 mM $MgCl_2$, 10 mM EGTA, 2 mM Na_2-ATP, 0.5 mM Na_2-GTP. Alternatively, the following combination can be used as intracellular solution: 130 mM KCl, 5 mM NaCl, 0.4 mM $CaCl_2$, 1 mM $MgCl_2$, 10 mM HEPES, and 11 mM EGTA. The pH is adjusted to 7.3 using KOH. Optimal osmolarity of intracellular solution is between 260 and 280 mosm. After preparation, the solution must be filtered (0.2 μm) and stored at −20 °C. A microloader tip is used to fill the pipette.

2.3.2 Patch Clamp Recordings

1. **Anti-vibration air floating table** to isolate microscope and patch clamp headstage and recording pipette from ambient vibrations which is critical for obtaining and preserving a stable seal between the electrode and the cell (*see* **Note 5**).

2. **Microscope and Camera** for simultaneous visualization of cell and patching electrode (*see* **Note 6**).
3. **Faraday cage** to shield from surrounding artifacts (*see* **Note 7**).
4. **Micromanipulator** to control the position of the pipette in *X*, *Y*, and *Z* axes.
5. **Headstage and pipette holder** to attach patch pipette and transfer the electrical signal from the recording pipette to the amplifier (*see* **Note 8**).
6. **Pressure controller and sensor** to apply positive and negative pressures through the recording pipette at different steps of the recording (*see* **Note 9**). A pressure controller tubing is connected to the patch clamping pipette (Fig. 3). The other end of the pressure controller tubing is connected to a 5 or 10 mL syringe through a three-way valve. After applying positive or negative pressures by syringe, the valve is closed to provide a constant pressure in the patch clamp pipette.
7. **Amplifier and data interface** to measure the subtle changes in membrane potential and current flow. Amplifier is coupled to the digitizer to convert the analog data into digital (AD) data for further processing. Other experimental data like command potential is converted from digital to analog (DA) to be delivered to the pipette (*see* **Note 10**).
8. **Perfusion system**: The perfusion system consists of an extracellular solution reservoir, tubings, a three-way valve, an air bubble trapper, a flow controller, and a carbogen (95% O_2 and 5% CO_2) bubbling system (optional) plus a draining system (*see* Fig. 3 and **Note 11**).
9. **Computer interface**: Software modules for patch clamp include a controlling, a recording, and an analyzing software (*see* **Note 12**).
10. **Light source and path for optogenetics**: The required optical setups for optical stimulation are described in Subheading 2.5. In general, LED or laser light sources with appropriate wavelengths, depending on optogenetic actuator (e.g., 470 nm for channelrhodopsin-2 (ChR2)), are guided through the objective or an optic fiber to the sample (*see* **Note 13**).
11. **Glass capillary puller and glass capillaries**: A programmable pipette puller system (e.g., Sutter Instruments) to generate patch pipettes with reproducible parameters such as pipette resistance must be used. Required parameters for the glass capillaries for whole-cell recording as well as the pipette pulling parameters are provided within the pipette puller user manual. An optimal pipette for cultured human stem cell-derived neurons has a 2–3 μm tip, a 3–4 mm taper with a sharp cone angle and is generated using a thick-walled Borosilicate glass and a

2.5 or 3.0 mm box filament (*see* **Notes 14** and **15**). Alternatively, pre-pulled pipettes can be acquired from specialized vendors.

2.4 MEA Electrophysiology

1. **MEA chips**: Standard MEA chips (e.g., Multichannel Systems) composed of more than 60 electrodes with different sizes (10–30 μm), configurations (square, rectangular or other), and electrode pitch (100–500 μm) (Fig. 5 and **Note 16**).
2. **Headstage**: This unit amplifies and digitizes the recorded signal, hosts the MEA chip, and controls the temperature during the recording period (Fig. 5).
3. **Interface board**: Includes several analog and digital inputs and outputs and receives data from the headstage. The interface board includes a digital signal processor with the capacity to be freely programmed for real-time signal detection and designing feedback commands. Signals from the optical light source (timing of the light activation) can be logged within the recording files generated from the interface board (Fig. 5).
4. **Data acquisition system**: Commercially available software (provided by supporting company, e.g., Multichannel Systems) is used for visualization of data simultaneously from all recording electrodes as well as analog or digital channels. The user can adjust the sampling rate (*see* **Note 17**).
5. **Analyzing software**: Recorded extracellular activity is detected online or offline and extracted and used for evaluation of the data based on signal waveform and timestamps (*see* Subheading 3.4.1).
6. **Light source and path for optogenetics**: The same device that is used for stimulation during patch clamp recording is used for the MEA system (Fig. 3, *see* Subheading 2.5 and **Notes 13** and **18**).

2.5 Optogenetic Stimulation Device

1. **Wide-field stimulation (basic setup)**: A fluorescence microscope equipped with the *adjusted filter sets* and an *adequate light source*. Often, a collimated light source is connected through direct coupling or via a light guide (or optic fiber) to the back epi-fluorescence port [19, 20] (Fig. 1).
2. **Patterned stimulation:** Simultaneous stimulation of different cells in parallel is achieved by modulating the light intensity or the phase of the light beam. Required components are collimated light from a laser diode (e.g., **450** nm), a spatial light modulator (SLM) that offers binary phase modulation, and a linear polarizer (Fig. 2) [21, 22].

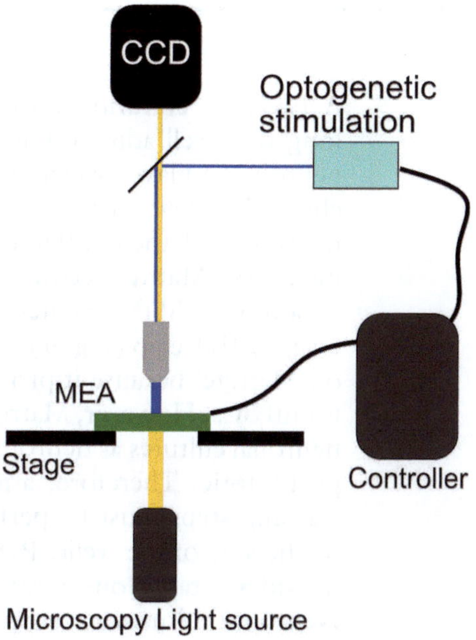

Fig. 1 Setup for wide-field optogenetic stimulation of hiPSC-derived neuronal cultures. An upright microscope configuration is shown. The light for optogenetic stimulation is guided through the light path normally used for fluorescent stimulation. This system can be combined with MEA electrophysiology or patch clamp recordings. *CCD* charge-coupled device (camera), *MEA* multi-electrode array

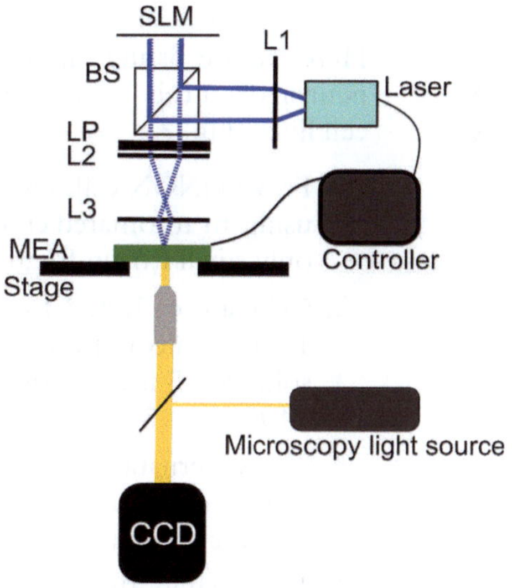

Fig. 2 Scheme of patterned optogenetic stimulation setup with single-cell resolution based on holographic pattern generation, adopted from [22]. *SLM* spatial light modulator, *L1–L3* lenses, *LP* linear polarizer, *MEA* microelectrode array

3 Methods

A two-step culturing protocol is applied to facilitate stable and long-term cell adhesion and thereby survival of stem cell-derived neurons on glass coverslips for patch clamp recordings or in MEA chips. The first step includes the initial culture of the inducible neurons and their differentiation (mTeSR1 + doxycycline treatment) on Matrigel-coated well plates. The second step requires re-seeding of differentiated neurons on a PDL-laminin-coated coverslip/MEA chip (Fig. 4). Preliminary differentiation is performed on Matrigel because it provides an appropriate microenvironment for hiPSCs. However, Matrigel coating is not optimal for long-term neuronal cultures as neurons tend to detach and die after becoming postmitotic. Therefore, after differentiation, media changes and washing steps must be performed with care by adding the media to the side of the wells. PDL-laminin coating, on the other hand, provides a stable long-term substrate and enhances the attachment. Neuronal cultures are supplemented with astrocyte cultures (e.g., primary rat astrocytes), the so-called Banker culture system [6, 23], to boost neuronal network maturation and improve its long-term survival. In addition, for long-term cultures, only half of the medium is exchanged with fresh maturation medium (complete BrainPhys medium) once per week. Evaporated water must be compensated [6]. Alternatively, an astrocyte culture in a separate dish is used to harvest the astrocyte-conditioned neuronal media and mix it (1:2) with fresh complete BrainPhys media for maturation of the hiPSC-derived neurons [18].

3.1 Neuronal Cultures for Patch Clamp Recordings

Here we are listing all steps for functional recordings of iNGN neurons, but this protocol is adaptable for other inducible neuronal cell lines (Fig. 3).

1. Frozen iNGN cells are thawed within a warm water bath or by using an automated cryo-thawing device (e.g., BioCision) until only small frozen fragments remain.
2. Cells are collected from the cryotube and transferred into a 15 mL falcon tube. PBS without calcium and magnesium solution (5 mL) is added, and cell suspension is centrifuged at 1400 rpm (359 × g) for 4 min.
3. The supernatant is aspirated, and the cell pellet is resuspended in 1 mL mTeSR1 + ROCK inhibitor solution. About 100,000 cells are seeded per well of a Matrigel-coated six-well plate.
4. 1 day later, the medium is exchanged to mTeSR1 without ROCK inhibitor.
5. The medium is changed every day until 80% confluency is reached.

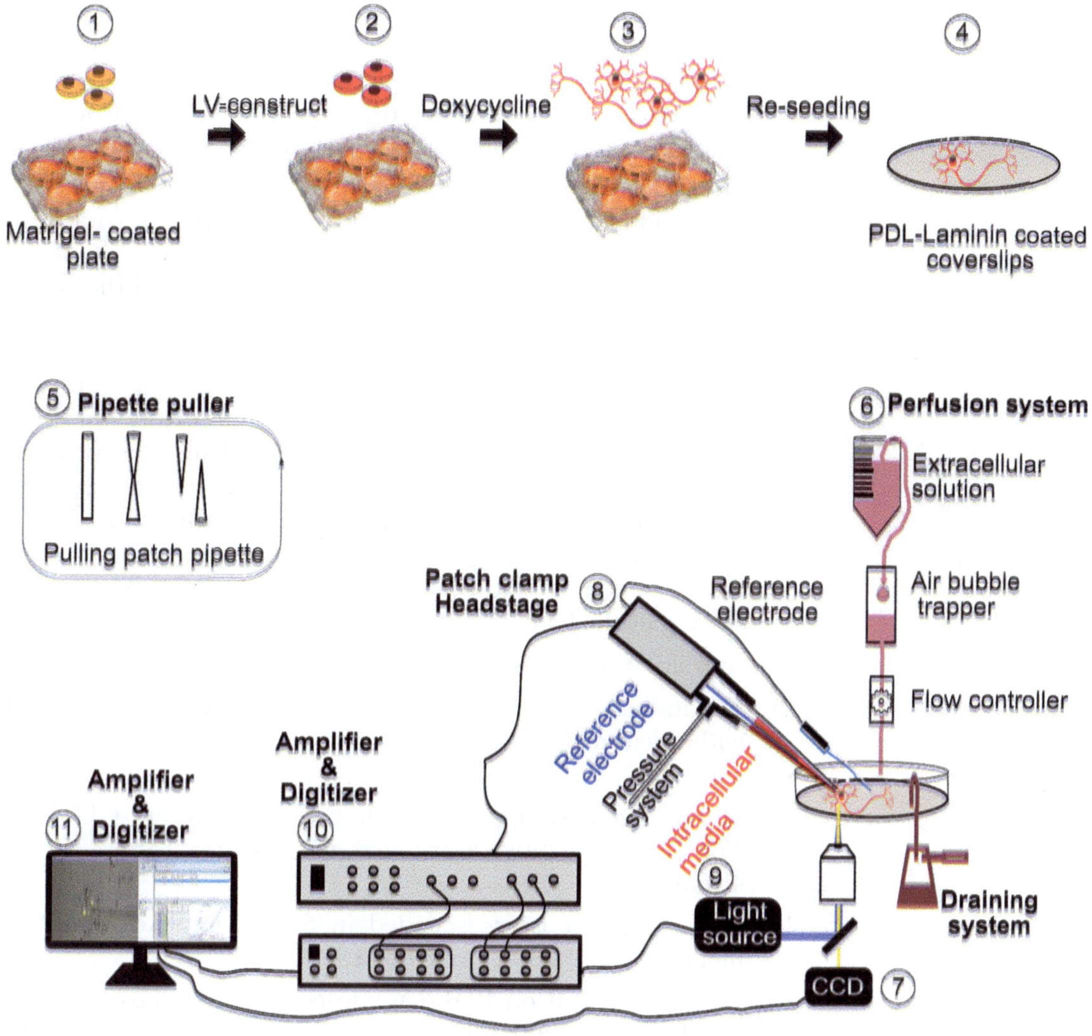

Fig. 3 Preparation of hiPSC-derived neurons for patch clamp recordings and details of the patch clamp recording setup. After transducing iNGN cells with lentiviral particles bearing the genetic information for the optogenetic actuators, the neuronal induction process is started by doxycycline treatment for 5 days. Differentiated neurons are re-seeded on a PDL-laminin-coated coverslips and kept in complete BrainPhys medium for long-term maturation. The Patch clamp setup is composed of perfusion and draining systems, headstage, amplifier and digitizer, and PC for visualization and recording. The light source for optogenetic stimulation is connected to digitizer for synchronization of the light pulses and recorded data

6. Cells are passaged (detached by 2 min incubation in 1 mL TrypLE Express, centrifuged in 1400 rpm (359 × *g*) for 4 min, and the cell pellet is re-suspended in mTeSR1 + ROCK inhibitor), and seeded again with higher concentrations (300,000 cells per well of a Matrigel-coated six-well plate).
7. 1 day after seeding, the medium is changed to mTeSR1. Lentiviral transduction of optogentic tools is performed as previously described [6, 18] (*see* **Note 1**).

8. The next day, the cells are washed twice with PBS without calcium and magnesium, then mTeSR1 + doxycycline is added from 1 dpi for 5 consecutive days to differentiate the hiPSC to the neurons. When the optogenetic tool is fused to a fluorescent reporter, its expression should be clearly detectable 48 h post lentiviral transduction.
9. PDL and laminin coated coverslips are prepared in parallel to the neuronal culture in Matrigel-coated plates. 50 μL of 0.1 mg/mL PDL is added directly to the center of the coverslips and incubated overnight. Coverslips are washed (3×) with sterile distilled water and dried under the culture hood. 1 mg/mL laminin stock solution is diluted to 0.05 mg/mL with complete BrainPhys culture medium (1:20, store for max. 1 week at 4 °C) and 60 μL of the acquired solution is added to the 12 mm cover slips that have already been coated with PDL. The laminin-coating requires 3 h incubation time (37 °C, 100% relative humidity, ideally in the incubator). The solution should be removed just before adding the cells. It is of importance that the laminin-coated cover slips never dry.
10. At 5 dpi, the pre-used culture medium is collected and stored.
11. The induced neurons are gently washed once with warm PBS with calcium and magnesium.
12. The induced neurons are detached by 5 min incubation with 500 μL Accutase and collected in pre-used medium + 1% BSA. The cells are gently centrifuged at 1400 rpm (359 × *g*) for 4 min.
13. The cell pellet is resuspended in fresh complete BrainPhys medium + 0.2% BSA (*see* **Note 19**).
14. Cells are seeded on PDL-laminin-coated coverslips (between 10,000 and 30,000 cells per coverslip). The cells are kept in the incubator.
15. After 2 h of incubation, pre-warmed conditioned complete BrainPhys medium is added (*see* **Note 20**).
16. Half of the complete BrainPhys medium is exchanged with conditioned complete BrainPhys medium once every 3–7 days (*see* **Note 20**).

3.2 Patch Clamp Electrophysiology

Intracellular currents depending on optogenetic excitation or inhibition at the single-cell level as well as other biophysical properties of neural cells like membrane capacitance, densities of ion channels, shape of induced action potential, and response to the light stimuli can be extracted using patch clamp recordings. There are several specific configurations for cell-attached patch clamp measurements available. They include whole-cell patch clamp, inside-out patch, outside-out patch, and perforated patch. For details, please refer to

[24]. Here, we focus on the whole-cell patch clamping that allows the measurement of intracellular currents as a response to optogenetic stimulation. The measurements can be performed in voltage-clamp or current-clamp modes. Voltage-clamp measures membrane ion currents by compensating the changes in ion current through a negative feedback to keep the membrane potential at the set level. In current clamp, changes in membrane potential in response to the known injected current or applied light pulses are measured [24].

3.2.1 Performing Patch Clamp Recordings and Optogenetic Stimulation

1. Start the patch clamp equipment and microscope.
2. Start the perfusion system and make sure that the perfusion speed and draining is adjusted well (*see* **Note 11**).
3. Fabricate capillary glass pipettes with a pipette resistance around 4 MΩ (pipette resistance is measured when filled with internal solution, *see* **Note 17**).
4. Place the coverslip with the neuronal culture (Fig. 3, *see* Subheading 3.1) in the recording chamber and make sure that the coverslip is fixed well and is not moving.
5. Fill the pipette with internal solution and load it into the pipette holder. Often tapping the pipette will help to remove trapped air bubbles. The internal solution must cover at least one-third of the recording electrode.
6. Apply positive pressure through the side valve of the pipette holder before inserting the pipette into the extracellular solution (*see* **Note 9**).
7. Use the microscope to find a neuron for patching.
8. Using the micromanipulator and reading the *X*, *Y*, *Z* positions guide the pipette in the bath solution toward the to-be-recorded neuron. The positive pressure inside the pipette should be kept constant during the guiding process by closing the three-way valve.
9. As the pipette is getting closer to the neuron, set the micromanipulator speed to the finest level.
10. At this step, the amplifier is set to the voltage-clamp mode.
11. The pipette current needs to be offset corrected showing that no current is passing through the pipette (controlled by software in new amplifier versions).
12. Before reaching the cell, a seal test is performed by applying a 10 mV test pulse at 100 Hz. The protocol of the pulse test should have already been generated and saved. Based on Ohm's law, applying the pulse test will reveal the pipette resistance.

13. Depending on the recorded amplitude of the rectangular current, the resistance of the pipette is calculated. This value is needed for further analysis.
14. Keep the test pulse running and guide the pipette toward the cell until a tiny dimple on the cell membrane is visible which is wobbling with low amplitude in the phase of the test pulse.
15. Release the positive pressure by opening the three-way valve and simultaneously apply a strong but sharp suction to the pipette to form a giga seal. It is demonstrated by a sharp decrease in current amplitude until it reaches to the zero level.
16. To open the cell membrane, apply a sharp but light suction (negative pressure) through the pipette by a syringe which has been attached to the end of the pressure system through three-way valve (Fig. 3). A sharp rise followed by slow decay of current response to the test pulse confirms the opening of the cell membrane. Now, membrane capacitance and other electrical parameters can be measured. From this step on, depending on the intended experiment, the mode can be kept at voltage clamp or changed to current clamp.
17. Voltage clamp mode can be selected for recording excitatory and inhibitory postsynaptic currents (EPSCs and iPSCs) by holding the neuronal membrane potential at −70 mV. This experiment can be performed in the presence of synaptic blockers to distinguish between the inhibitory and excitatory currents.
18. To measure photo-induced currents and to separate them from voltage-gated sodium currents, Tetrodotoxin (1 μM) is applied to the bath solution to block the sodium current through voltage-gated sodium channels. The timing of the light pulse (optogenetic stimulation) and the recording should be precisely controlled and synchronized by appropriate data transfer between amplifier and light source. This is important for the analysis of the data afterward.
19. Current clamp is used for studying the cell behavior (changes in membrane potential) in response to injected currents or optical stimulation. In this mode, passive membrane properties and action potential features are studied by applying different current levels and measuring the change in membrane potential (Vm) caused by the applied current.
20. Optogenetic stimulation is applied in different light intensities, pulse widths, and intervals, which allows to compare cell behaviors with respect to these parameters. All protocols for the optical stimulation can be developed further and modified during the experiment (*see* **Notes 21** and **22**).

3.3 Banker Culture for MEA Electrophysiology

The Banker culture system enables the co-culturing of hiPSC-derived neurons and astrocytes on MEA chips. The neurons are attached to the MEA surface to cover the electrodes and the astrocytes are separated on paraffin-dot coverslips (*see* **Note 6**). The astrocytes are cultured on separate paraffin-dot coverslips, flipped, and placed on top of the neuronal culture. Paraffin-dots act as spacer between two cultures. The astrocyte-secreted molecules improve neuronal network survival and maturation [6].

1. Follow all steps in Subheading **2.1** until the end of **step 8**.
2. Parallel to the neuronal cultures, astrocyte cultures are prepared on paraffin-dot coverslips (*see* **Note 4**) coated with PDL-laminin.
3. Thaw astrocytes, mix them with astrocyte medium, and centrifuge at 1400 rpm for 5 min. Remove supernatant, resuspend the cell pellet in fresh astrocyte medium, and seed 20,000 cells on each paraffin-dot coverslip (*see* **Note 4**).
4. Incubate astrocytes on paraffin-dot coverslips for 2 h. Add pre-warmed astrocyte medium.
5. Change the astrocyte medium to complete BrainPhys medium 3 days before re-seeding the neurons on MEAs to expose and adapt astrocytes to the complete BrainPhys medium.
6. Wash and wet-autoclave (120 °C for 20 min) MEA chips 2 days before seeding iNGN cells.
7. After drying, plasma treatment (2 min, 0.3 mbar, and 50% power) increases the MEA chips' hydrophilicity (*see* **Note 23**).
8. Immediately after plasma treatment, coat MEAs with 50 μL of 0.1 mg/mL PDL (only in center) and incubate overnight (*see* **Note 23**).
9. Wash MEAs the day after PDL coating with sterile distilled water (3×).
10. Dry and coat the MEAs with 50 μg/mL laminin and incubate (37 °C, 100% RH) for 3 h before seeding differentiated neurons. Laminin coating should never dry.
11. At 4 dpi, potential undifferentiated iNGN cells within Matrigel-coated well plate must be removed by Ara-C treatment (1:500) for 1 day. This step is important as otherwise, remaining hiPSC will overgrow the neurons.
12. At 5 dpi collect pre-used medium from the differentiated neuronal cultures within Matrigel-coated well plate.
13. Wash the culture with warm PBS without calcium and magnesium.

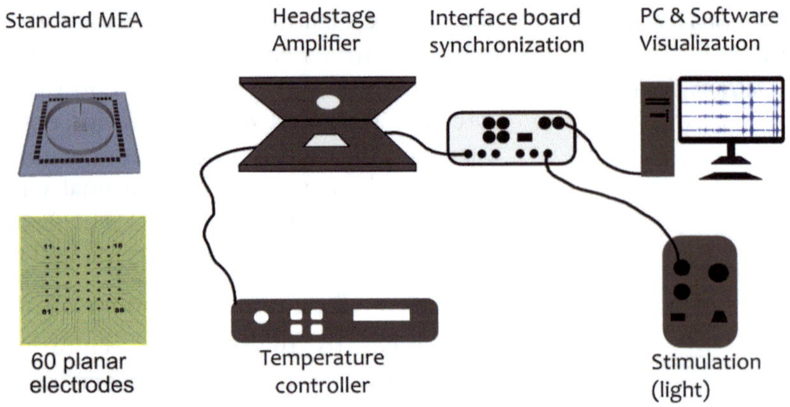

Fig. 4 MEA chip and setup. Left: A standard MEA chip with 8 × 8 electrode configuration. Right: MEA recording setup including headstage and temperature controller, interface board and acquisition system. Light source is connected to the interface board for synchronization of the data

14. Detach iNGN cells using 5 min incubation with 500 μL Accutase and collect them in pre-used medium from **step 12** + 1% BSA and centrifuge at 1400 rpm for 4 min.
15. Resuspend cell pellet in fresh complete BrainPhys medium + 0.2% BSA.
16. Gently drain laminin coating from MEAs.
17. Seed iNGN cells immediately on PDL-laminin coated MEAs (around 50,000–75,000 cells, in 50 μL per MEA). The iNGN cells should be placed in the center covering the electrode area.
18. Incubate seeded neurons for 2 h.
19. Take an astrocyte culture on a paraffin-dot coverslip, flip astrocyte culture, and place it on top of the neuronal culture on the MEA (Fig. 4).
20. Add pre-warmed complete BrainPhys medium (1.5 mL) to the MEA and leave the culture in the incubator.
21. Change medium every 3–7 days depending on the cell density or medium consumption and evaporation (*see* **Note 24**).

3.4 MEA Electrophysiology

Even though MEA technology offers simultaneous stimulation and recording from the same electrodes with high temporal resolution [7, 25], combining it with optogenetic stimulation results in a higher spatial resolution and decreased stimulation-related noise levels in the recording electrodes [6, 22]. In addition, optogenetics facilitates the selective stimulation of small neuronal population or even single neurons that are expressing the particular optogenetic actuator. Therefore, it increases the flexibility to control neuronal cell activity [26]. MEA electrophysiology is an optimal option for long-term network electrophysiology because one can record from the same chip over time [27–29] (Fig. 5).

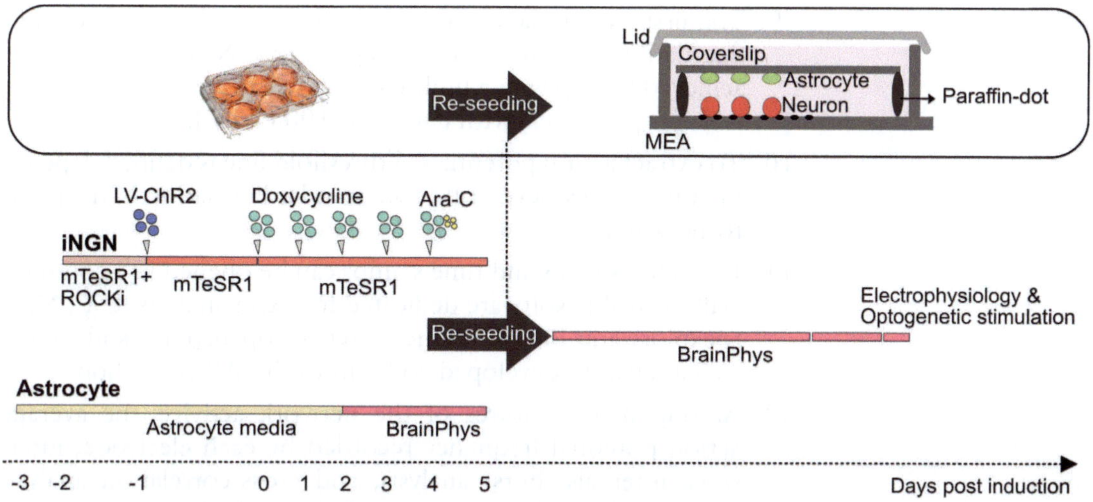

Fig. 5 Preparation of hiPSC-derived neuronal culture for long-term MEA electrophysiology and optogenetic stimulation. iNGN neurons and astrocytes are prepared in parallel. iNGN cells are treated with lentiviral particles transducing the genes for optogenetic actuators. Then neuronal induction is started with mTeSR1 + doxycycline. Ara-C is added to remove un-differentiated cells. Re-seeding of neurons is performed by detaching the cells using Accutase and seeding them on PDL-laminin coated MEA substrate. Astrocytes on coverslips are flipped and placed on top of the neuronal culture. The medium is changed to complete BrainPhys medium. Electrophysiology and optogenetic stimulation are performed over time under sterile conditions

3.4.1 Recording Network Spontaneous Activity

1. Plan the recording experiment ahead to minimize the time the MEA culture will stay outside of the incubator (*see* **Note 24**).
2. Turn on the recording software and temperature controller and adjust the temperature of the heating plate above 37 °C.
3. Take the MEA outside of the incubator and clean the contact pads of the MEA chip with pure ethanol to decrease the noise level.
4. Place the MEA on the amplifier. Make sure that the MEA has been positioned correctly and internal reference electrode is in the correct orientation.
5. Run the recording software and check the noise level. MEAs with smaller electrode diameter show higher noise levels than MEAs with larger electrodes. The noise levels in general should be between ±10 and ±40 μV in Titanium nitride (TiN) electrodes.
6. Record the raw signal for 10–15 min (*see* **Note 25**).
7. Release the MEA and return the chip into the incubator before performing other experiments or medium exchange.
8. The recorded raw signal can be replayed offline for extracting extracellular activity and timestamps of the action potentials.

9. The first step in data analysis is to filter out the low frequency fluctuations in the baseline signal (*see* **Note 26**). This is achieved by applying a high-pass filter (e.g., second-order Bessel high-pass filter) with cut-off at 100 Hz [7].
10. To extract action potentials, a threshold line is defined depending on the noise level (often ±5 standard deviation of the peak-to-peak noise).
11. Extracted signals and time stamps can be opened in a commercially available software dedicated for extra analysis (e.g., Neuroexplore and Plexon offline sorter) or opensource software as well as custom developed codes in MATLAB or Python.
12. Among many features of the network activity, the average action potential frequency recorded by each electrode, interspike intervals, burst analysis, and cross-correlation analysis between electrodes are extracted to monitor network activity over time and map the network functional connectivity patterns [7, 28].

3.4.2 Recording Network Optogenetic-Evoked Activity

Before starting the optogenetic stimulation, follow all first eight steps of recording spontaneous activity (previous section) at baseline to compare light-stimulated conditions with baseline activity. Afterwards:

1. Select the appropriate wavelength (narrow spectrum with a filter if necessary) and put it to the predetermined intensity. The intensity of the light needs to be measured at MEA surface beforehand. Therefore, a control MEA filled with medium is placed under the objective, and the light intensities are measured with a power meter. Later, these values are used for applying the desired light intensity on MEA surface (*see* **Note 22**).
2. The duration of each light pulse, the intervals between pulses and the total number of pulses must be determined before the experiment. A complex protocol composed of several episodes of light stimuli with different pulse widths, intervals and numbers is used to check the network activity response to different stimulation parameters.
3. Make sure that the recording software logs the timings of the light pulses.
4. First run the recording software and record for a few seconds. This baseline recording is important for analysis and offset correction later on. Then turn on the designed protocol (*see* **Note 18**).
5. After recording the optogenetic-evoked activity, record baseline activity for a few minutes. This is important to check if the

stimulation has an influence on the whole network activity for an extended period.

6. Extract the activity in baseline and optogenetic-evoked activity with the same filter and threshold parameters. The steps are the same as for the spontaneous activity.
7. After extracting the timestamps of the action potentials, they can be used together with the recorded timings of the stimulus to plot and analyze the evoked activity of the network in conjunction with stimuli.
8. Following parameters can be extracted from the acquired data: Frequencies and latencies upon optogentically evoked activity, the maximum firing rate in response to light stimulation and to different pulse widths and frequencies.

4 Notes

1. **Lentiviral transduction**: Lentiviral particles are added to the hiPSC culture 1 day after seeding. After 24 h incubation, the medium is drained, and the cells must be washed with PBS without calcium and magnesium. Subsequently, the neuronal induction is started by doxycycline treatment. Additional PBS washing steps are needed after 48 h to remove potential remaining lentiviral particles [18].
2. **Lentiviral transductions**: In case lentiviral particles are required, production and transduction require biosafety-level 2 facilities and experienced researchers (for more details, *see* [18]).
3. **Astrocytes**: Astrocytes are quite sticky to the cell culture well surface. Therefore, they should be handled carefully. All parts of the pipette tips and falcon tubes that will be in contact with astrocytes during the experiment should be pre-exposed to the medium. This will decrease their attachment to the pipette tip or falcon tube.
4. **Preparation of the paraffin-dot coverslips**: Autoclave low melting paraffin in a 200–500 mL bottle and melt it on a heating plate (up to 250 °C) under a laminar flow hood. Use a 200 μL pipette tip attached to an aspiration pipette and soak in the melted paraffin. Place three small droplets of paraffin on three corners of the coverslips to create paraffin dots that will be used as spacer between neuron and astrocyte cultures.
5. **Air table:** An air table equipped with magnetic breadboard tabletop with metric or imperial tapped holes are suitable for firm installation of the microscope and other patch clamp equipment.

6. **Microscope and camera:** Depending on experimental setup, both upright and inverted microscopes can be used for patch clamp. For cell culture systems, mainly the inverted microscope configuration is used to provide enough space for micromanipulators and recording electronics. For recording from thick tissue cultures, upright microscope is required to visualize the sample from top. A good-quality camera attached to the microscope is always helpful on obtaining good patch clamp results.
7. **Faraday cage:** All parts of the recording instruments and Faraday cage should be grounded correctly. In case an unknown noise source is detected in the recordings, unplug attached equipment one by one and try to isolate the source of the noise. Following equipment can be the source of noise: light source of the microscope or optogenetic light source, attached computer screens, perfusion pump, digital oscilloscope, instrument power supplies as well as other large equipment in the vicinity of the setup.
8. **Patch clamp headstage:** Each headstage should be used with its own corresponding amplifier.
9. **Positive pressure in patch pipette:** Positive pressure in the pipette before entering into the bath solution prevents a potential contamination of the pipette tip.
10. **Amplifier and Digitizer:** A high-frequency sampling rate by a digitizer is required for recording precise patch clamp data. In addition to the data channels, digitizer offers trigger inputs. They are used for timing signals that synchronize optical stimulation and patch clamp recording.
11. **Perfusion system:** For simple experiments, only one container including the extracellular solution connected to the carbogen (95% O_2 and 5% CO_2) bubbling system is sufficient. For testing the effect of chemical compounds like excitatory and inhibitory synapse blockers, additional containers and perfusion roots need to be added. For a stable recording, an equivalent rate of perfusion and draining is required. Generally, the perfusion speed for the neuronal cell culture is around 1.5 mL per min. Higher speeds of perfusion (more than 2 mL per min) can induce mechanical disturbances on pipette and thereby noise.
12. **Patch clamp PC:** Normally two separate screens are required: one for visualization of the neuronal cell during patching and another for controlling the software and to visualize signal.
13. **Light source for optogenetic stimulation:** LED and laser sources are most suitable for optogenetic stimulation. Stimulation frequency and intensity of the light source as well as exact timing of ON/OFF periods can be adjusted using proper devices (filters, shutters, etc.) and custom-made codes or commercially available software. Also, patch clampPatch clamp

setups allow to control the stimulation through the digitizer. The abovementioned parameters can also be programmed through controlling a shutter placed between the light guide; however, one has to consider that the minimum ON/OFF timing can be limited because it needs time to open and close the shutter mechanically. In case the light source does not produce narrow wavelength, it can be additionally narrowed down using proper set of bandpath filter sets.

14. **Pulling patch clamping pipettes:** Reproducible resistance and shapes of pulled pipettes significantly improve the data quality. To fabricate symmetrically pulled pipettes, it is important to place glass capillaries exactly in the middle of the heating filament.

15. **Patch pipettes:** Lower resistance pipettes give loose sealing and higher resistance pipettes give tight and stable sealing; however, obtaining a seal with higher resistance pipettes is more challenging.

16. **Standard MEA chips**: Gold- or titanium-based planar electrodes of MEA chips are embedded in a glass substrate that make MEAs transparent. They can be used both with upright and inverted microscopes. A transparent Teflon lid is used to prevent medium evaporation when the chips are outside of the incubator for microscopy and electrophysiology recordings.

17. **Giga seal:** In case there remains a current after seal generation, the seal is not tight enough. Often initial seals of more than 400 MΩ recover in few minutes to more than 1 GΩ, especially if a negative potential of around −60 to −70 mV is determined for the voltage clamp. Consider that still the cell membrane has not been opened in this phase.

18. **Light stimulation setup (light source):** Well suited for full-field stimulation is a LED as a stimulation source (e.g., pE-300 ultra, CoolLED Ltd.), since it is easy to use and has a long lifetime. Wavelength filters have to be chosen according to the optogenetic tools used. The light source should be adjustable and electronically controllable.

19. **Resuspending the differentiated neurons:** Because differentiated neurons have already developed branches and connections, after centrifugation, the cell pellet needs to be treated gently for dissociating the neurons.

20. **Conditioned complete BrainPhys medium:** Primary rat astrocyte cultures are prepared on PDL-laminin-coated coverslips in parallel to the neuronal culture. Conditioned medium is prepared by mixing medium derived from astrocyte culture and fresh complete BrainPhys medium (1:2).

21. **Light stimulation setup (parallel imaging/stimulation):** If imaging should be done in parallel to stimulation, one has to consider possible interference of the two light sources. The additional illumination can cause stimulation when it is not intended. A cross-stimulation is also possible when multiple opsins are used. The thresholds and potential crosstalk must be checked before an experiment.
22. **Light stimulation setup (light intensity):** For wide-field illumination in a microscope setup, the illumination area is equal to the field of view. Thus, this is linked to the magnification of the objective used. Smaller magnification means larger illumination area but also less light intensity. On the other hand, a higher magnification objective leads to a smaller area for stimulation with a high light intensity. Thus, one always has to consider the light intensity per area (usually given in mW/mm^2) for each experiment. This value has to reach a certain threshold in order to activate the opsins reliably. A tunable light source is necessary. For example, typical values for ChR2 are in the range of 1–5 mW/mm^2 [31]. Intensity should be measured with a power meter before the experiment and divided by the area. Attention with patterned stimulation, since the illuminated area is smaller than the field of view of the objective.
23. **MEA chip plasma treatment:** Plasma activation is maximally stable for 10–20 min at ambient conditions. Therefore, PDL treatment should be performed immediately after plasma treatment.
24. **Changing medium for MEA cultures:** Consider that medium change temporarily affects the baseline network activity because of changes in the cellular microenvironment. This effect can last for 1 day. Therefore, the best way to deal with this issue is to do the medium exchange after a recording session. It is also helpful to perform microscopy imaging after a recording session to minimize the effect of environmental disturbances on network function [30].
25. **Data acquisition in MEA systems:** Normally data are acquired at 25 kHz, and it is better to record raw data even though the stored file size could be several gigabytes for 15 min of recording. For very long recordings, data can be analyzed and saved as extracted action potentials which decreases the file size to a few megabytes.
26. **Filtering MEA data:** In applying the high-pass filter for MEA data, consider that adjusting the cut-off to higher values can distort the signal shape.

Acknowledgments

The Volkswagen Foundation (Freigeist—A110720), the European Research Council (ERC-StG 678071—ProNeurons), and the Deutsche Forschungsgemeinschaft (EXC-2151-390873048—Cluster of Excellence—ImmunoSensation^2 at the University of Bonn and SPP2127) support VB. JS acknowledges the support by the Joachim Herz Stiftung.

References

1. Busskamp V, Lewis NE, Guye P et al (2014) Rapid neurogenesis through transcriptional activation in human stem cells. Mol Syst Biol 10:760. https://doi.org/10.15252/msb.20145508
2. Taylor CJ, Bolton EM, Bradley JA (2011) Immunological considerations for embryonic and induced pluripotent stem cell banking. Philos Trans R Soc Lond Ser B Biol Sci 366:2312–2322. https://doi.org/10.1098/rstb.2011.0030
3. Lam RS, Töpfer FM, Wood PG et al (2017) Functional maturation of human stem cell-derived neurons in long-term cultures. PLoS One 12:e0169506
4. Silva MC, Haggarty SJ (2019) Human pluripotent stem cell-derived models and drug screening in CNS precision medicine. Ann N Y Acad Sci 1471(1):18–56. https://doi.org/10.1111/nyas.14012
5. Engle SJ, Blaha L, Kleiman RJ (2018) Best practices for translational disease modeling using human iPSC-derived neurons. Neuron 100:783–797. https://doi.org/10.1016/j.neuron.2018.10.033
6. Klapper SD, Sauter EJ, Swiersy A et al (2017) On-demand optogenetic activation of human stem-cell-derived neurons. Sci Rep 7:14450. https://doi.org/10.1038/s41598-017-14827-6
7. Habibey R, Latifi S, Mousavi H et al (2017) A multielectrode array microchannel platform reveals both transient and slow changes in axonal conduction velocity. Sci Rep 7(1):8558. https://doi.org/10.1038/s41598-017-09033-3
8. Habibey R, Sharma K, Swiersy A, Busskamp V (2020) Optogenetics for neural transplant manipulation and functional analysis. Biochem Biophys Res Commun 527(2):343–349. https://doi.org/10.1016/j.bbrc.2020.01.141
9. Zhang K, Cui B (2015) Optogenetic control of intracellular signaling pathways. Trends Biotechnol 33:92–100. https://doi.org/10.1016/j.tibtech.2014.11.007
10. Klapoetke NC, Murata Y, Kim SS et al (2014) Independent optical excitation of distinct neural populations. Nat Methods 11:338–346. https://doi.org/10.1038/nmeth.2836
11. Fenno L, Yizhar O, Deisseroth K (2011) The development and application of optogenetics. Annu Rev Neurosci 34:389–412. https://doi.org/10.1146/annurev-neuro-061010-113817
12. Hooks BM (2018) Dual-channel photostimulation for independent excitation of two populations. Curr Protoc Neurosci 85:e52. https://doi.org/10.1002/cpns.52
13. Latifi S, Mitchell S, Habibey R et al (2020) Neuronal network topology indicates distinct recovery processes after stroke. Cereb Cortex 30(12):6363–6375. https://doi.org/10.1093/cercor/bhaa191
14. Daadi MM, Klausner JQ, Bajar B et al (2016) Optogenetic stimulation of neural grafts enhances neurotransmission and downregulates the inflammatory response in experimental stroke model. Cell Transplant 25:1371–1380. https://doi.org/10.3727/096368915X688533
15. Weitz AJ, Lee JH (2016) Probing neural transplant networks in vivo with optogenetics and optogenetic fMRI. Stem Cells Int 2016:8612751. https://doi.org/10.1155/2016/8612751
16. Weick JP, Johnson MA, Skroch SP et al (2010) Functional control of transplantable human ESC-derived neurons via optogenetic targeting. Stem Cells 28:2008–2016. https://doi.org/10.1002/stem.514
17. Steinbeck JA, Choi SJ, Mrejeru A et al (2015) Optogenetics enables functional analysis of human embryonic stem cell-derived grafts in a Parkinson's disease model. Nat Biotechnol 33:204–209. https://doi.org/10.1038/nbt.3124

18. Sauter EJ, Kutsche LK, Klapper SD, Busskamp V (2019) Induced neurons for the study of neurodegenerative and neurodevelopmental disorders. In: Ben-Yosef D, Mayshar Y (eds) Fragile-X syndrome: methods and protocols. Springer, New York, pp 101–121
19. Lee S, George JH, Nagel DA et al (2019) Optogenetic control of iPS cell-derived neurons in 2D and 3D culture systems using channelrhodopsin-2 expression driven by the synapsin-1 and calcium-calmodulin kinase II promoters. J Tissue Eng Regen Med 13: 369–384. https://doi.org/10.1002/term.2786
20. Renault R, Sukenik N, Descroix S et al (2015) Combining microfluidics, optogenetics and calcium imaging to study neuronal communication in vitro. PLoS One 10:e0120680
21. Schmieder F, Habibey R, Büttner L, et al (2019) Optogenetic investigation of in vitro human iPSC-derived neuronal networks (conference presentation). In: Proceedings of SPIE
22. Schmieder F, Klapper DS, Koukourakis N et al (2018) Optogenetic stimulation of human neural networks using fast ferroelectric spatial light modulator-based holographic illumination. Appl Sci 8(7):1180
23. Kaech S, Banker G (2006) Culturing hippocampal neurons. Nat Protoc 1:2406–2415. https://doi.org/10.1038/nprot.2006.356
24. Molleman A (2003) Patch clamping: an introductory guide to patch clamp electrophysiology. Wiley, New York
25. Habibey R, Golabchi A, Blau A (2015) Microchannel scaffolds for neural signal acquisition and analysis. In: Londral AR, Encarnação P, Rovira JLP (eds) Neurotechnology, electronics, and informatics. Springer International Publishing, Cham, pp 47–64
26. Maybeck V, Schnitker J, Li W et al (2016) An evaluation of extracellular MEA versus optogenetic stimulation of cortical neurons. Biomed Phys Eng Express 2:55017. https://doi.org/10.1088/2057-1976/2/5/055017
27. Wilk N, Habibey R, Golabchi A et al (2016) Selective comparison of gelling agents as neural cell culture matrices for long-term microelectrode array electrophysiology. OCL 23(1): D117. https://doi.org/10.1051/ocl/2015068
28. Habibey R, Golabchi A, Latifi S et al (2015) A microchannel device tailored to laser axotomy and long-term microelectrode array electrophysiology of functional regeneration. Lab Chip 15(24):4578–4590. https://doi.org/10.1039/c5lc01027f
29. Latifi S, Tamayol A, Habibey R et al (2016) Natural lecithin promotes neural network complexity and activity. Sci Rep 6:25777. https://doi.org/10.1038/srep25777
30. Saalfrank D, Konduri AK, Latifi S et al (2015) Incubator-independent cell-culture perfusion platform for continuous long-term microelectrode array electrophysiology and time-lapse imaging. R Soc Open Sci 2(6):150031. https://doi.org/10.1098/rsos.150031
31. Yizhar O, Fenno LE, Davidson TJ et al (2011) Optogenetics in neural systems. Neuron 71: 9–34. https://doi.org/10.1016/j.neuron.2011.06.004

INDEX

Valentin Gordeliy (ed.), *Rhodopsin: Methods and Protocols*, Methods in Molecular Biology, vol. 2501, https://doi.org/10.1007/978-1-0716-2329-9,

MIX
Papier aus verantwortungsvollen Quellen
Paper from responsible sources
FSC® C105338

If you have any concerns about our products,
you can contact us on
ProductSafety@springernature.com

In case Publisher is established outside the EU,
the EU authorized representative is:
Springer Nature Customer Service Center GmbH
Europaplatz 3, 69115 Heidelberg, Germany

Printed by Libri Plureos GmbH
in Hamburg, Germany